Sustainable Engineering Forum 2014

Core Programming Area at the 2014 AIChE Spring Meeting and 10th Global Congress on Process Safety

New Orleans, Louisiana, USA
30 March – 3 April 2014

ISBN: 978-1-63439-080-4

Printed from e-media with permission by:

Curran Associates, Inc.
57 Morehouse Lane
Red Hook, NY 12571

CURRAN ASSOCIATES INC. proceedings.com

Some format issues inherent in the e-media version may also appear in this print version.

Printed by Curran Associates, Inc. (2014)

AIChE
3 Park Avenue
New York, NY 10016-5991

Phone: (203) 702-7660
Fax: (203) 775-5177

www.aiche.org

Additional copies of this publication are available from:

Curran Associates, Inc.
57 Morehouse Lane
Red Hook, NY 12571 USA
Phone: 845-758-0400
Fax: 845-758-2634
Email: curran@proceedings.com
Web: www.proceedings.com

Sustainable Engineering Forum 2014

Core Programming Area at the 2014 AIChE Spring Meeting and 10th Global Congress on Process Safety

New Orleans, Louisiana, USA
30 March – 3 April 2014

ISBN: 978-1-63439-080-4

AIChE

TABLE OF CONTENTS

354012 Possible vs. Practical: Engineers Must Lead the Development of Practical Technologies

Abstract:

Chemical Engineering is need today more than ever. There is a strong societal desire for sources of energy and feedstocks that can sustainably support a high standard of living for all. Thermodynamic, material and economic challenges stand in the way. Most of the general public does not have the basic understanding of the energy flows, chemical transformations, and the scale of the systems that support our current way of life. These concepts are fundamental to chemical engineering. Chemical engineers should be on the vanguard of separating hype from truly practical options yet occationally we let society down. This talk will present several examples where the fundamentals of chemical engineering can be used to access the practicality of proposed solutions to energy and water issues, as well as explain why the venture capital model struggles in the materials market.

352423 Preparation of Cellulose Nanowhiskers and Their Applications on Liquid Fuels Production

Abstract:

In this study, cellulose nanowhiskers (CNWs) were prepared from raw biomass materials and used as a catalyst support to synthesize iron-based nano catalysts for converting syngas to liquid hydrocarbons. The biomass materials used for preparing CNWs included cotton linters, kenaf bast, and switchgrass. The yields and physicochemical properties of CNWs prepared from these materials were determined and compared. Iron nano catalysts supported on CNWs were synthesized and evaluated for their activity on syngas conversion to liquid hydrocarbons via Fischer-Tropsch synthesis. CNWs-supported iron catalysts showed very promising activity on CO conversion and synthesis of liquid hydrocarbons. Moreover, compared to nanofibers, CNWs-supported iron catalysts had higher stability and higher selectivity towards liquid hydrocarbons. Our study demonstrated the feasibility of developing potential cost-effective nano catalysts for liquid fuels production from domestic sources.

352551 Computational Design of Catalysts and Chemical Modeling of FTS from Syngas

Abstract:

Computational Design of Catalysts and Chemical Modeling of FTS from Syngas

Suraj Gyawali, Purnima Kharidehal, Fernando Soto and Dr. Daniela Mainardi

Institute for Micromanufacturing, Chemical Engineering,

Louisiana Tech University

Fischer -Tropsch Synthesis (FTS) is a surface catalyzed polymerization reaction that synthesize hydrocarbons by hydrogenation of carbon monoxide. During the reaction, carbon monoxide is first adsorbed on the surface of the catalyst, and is hydrogenated to produce CH_X monomers. These CH_X monomers then propagate to form a wide product range of long chain hydrocarbons. Catalysts used in this reaction are *3-d* transition metals like Cobalt, Iron, Nickel, and Ruthenium. These catalysts play an important role in product selectivity, since the products can be olefins, paraffin, and/or alcohol. Currently, FTS has gained a lot of attention in the energy industry because it is also a promising method of producing clean transport fuels with exceptionally low aromaticity and lack of sulphur content. Controlling product selectivity and overcoming catalyst deactivation has been a challenge in order to make this technology commercially available. An optimal design of a catalyst, which can make the FTS commercially viable, includes the study of CO adsorption and the reaction mechanism of FTS on the surface of the catalyst. In this work, stepped surface and alloyed nano-cluster catalysts based on Co, Fe, Ni, and Ru have been studied using electronic structure methods. Alloyed nano-cluster and stepped catalysts are attractive because they have different catalytic properties than either of the parent metals or the flat surfaces. Therefore, Density Functional Theory has been used to understand the catalytic activity in Fischer-Tropsch reaction, which includes the geometry optimization of the stepped and alloyed nano-cluster catalysts, study of CO adsorptions, CO dissociations, and the reaction mechanism on the surface of the different catalysts.

352675 Study of the Interaction of Alkali Metals with Cobalt-Based Bimetallic Catalysts Using Density Functional Theory

Abstract:

The conversion of syngas to liquid hydrocarbons via Fischer-Tropsch synthesis (FTS) is considered as an attractive emerging technology in clean energy for the present century. Despite the experimental and theoretical work, several mechanistic details for FTS remain unclear and unpredictable. To exemplify, the CO dissociation pathways remain largely unresolved [1-4]. Thus, FTS requires detailed atomic studies of the hydrocarbon formation sequence. For chemical reactions, reactant conditions and sources (i.e. Natural gas, coal, biomass) are relevant. However, in FTS, the reaction kinetics can improve with a suitable catalyst. Hence, appropriate catalysts reduce costly investments in the production or purification of syngas. Commonly used catalysts for the FTS process are 3–*d* transition metals such as Co, Fe, and Ru. Typically, high–surface–area supports (i.e. Silica, alumina, or zeolites) [5] are used for these catalysts. Consequently, the ability to control or engineer the catalyst surface provides leverage for controlling reaction pathways and product distribution. Indeed, catalysis already play a key role in the continuing development of clean energy conversion processes, which is essential to our standard of living [6]. Particularly, bimetallic catalysts show an increased reactivity in FTS for fuel production. Furthermore, alkali metals and transition metals have strong promoting effect in catalysts. However, the exact role and effectiveness of promoters in bimetallic catalysts is not fully elucidated. In order to increase catalyst selectivity and efficiency, it is necessary to investigate the surface structure, energies and electronic structure of catalysts and promoters using atomistic simulations. Therefore, in this research, the binding and interaction of alkali and transition metals (Li, Na, K, Rb and Mn) with the cobalt-based (113) surface slabs have been computed using density functional theory (DFT).

[1] M. Ojeda, R. Nabar, A.U. Nilekar, A. Ishikawa, M. Mavrikakis, E. Iglesia, Journal of Catalysis, 272 (2010) 287-

297.

[2] M. Zhong-Yun, C.F. Huo, X.Y. Liao, Y.W. Li, J. Wang, H. Jiao, Journal of Physical Chemistry C, 111 (2007) 4305-4314.

[3] J. Cheng, X.Q. Gong, P. Hu, C.M. Lok, P. Ellis, S. French, Journal of Catalysis, 254 (2008) 285-295.

[4] O.R. Inderwildi, S.J. Jenkins, D.A. King, Journal of Physical Chemistry C, 112 (2008) 1305-1307.

[5] S.N. Rashkeev, M.V. Glazoff, The Journal of Physical Chemistry C, 117 (2013) 4450-4458.

[6] J.J. Spivey, Catalysis Today, 100 (2005) 171-180.

SUMMARY

A study has been performed to investigate the feasibility of using a heat pump on the Isobutylene Concentration Column (IBC). Three heat pump options were considered: 1.) Overhead Vapor Recompression, 2.) Reboiler Liquid Flashing, and 3.) Heat Integrated Distillation. The most cost effective heat pump option for the IBC was determined to be the heat integrated distillation option. Note that in this option the rectifying section is operated at a higher top pressure (i.e. 76.4 psig) than the stripping section which operates at a top pressure of 30.6 psig. Compressor horsepower was estimated at 5850 Hp. Total Installed Cost (TIC) for this option was estimated at **$ 21.8 MM**. Internal Rate-of-Return (IRR) was estimated at **11.0 %**; while the payback period was estimated to be **5.6 yrs.** Net Present Value (NPV) was estimated at **$ 834.22 K**. While the TIC for this project is high, it is recommended that it be considered as one component of several options that may reduce the estimated $ 50 MM cost of a new boiler or eliminate it altogether.

BACKGROUND

The IBC is a significant user of both steam and cooling water which supplies multiple operating units with feedstock at the TPC Group Houston production facility. Note that the IBC is part of the MTBE unit and although currently idle, is scheduled to come online with the startup of the Phoenix project in mid-to-late 2014. It should be pointed out that at design operating rates IBC will consume on the order of 100 MPPH of steam and 13,200 GPM of cooling water.

The high steam and cooling water flows associated with operation of the IBC are due to the difficulty in separating isobutylene from isobutane, which are close boiling components with a low relative volatility. The difficulty of this separation is also reflected by the fact that 140 trays and a reflux ratio of 16.5 are required for this separation. At the time IBC was built, the large number of trays required for this separation could not be accommodated in a single column, hence two columns were required.

Feed to the IBC will be provided by via Tank No.3 (i.e. Isobutylene & Isobutane MTBE Rxr Feed). Note that the overhead product from the IBC is the make-up feed to the MTBE “C” reactors along with methanol.

The reactor effluent from the MTBE “C” reactors are fed to the MTBE Separation Tower. Overhead product from the MTBE Separation Tower is recycled back to the MTBE “C” reactors in order to convert unreacted isobutylene to MTBE. However, a small slipstream from the MTBE Separation Tower overhead product is also purged to the Dehydro Feed Sphere to convert unreacted isobutane to isobutylene.

The MTBE Separation Tower bottoms are the MTBE product and it is cooled and provides supplemental reboil heat to the IBC via auxiliary reboilers before proceeding to the MTBE Storage Tank. At present the MTBE Separation Tower bottoms cooler is the only equipment in place that is used to offset the large demand for steam by the IBC reboilers.

The liquid side-draw product from tray No. 126 of the IBC consists of approximately 88 wt. % isobutylene and is used as feedstock to the polyisobutylene units via the Isobutylene Concentrate Sphere.

Finally, the bottoms product from the IBC which consists of low purity 1, 3 butadiene is recycled to the Prefractionation Tower for recovery of 1, 3 butadiene via the DMF and furfural extraction trains.

The use of a heat pump on the IBC would “free-up” the approximately 100 MPPH of steam for use elsewhere within HNO or could be used to improve existing plant steam systems reliability.

DISCUSSION

The basic concept of a vapor recompression (i.e. heat pump) distillation system is to upgrade the heat content of the columns overhead vapor via compression to supply the heat required at the reboiler.

Note that while conventional distillation systems have a separate reboiler and a separate condenser, a heat pumped distillation column has a combined "condenser-reboiler". In such an arrangement column overhead vapor is first sent to a compressor to raise its temperature and pressure. The heat content of the compressed overhead vapor stream is utilized to provide heat duty to reboil the column bottoms. In the process of giving up heat to reboil the column bottoms the overhead vapor is condensed and part of the condensed overhead stream is sent back to the column as reflux.

Note that in the case of the IBC there are three alternatives to consider with respect to a heat pump for this distillation system: 1.) Overhead Vapor Recompression, 2.) Reboiler Liquid Flashing, and 3.) Heat Integrated Distillation Columns. Another option not considered for the IBC involves using an external working fluid in a closed loop refrigeration cycle where heat is removed from the compressed vapors in the column reboiler and heat is supplied via the column condenser which acts as the evaporator.

The basic design for an Overhead Vapor Recompression Heat Pump is shown below in Figure No. 1:

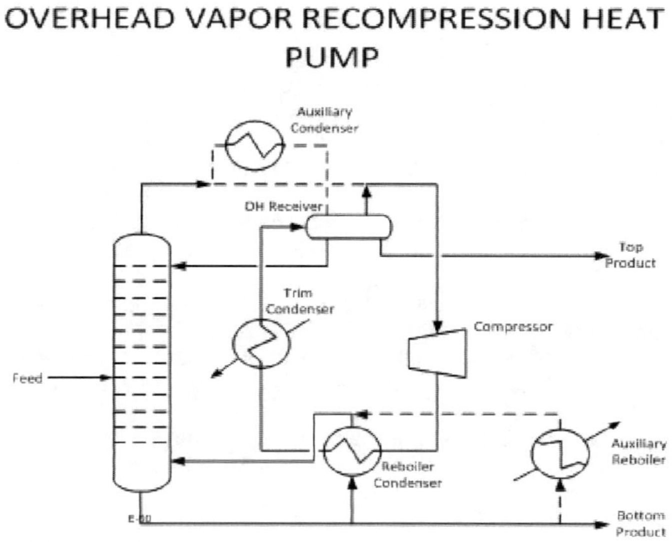

Figure No. 1: Overhead Vapor Recompression Heat Pump

An alternative to a vapor recompression heat pump involves dropping the pressure of the bottoms stream to provide cooling via the Joule-Thompson effect and then using the cooled bottom stream to condense the overhead vapor from the column. In the process of condensing the column overhead vapor the partially flashed bottoms stream picks up heat and is fully vaporized. The vaporized bottoms stream is then compressed and is sent back to the bottom tray of the column as reboil vapor. Note that this scheme is referred to as a Reboiler Liquid Flashing Heat Pump. The basic design for a Reboiler Liquid Flashing Heat Pump is shown below in Figure No.2:

REBOILER LIQUID FLASHING HEAT PUMP

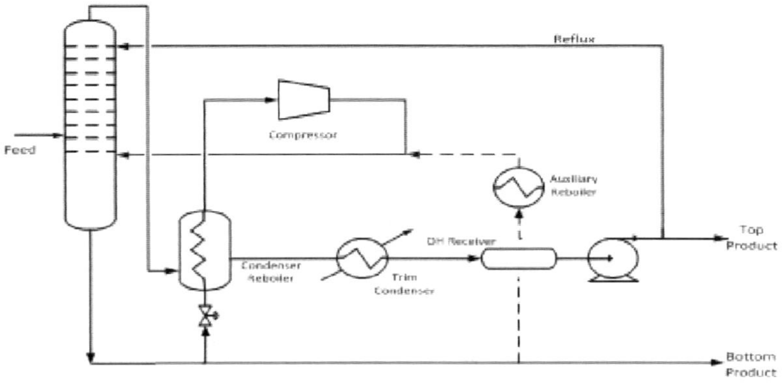

Figure No. 2: Reboiler Liquid Flashing Heat Pump

Finally a third option for a heat pump on the IBC involves operating the rectifying section at a higher operating pressure than the stripping section. A compressor is used to raise the pressure of the overhead vapor from the stripping section to the pressure of the rectifying section. The basic design for a Heat Integrated Distillation Column is shown below in Figure No. 3:

HEAT INTERGRATED DISTILLATION

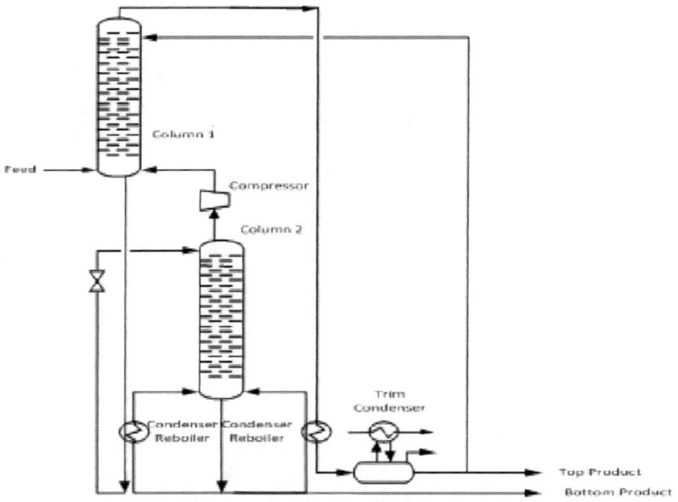

Figure No. 3: Heat Integrated Distillation

A common feature of all heat pump options is the need for a trim condenser. A trim condenser is needed to balance the heat input generated by the compressor. Another common feature is the need for auxiliary reboil and condensing capability required during startup and/or process upsets. Note that in the case of the IBC the existing steam reboilers and water-cooled condensers would act as the auxiliary reboilers and condensers.

The path forward in evaluating the various heat pump options for the IBC involved first building and validating a simulation for the existing configuration using lab and historical operating data for operating conditions and stream compositions. The next step involved simulation of the IBC using the feed rate as set by the Phoenix material balance in order to generate expected product flows, compositions etc. Note that this simulation served as the reference or "Base Case" simulation.

Following that step each heat pump configuration was simulated using the same feed rate, composition and operating conditions as listed in the base case simulation while being careful to ensure that all overhead, side draw, and bottoms flows, compositions, and temperatures closely matched those in the base case simulation.

The final step involved comparing each heat pump option against its required utilities (i.e. compressor power, cooling water etc.) and its installed cost for the new compressor and total minimum required heat transfer surface.

An important concept used to determine the applicability of heat pumps in distillation service is known as "Temperature Lift". Temperature lift refers to temperature difference between the vapor leaving the top tray of the column and the bottoms product leaving the reboiler. Note that a high value of temperature lift translates into high compressor horsepower and hence operating cost since the gas must be compressed to a higher pressure in order to generate the higher temperature and hence driving force for heat transfer.

Conversely a low value of temperature lift translates into relatively low compressor horsepower. In general, heat pumps are not recommended for temperature lifts greater than about 30 °F. Close boiling mixtures such as ethylene – ethane and propylene-propane have low temperature lifts. Hence, heat pumps are commonly applied in distillation systems separating these components. In the case of the IBC the temperature lift is 24.9 °F, which indicates that a heat pump may be applicable for this separation.

RESULTS:

After running and converging a simulation of the existing operating configuration for the IBC the next step was to validate the model using historical operating data and lab analysis. It should be noted that the thermodynamic property method selected for this simulation was the UNIFAC model. The UNIFAC model was selected based on binary interaction parameters for isobutane and isobutylene which resulted in the best fit with plant operating data. The UNIFAC thermodynamic property method was also used by JACOBS Engineering in preparing the material and energy balance for the Phoenix project.

As an initial check on the simulation, the material & energy balances for the IBC generated by JACOBS Engineering as part of the overall Phoenix project material & energy balance were reproduced by the simulation model for the IBC.

Validation of the simulation model utilized historical operating data for 8/13/01 and was provided by the TPC operations department. Lab analysis for the IBC feed, overhead, side draw, and bottoms products were also obtained for the same date. The IBC simulation was then re-run using historical operating data as input and a fixed reboiler duty of 67.9 MM Btu/hr as estimated from the historical operating data.

The simulation output was used to calculate all IBC product flows, compositions, temperature profiles, pressures, condenser duty as well as reflux ratio. A comparison of the simulation results with operating data is given below in Table I:

Table I: IBC Operating Data vs. Simulation Values

Stream Desc.	Plant Data Flow (bbl/hr)	Simulation Flow (bbl/hr)	Press. - Oper.[1] (psig)	Press. - Sim. (psig)	Reboiler Duty - Oper.[1] (MM Btu/hr)	Reboiler Duty - Sim. (MM Btu/hr)	Reflux Ratio - Oper.	Reflux Ratio - Sim.
Feed[2]	348.38	348.42	N/A	95.30	67.90	67.90	17.27	13.95
OH Prod.[3]	189.58	183.52	76.40	76.40	N/A	N/A	N/A	N/A
Side Draw Prod.[4]	170.15	169.75	N/A	86.82	N/A	N/A	N/A	N/A
Bottoms Prod.[5]	4.00	4.39	87.99	87.99	N/A	N/A	N/A	N/A

Notes:

1.) N/A indicates no instrument tap available or is not applicable.
2.) Numbers in blue represent simualtion values for IBC Feed, reboil duty & reflux ratio.
3.) Numbers in green represent simualtion values for IBC OH Product.
4.) Numbers in red represent simualtion values for IBC Side Draw Product.
5.) Numbers in purple represent simualtion values for IBC Btms Product.

It should be pointed out that the simulation assumed a murphree tray efficiency of 70 % across all 140 trays in the IBC. The largest discrepancy between actual operating data and simulation results occurs for the values of reflux ratio as estimated by the simulation. The simulation was run with alternative values of tray efficiency but the results listed in Table I represent the best fit that could be obtained with the model.

In setting up the simulation model note that both 1.) Bottoms Flow and 2.) Reboiler Duty was fixed. Another way to specify the inputs would be to specify bottoms flow and reflux ratio instead of reboiler heat duty. However when this set of specifications is used the calculated reboiler duty is near 80 MM Btu/hr as opposed to the observed value of 67.9 MM Btu/hr.

A comparison between the actual vs. simulation temperature profile for the trays in the IBC is given below in Table II:

Table II: IBC Operating vs. Simulation Temperature Profile

Tray No.	Temp. - Oper. (° F)	Temp. - Sim. (° F)
Top Tray	115.41	115.90
Tray # 17	118.46	118.03
Tray # 39	121.19	120.24
Tray # 71	124.48	123.37
Tray # 107	128.94	127.53
Tray # 136	135.80	134.10
Bottom Tray	140.27	135.08

It will be noted that the largest discrepancy between actual operating values and simulation values occurs for the bottom tray temperature, where the difference in temperature is about 5 ° F. Other than the bottoms tray temperature there is reasonably good agreement between the temperature profile predicted by the simulation and actual values. Once again running the simulation using alternate values for tray efficiency did little to improve the fit.

The comparison between actual product compositions vs. simulation values for the IBC is given below in Table III:

Table III: IBC Simulation vs. Plant Data for Feed & Product Compositions

Component	Formulae	Feed wt. %	OH Prod. wt. %	Side Draw wt. %	Bottoms wt. %
Propane	C_3H_8	0.326 / 0.326	0.58 / 0.64	0.0 / 0	0.0 / 0
Propene	C_3H_6	0.10 / 0.10	0.17 / 0.20	0.0 / 0	0.0 / 0
Iso-Butane	C_4H_{10}	47.19 / 47.29	88.79 / 82.87	7.30 / 11.05	0.88 / 2.80
N-Butane	C_4H_{10}	1.13 / 1.13	0.01 / 0	1.63 / 2.21	8.98 / 5.62
Iso-Butylene	C_4H_8	49.51 / 49.61	12.41 / 15.82	88.74 / 84.32	68.25 / 81.24
1-Butene	C_4H_8	0.495 / 0.496	0.03 / 0.29	0.95 / 0.71	0.96 / 0.56
Trans-2-Butene	C_4H_8	0.53 / 0.53	0.0 / 0	0.84 / 0.99	7.89 / 4.48
Cis-2-Butene	C_4H_8	0.41 / 0.41	0.0 / 0	0.43 / 0.71	9.67 / 5.50
1,3 Butadiene	C_4H_6	0.10 / 0.1	0.02 / 0.10	0.12 / 0.01	0.15 / 0.01
N-Pentane	C_5H10	0.0 / 0	0.0 / 0	0.0 / 0	0.0 / 0

Notes:

1.) Numbers in **black** represent lab data for IBC Feed, OH, Side Draw, & Btms comp
2.) Numbers in blue represent simualtion values for IBC Feed.
3.) Numbers in green represent simualtion values for IBC OH Product.
4.) Numbers in red represent simualtion values for IBC Side Draw Product.
5.) Numbers in purple represent simualtion values for IBC Btms Product.

Here the largest discrepancy is seen to occur for the composition of isobutylene in the bottoms flow, where the simulation predicts about 81 wt. % isobutylene against only 68 wt. % isobutylene for the lab sample of the column bottoms. Concentrations for N-Butane, 1-Butene, Cis-2-Butene & Trans-2-Butene and 1, 3 Butylene are correspondingly lower than actual lab values as well.

Overall it was decided that the fit between the simulation and actual operation was reasonably accurate to justify proceeding to evaluate heat pump options for the IBC.

After validating the IBC simulation model with historical operating data, the next step was to ramp up the feed rate to match the design operating feed rate as listed in the Phoenix project Material & Energy balance as developed by JACOBS Engineering. Note that this simulation would serve as the "base case" or reference model. Note also that the feed composition to the IBC in the base case model is based on actual operating data.

Simulation models were developed for each of the three heat pump options for the IBC using the same feed rate and composition as in the base case while being careful to ensure that all product compositions and flows were the same as in the base case simulation. In certain cases due to the change in tower operating pressure with a particular heat pump option product temperatures and pressures will not match with the base case simulation. In all three simulation cases as well as the base case the total pressure drop across the IBC was assumed to be 15 psi, that is the same as for the Phoenix Material & Energy balance as developed by JACOBS Engineering.

Each heat pump simulation was ran across a fairly wide range of tower operating pressure in order to at least bracket where the optimum operating pressure was likely to be. In this case the optimum tower operating pressure results in the lowest utility consumption that is compressor horsepower as well as minimum trim condenser duty.

In the case of the overhead vapor recompression and reboiler liquid flashing heat pump option the optimum tower operating pressure was found to occur for a tower top pressure near 50 psig.

In the case of heat integrated distillation columns the optimum operating pressure for the rectifying section was found to be essentially the same as the current operating top pressure (i.e. 76.4 psig) whiles the optimum operating pressure for the stripping section was found to occur at approximately 30.6 psig.

Also for the purpose of comparison the heat integrated distillation heat pump simulation was also ran with the IBC rectifying section operating with a top pressure of 76.4 psig (i.e. normal operating pressure) and the IBC stripping section operating with a top pressure near 50 psig.

Finally in order to allow for accurate screening of the three alternative heat pump options tower operating pressure was adjusted so that the compressor power was the same for all three options. Next the minimum new heat transfer surface for each option was calculated based on the duty for each respective exchanger and the Minimum Temperature Difference (MTD).

The results are listed below in Table IV:

Table IV: IBC Comparison of Heat Pump Options

Option No.	Desc.	Comp. Power[1] (Hp)	Reb. / Cond. Duty (MM Btu/hr)	Trim Cond. Duty (MM Btu/hr)	Reb. / Cond. MTD (° F)	Trim Cond. MTD (° F)	Reb. / Cond. Surface[2] (sq. ft)	Trim Cond. Surface[3] (sq. ft)	Total New Heat Exch. Area (sq. ft)	Comments[4,5]
I	1D-506 OH Vapor Recompression	5511	104.82	-13.59	22.3	20.05	104505	11757	116262	1D-506 Top Press. = 50.3 psig 1D-507 Btms Press = 65.3 psig Trim Cond. Uses CT water.
II	1D-507 OH Vapor Recompression	2952	88.94	N/A	3.4	N/A	>800759	N/A	>800759	1D-506 Top Press. = 76.4 psig 1D-507 Top Press. = 50.3 psig Trim Cond. Uses CT water.
III	1D-507 OH Vapor Recompression	5517	82.18	-13.69	22.95	22.5	77219	11055	88274	1D-506 Top Press. = 76.4 psig 1D-507 Top Press. = 30.55 psig Trim Cond. Uses CT water.
IV	1D-507 Reboiler Liquid Flashing	5517	93.3	-13.77	20.7	21.9	104471	11791	116262	1D-506 Top Press. = 50.3 psig 1D-507 Btms Press = 65.3 psig Trim Cond. Requires chilled water

Notes:

1.) Assume compressor polytropic efficiency = 75 %, Mechanical Efficiency = 95 %
2.) Combined Reboiler / Condenser heat exchange surface based on minimum req'd.
3.) Trim Condenser heat exchange surface based on minmimum req'd.
4.) All scenarios assume a Delta P across IBC of 15 psi.
5.) IBC Btms pressure = 65.3 psig for Option No. 2

As can be seen from Table IV the heat integrated distillation column (i.e. Option II) with the IBC rectifying section operating at 76.4 psig and the IBC stripping section operating near 50.3 psig resulted in the lowest compressor horsepower input. However, the heat exchange surface required to implement this option is prohibitively large. The reason for the exorbitant amount of heat exchange surface is due the very small mean temperature difference.

It can also be seen from Table IV that the reboiler liquid flashing heat pump (i.e. Option IV) option results in the same amount of new heat exchange surface as the vapor recompression heat pump (i.e. Option I) despite the fact that the duty of the reboiler-condenser unit for this option is around 9 % lower than for the vapor recompression heat pump option. The reason for this apparent discrepancy can be traced to the lower mean temperature difference for this option when compared to the vapor recompression heat pump option.

Also note that in order for the trim condenser associated with the reboiler liquid flashing option to have the same surface as the trim condenser for the overhead vapor recompression option chilled water at 60 ° F is required.

An inspection of Table IV reveals that the most cost effective option is the heat integrated distillation option with the IBC rectifying section operating at a top pressure of 76.4 psig and the IBC stripping section operating with a top pressure of 30.55 psig (i.e. Option III).

Note that for the purposes of comparison minimal pressure drop was assumed when calculating the minimum new amount of heat exchange surface required.

TEMA data sheets have been developed for both the combined reboiler and trim condenser units associated with the most cost effective heat pump option and have been optimized to reduce the amount of heat transfer surface by taking advantage of the maximum allowable pressure drop. These units were sized with a 10 % margin with respect to hot and cold liquid stream flow, duty and heat transfer surface.

Operating cost savings were estimated based on the long term natural gas price of $ 3.70 / MM Btu. Using this factor and the amount of heat required to raise steam at 750 # from saturated water at 230 ° F, the cost of steam can be estimated via DOE best practices. DOE best practices set the price of steam at 1.3 times the cost of natural gas required to supply the required heat. Using this approach the cost of steam was estimated at $ 5.44 /1000 lbs.

JACOBS Engineering estimated IBC steam consumption at 104.762 MPPH at the design operating rate. Consequently, annual steam cost associated with the IBC in its existing configuration was estimated at $ 4,992,118 assuming 8760 operating hours in a year using this steam consumption.

Operating cost for electric power required to operate the compressor was estimated based on a purchase price of $ 30.00 / MW*hr. Compressor power at design rates is 5850 Hp. Consequently, annual power cost for the heat pump option was estimated at $ 1,146,424.27 assuming 8760 operating hours in a year.

Utility savings associated with the heat pump option were estimated based on the difference between steam cost for the existing configuration and power cost associated with the heat pump option. Utility savings were estimated to be **$ 3,845,693.80** yearly. In the economic analysis a 2.4 % annual escalation on utility savings was applied.

The capital cost estimate for the IBC heat pump was prepared by the TPC Capital Projects Group. The basic approach used in preparing the cost estimate was to apply factors to the total cost of equipment in order to arrive at TIC cost. These factors were based on similar components from previously executed TPC project and late-stage estimates from EPC firms (i.e. JACOBS). The cost estimate includes about $ 750,000 of civil / structural work for foundations to support the new reboiler-condenser units. The Total Installed Cost was estimated at **$ 21.8 MM**.

After receiving the TIC estimate simple project economics were calculated based on the estimated annual cost savings and TIC. The Internal Rate-of-Return (IRR) was estimated at **11.0 %**, while project payback was estimated at **5.6 yrs**. Finally, Net Present Value was estimated at **$ 834,217**.

It should be stressed that while this project has a relatively small net present value, that does not mean that this project (as a single component of a larger group of projects) could not be more cost effective than spending $ 50 MM to replace an aging boiler.

Note that finding the best optimum steam generating capacity for the HNO over the next 5 years, (of which replacing aging boiler capacity) is a major objective of the HNO Steam Systems study.

CONCLUSIONS:

1.) A heat integrated distillation column heat pump is the most cost effective heat pump option for the IBC, as it requires the least amount of new heat exchange surface for a given compressor power.
2.) The optimum heat pump configuration is always a trade-off between compressor power and the amount of new heat exchange surface required.
3.) Both duty and Mean Temperature Difference (MTD) are important factors affecting heat exchange surface required.
4.) While the TIC for a heat pump on the IBC is high, it may still be a valid option when considered as one component of an overall program to look at the best option for replacing the older boiler.
5.) The optimum top pressure for the IBC stripping section in a heat integrated distillation column is near 30.6 psig.
6.) The optimum top pressure for the IBC rectifying section in a heat integrated distillation column is near 76.4 psig.
7.) Compressor horsepower at design rates for a heat pump on the IBC is approximately 5850 Hp.

RECOMMENDATIONS:

1.) An IBC heat pump option should be considered as part of a group of energy portfolio projects w / reduced economic hurdle criteria.
2.) Heat pump option for the IBC should be considered among a group of options that could reduce the size of a new replacement, or eliminate it altogether.

349258 Modeling of Phase Transitions during Supercritical Carbon Dioxide Migrating in a Leaky Pathway

Abstract:

Carbon dioxide sequestration in geological formations has emerged as one of the key technologies for reducing greenhouse gas emissions. Nevertheless carbon dioxide leakage from storage sites will lead to disastrous consequences for human beings. Owing to compressibility and buoyancy effects, carbon dioxide would possibly migrate upwards from these storage reservoirs to the atmosphere along potential leakage pathways, which are probably abandoned wells or geological faults. During leakage carbon dioxide transforms from the supercritical state to normal states under certain conditions that are influenced by the pressure and temperature of the surrounding formation. In the process of the phase transition, the dramatic change of properties of carbon dioxide can exert strong impacts over the carbon dioxide leakage process. Hence predicting phase change behaviors during carbon dioxide leakage along leaky pathways is of significant importance in the evaluation of potential risks of carbon dioxide storage in subsurface formations.

In this paper we established a two-dimensional mathematic model to describe flow and heat transfer of carbon dioxide migrating along a leaky pathway. The major objective is to investigate phase change behaviors and relative impact factors during carbon dioxide leakage through a vertical porous channel. In the mathematic model, on the basis of the features of carbon dioxide saturation line, we proposed a piecewise function to identify whether phase change occurs or not in the course of supercritical carbon dioxide flow in the porous channel. In the mass balance equation, a correction to the source term is adopted to describe the distribution of the phases. In the energy balance equation, the equivalent heat capacity method is employed to describe the energy change during phase change, in which the density, viscosity, thermal conductivity and other properties of carbon dioxide are expressed as properties of mixture of gas and liquid phases by piecewise functions.

The proposed model was solved by using a finite element method-based commercial software package COMSOL Multiphysics. The results indicate that phase change of carbon dioxide is mainly dependent upon the coupling effects of heat exchange between the surroundings and the leaky pathway and the temperature variation caused by carbon dioxide expansion process. The heat exchange depends on interactions between carbon dioxide flow and geological conditions, and the depressurization process of carbon dioxide determines the action of Joule-Thomson effect. Under different conditions of the formation pressure and temperature, the relevant properties of carbon dioxide would change dramatically due to phase change that carbon dioxide crosses over the saturation line. Consequently, accurately predicting the properties of carbon dioxide, such as density and viscosity, is the key to understand the behaviors of phase change process during carbon dioxide leakage. The model proposed in this paper can predict the phase change behaviors during carbon dioxide leakage along a leaky pathway, which will provide an analytical tool for the prediction of characteristics of carbon dioxide leakage and the assessment of potential leakage risks.

Application of Joukowsky Surge Pressure Equation to High Pressure Heat Exchanger Design

By
George Maliakkal & Fadi Wakim
Bechtel Corporation

Presented at AICHE Spring Meeting
New Orleans, LA
31 MARCH 2014

Abstract

Joukowsky equation is used to model the surge pressure peak due to tube rupture in shell and tube heat exchangers with vastly differing pressures between the high pressure (HP) gas side and low pressure (LP) liquid side.

This surge pressure, which can be several times the design pressure of LP side, travels with velocity of sound in the LP side liquid, for which conventional relief devices are not readily designed.

Exchanger parameters that determine the magnitude of this pressure surge are identified and analyzed.

The results can be applied to exchanger design to reduce the peak of the surge pressure to acceptable LP design pressures, potentially negating the need for quick acting relief devices.

WATER HAMMER PHENOMENA

JOUKOWSKY EQUATION FOR WATER HAMMER

- $\Delta p = \rho\, c\, \Delta v$
- Δp **pressure rise**
- ρ **density of liquid**
- **c sound velocity**
- Δv **change in velocity**
- **c =** $\sqrt{K/\rho}$
- **K bulk modulus, (same unit as pressure)**
- $\Delta p = Ps - P = \rho\sqrt{K/\rho}\,\Delta v = \sqrt{K\rho}\,\Delta v$
- **Ps surge pressure in line**
- **P initial pressure in line**

EXAMPLE: SURGE PRESSURE WHEN VALVE IS SUDDENLY CLOSED IN A WATER LINE

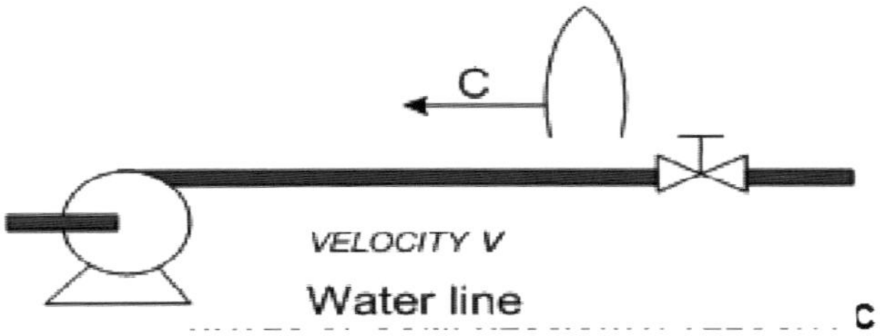

v =10 ft/s; Δv =10 ft/s for sudden closure;
K= 337067 psi; ρ = 60.11 lb/ft3

$$\Delta P = \frac{\sqrt{\overset{K}{337{,}067} * 144 * 32.17 * \overset{\rho}{60.11}} * \overset{\Delta V}{10}}{144 * 32.17} = 661 \text{ PSI}$$

BULK MODULUS K

K = dP/(dV/V) OR ***K = dP/(dρ/ρ)***

For water K~ 300,000 psi For HC liquids K 50,000 psi to 200,000 psi

$K = \rho c^2$

ρ density of liquid,
c sound velocity

K is K(T)

For gases K= P*γ

Example: K and "compressibility" for water

when compressed 10,000 psi shrinks 3%.
when compressed 1000 psi shrinks 0.3%

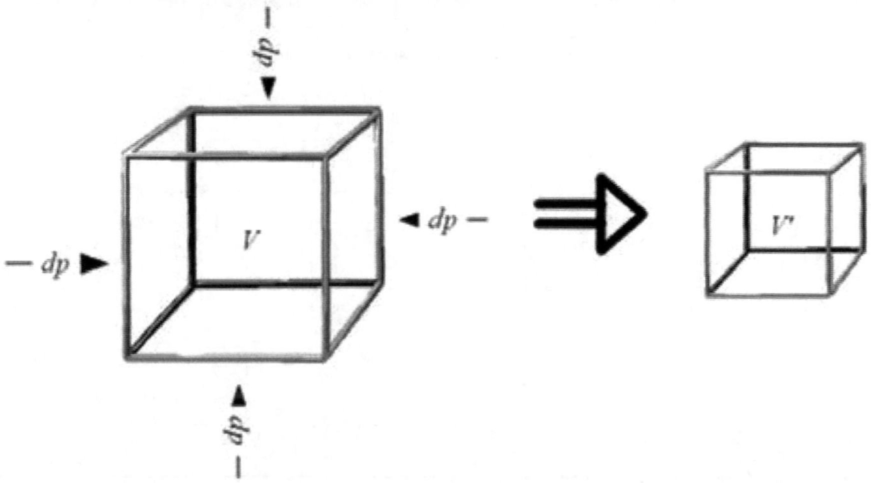

TUBE RUPTURE SCENARIO

Compare action of Water Hammer and Tube Rupture

Water Hammer: Stopping velocity induces pressure wave.
Tube Rupture: HP gas expansion causes change in velocity, which causes pressure surge.

Joukowsky Equation $\Delta p = \rho c \Delta V$ (1)

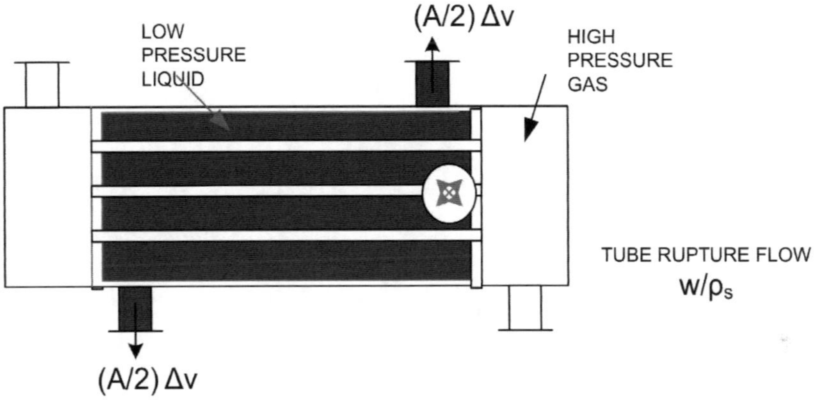

Continuity Equation $A \Delta v = w/\rho_s$ (2)

Gas Law Equation $P_s = \rho_s R T / M_w$ (3)

Surge Pressure Equation $P_s^2 - P_s P_L = (\rho c w R T)/(A M_w)$ (4)

In terms of Bulk Modulus (see Note) $P_s^2 - P_s P_L = (\mathrm{sqrt}(K\rho) w R T)/(A M_w)$ (5)

Note: K is effective bulk modulus of LP side. Ideally it is a composite of bulk modulus of the liquid medium and the enclosure (shell or tube material). Effect of bulk modulus of enclosure is omitted in this calculation. This omission results in a slightly higher (~ 2%) calculated surge pressure.

TUBE RUPTURE MODEL

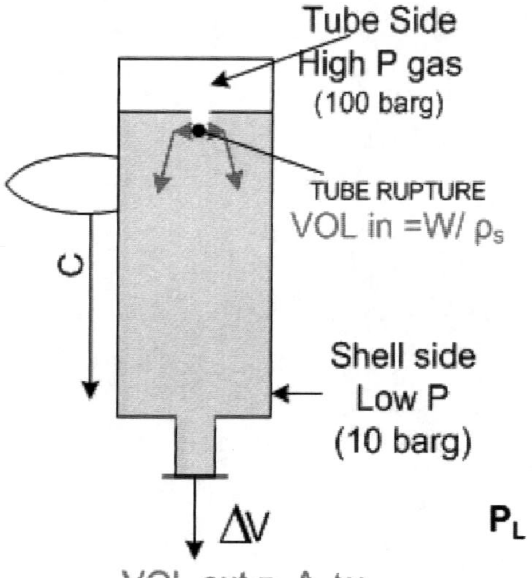

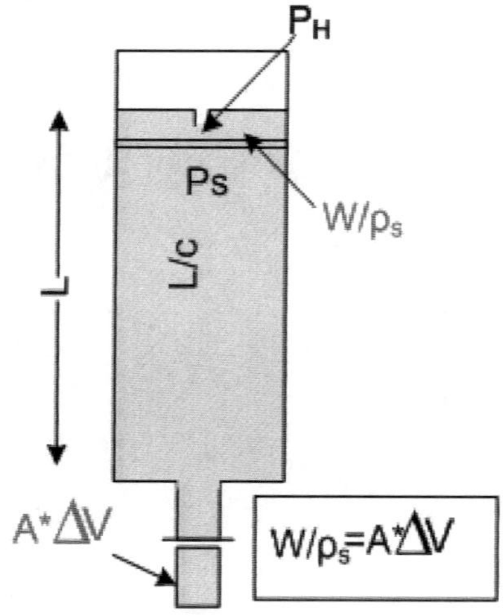

L/c=
20 feet/ 5000 ft/sec
=0.004 secs
=4 millisecond

- ρ_s **density of gas at** P_S
- **W tube rupture flow**
- P_S **surge Pressure =** $P_L + \Delta p$
- **A outlet nozzles area**
- Δv **change in velocity at nozzles**
- P_L **low P side operating pressure**
- ρ **liquid density**
- **K effective bulk modulus**
- **c velocity of sound in liquid**

Vol rate in (green shaded) = w/ρ_s
Vol rate out (blue block) = $A\,\Delta v$

TRIGGER FOR SURGE PRESSURE TRANSIENT ANALYSIS

- 1. The low-pressure stream is liquid
- 2. The high-pressure stream is gas, vapor, two-phase, or a liquid that vaporizes
- 3. High pressure side operating pressure is greater by 1000 psi (70 bar) over LP side pressure (verify against latest API 521)

Variables Influencing Surge Pressure

- $P_S - P_L = w/A\rho_s * \sqrt{K\rho}$

- "w" tube rupture flow α to (tube dia)^2
- "A" nozzle area α to (nozzle dia)^2
- "K" effective bulk modulus α 1 / $(1+f_g (K_s/(\gamma P_L-1)))$

K order of 20,000 bar; P_L order of 20 bar; for small volume fraction f_g of vapor in liquid (e.g. =0.01), K value can drop by as much as a factor of 10.

SURGE PRESSURE AS FUNCTION OF EXCHANGER TUBE SIZE

LP side	10 barg liquid
HP side	110 barg gas

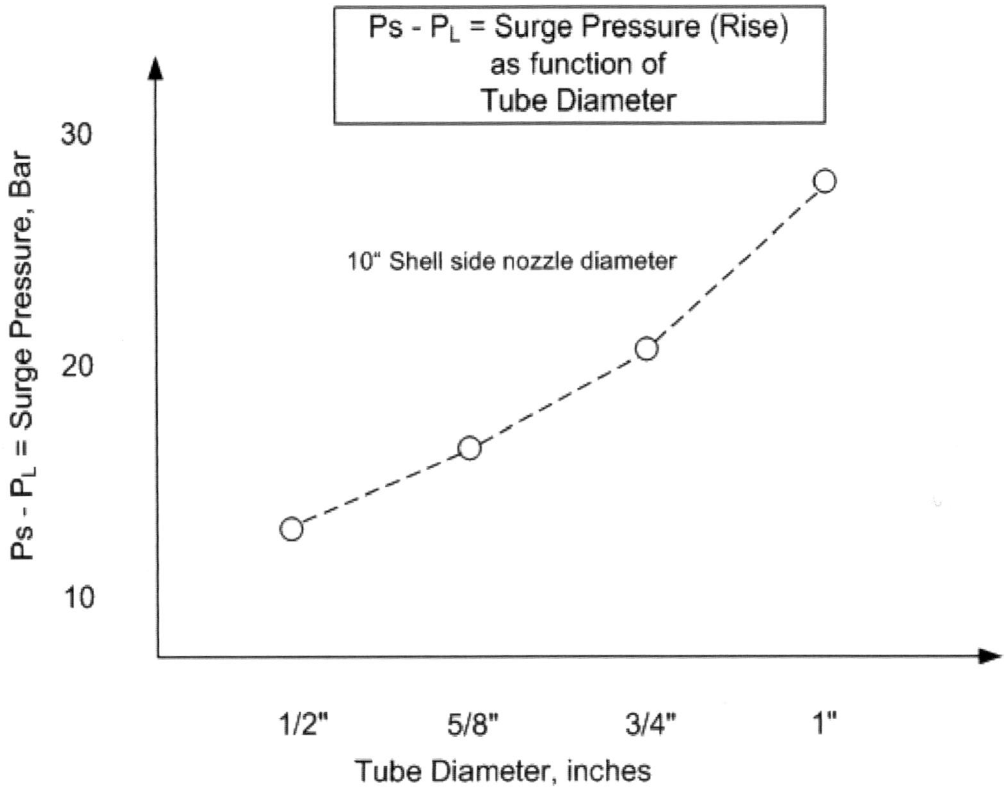

SURGE PRESSURE AS FUNCTION OF EXCHANGER SHELL-SIDE NOZZLE DIAMETER

LP side	10 barg liquid
HP side	110 barg gas

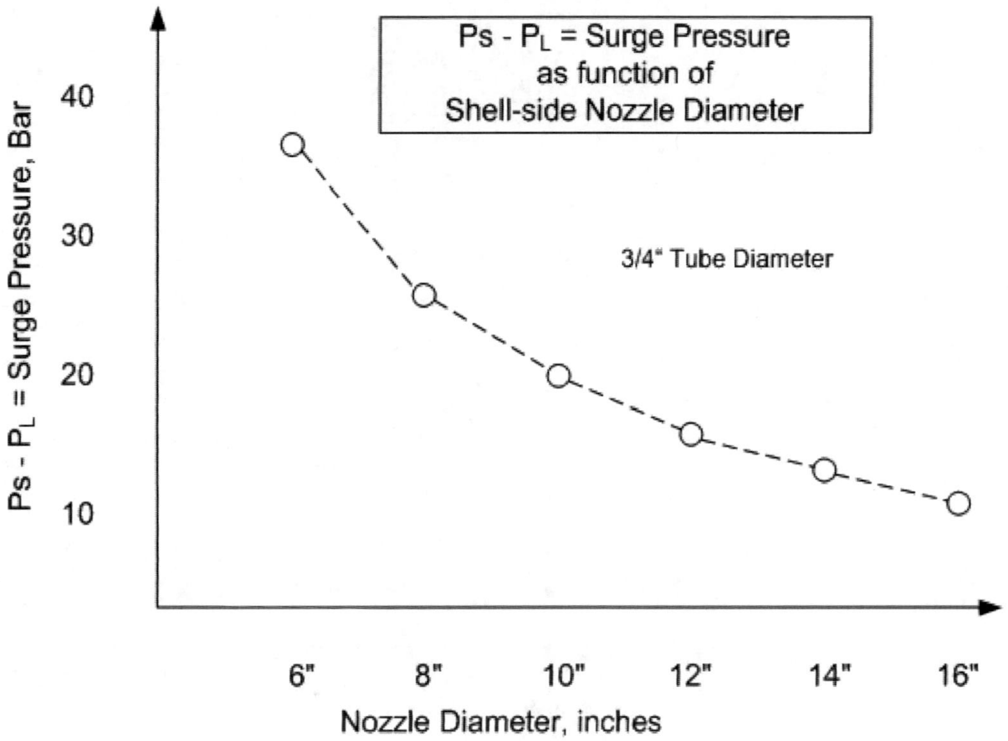

BULK MODULUS REDUCTION OF LP-SIDE LIQUID AS FUNCTION OF VAPOR FRACTION

LP side	10 barg liquid
HP side	110 barg gas

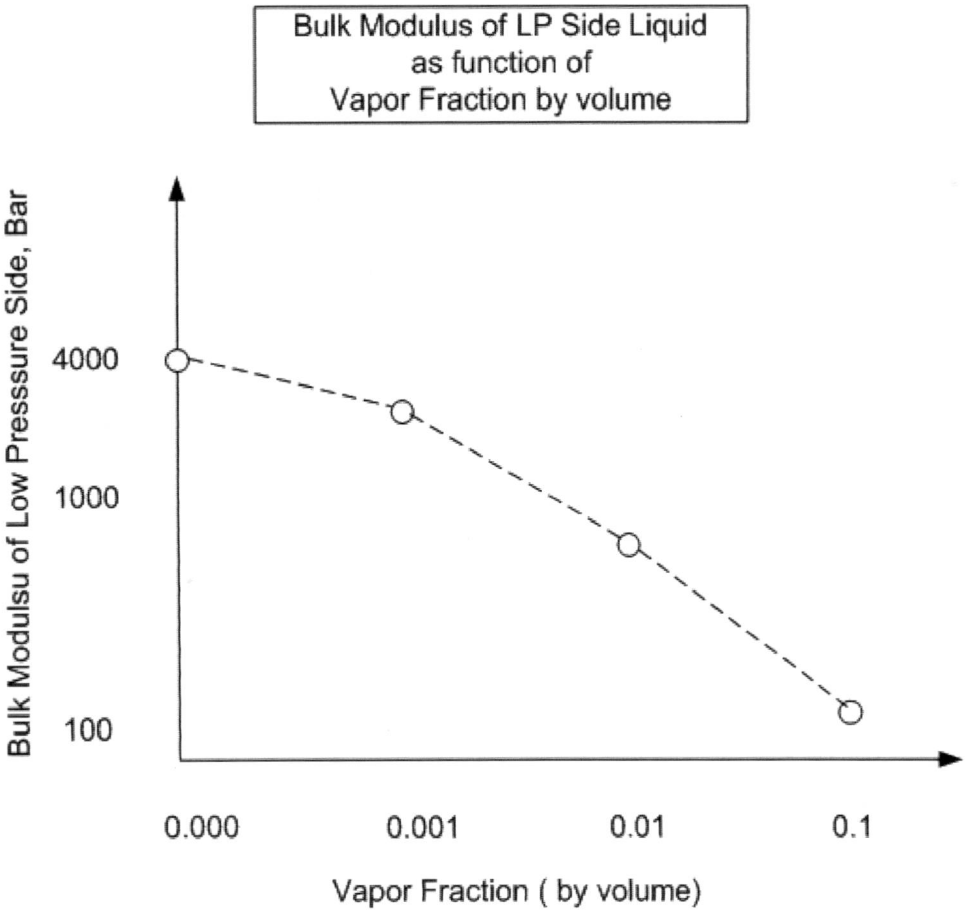

0.001 wt% vapor in this system was 0.01 vol%

SURGE PRESSURE AS FUNCTION OF VAPOR (GAS) FRACTION

LP side	10 barg liquid
HP side	110 barg gas

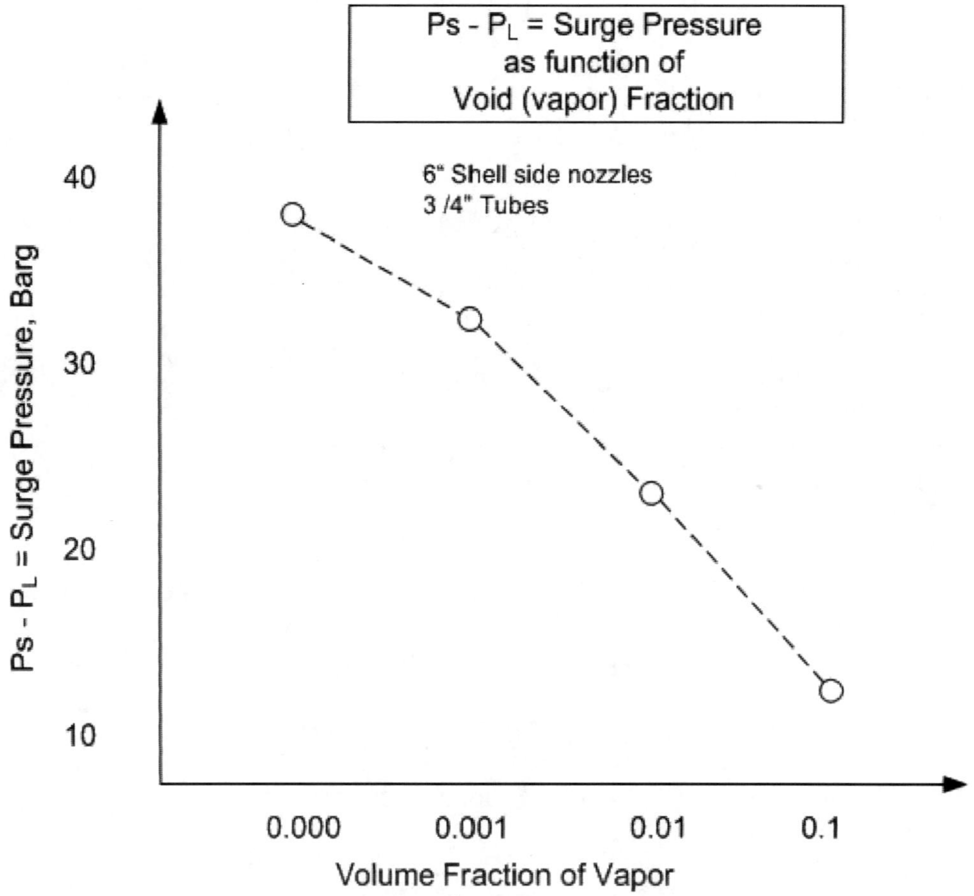

LP-SIDE
LIQUID AT BUBBLE POINT

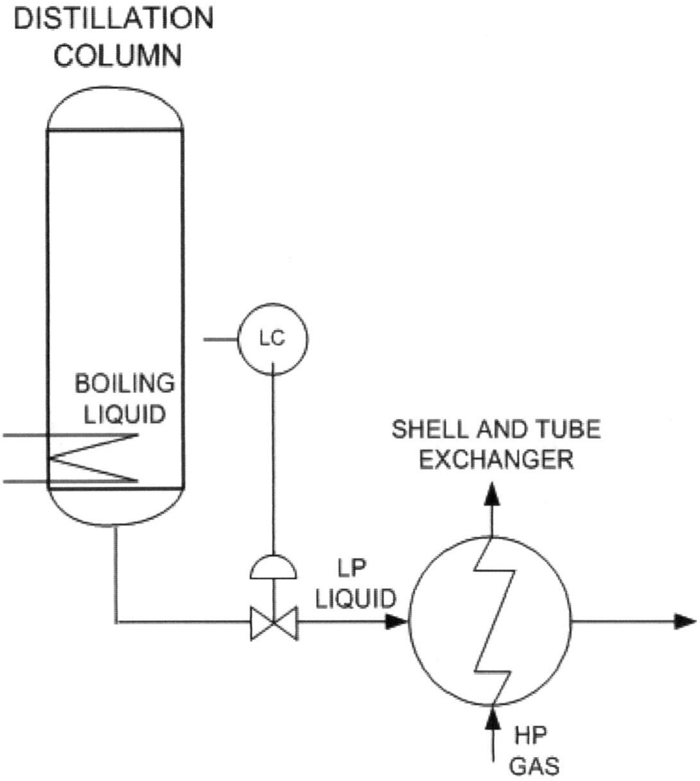

MINUTE AMOUNT OF VAPOR BUBBLES DUE TO PRESSURE DROP THROUGH CONTROL VALVE (ORDER OF 0.001 WT FRACTION) CAN RESULT IN BULK MODULUS REDUCTION BY FACTOR OF 10

LP-SIDE LIQUID
INJECTED WITH INERT OR FUEL GAS

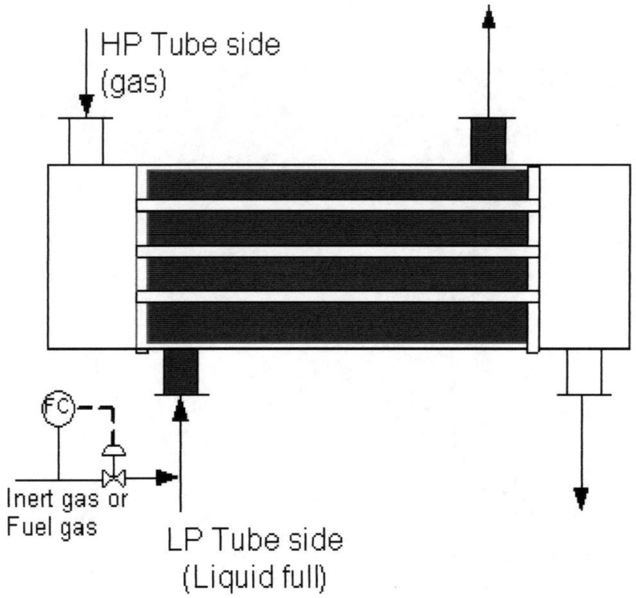

MINUTE AMOUNT OF INJECTED GAS (ORDER OF 0.001 WT FRACTION) CAN RESULT IN SIGNIFICANT REDUCTION IN SURGE PRESSURE; FOR EXAMPLE, 100,000 KG/HR LIQUID FLOW INJECTED BY 10 KG/HR NITROGEN CAN REDUCE SURGE PRESSURE FROM 36 BAR TO 20 BAR

INERT GAS FLOW RATE MAY BE MADE PROPORTIONAL TO TUBE SIDE LIQUID FLOW BY PROPER CONTROLLER ACTION

Reduce surge pressure by gas volume at outlet nozzle

LP side 10 barg liquid
HP side 110 barg gas

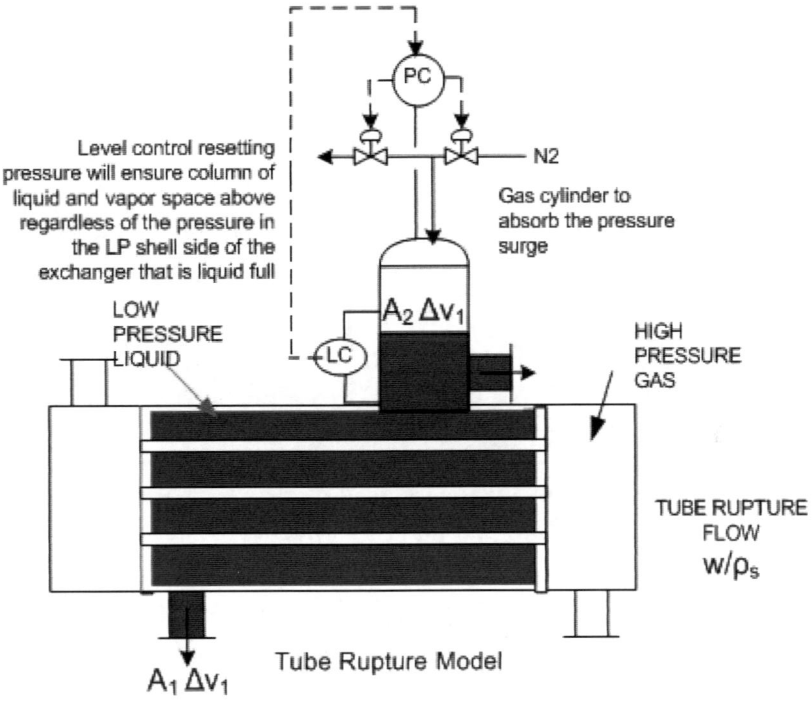

Tube Rupture Model

2 - 6 " nozzles gave surge ΔP of 13.9 bar
1-6" with 1-20" nozzle gave surge ΔP of 3.8 bar

Minimum level in the gas cylinder should be above the side (tubeside outlet) nozzle to prevent gas carryover into the tubeside liquid.

ASSUMPTION

- Joukowski water hammer equation can be adapted for surge pressure calculation in S/T exchanger tube rupture.

SUMMARY & CONCLUSIONS

- Pressure surge in systems with high pressure differentials (> 70 bar) between shell and tube design pressures are to be investigated for safety and mitigated per API 521

- Because the speed of pressure surge wave is at the velocity of sound of liquid medium, active mitigation of surge pressure is not readily accomplished by traditional relief devices.

- It is desirable to have passive methods to limit the pressure surge

- Passive methods are explored in this presentation in terms of exchanger geometry and exchanger fluid properties. These are tube size, shell nozzle size and vapor fraction in low pressure liquid.

- Reducing tube size results in reduction of surge pressure

- Increasing nozzle diameter results in reduction in surge pressure

- Decreasing bulk modulus of low pressure side liquid reduces surge pressure

- Bulk modulus of liquid side is reduced by presence of minute amounts of vapor (< 0.001 wt%) in flashing liquid or by injection of inert into LP liquid side.

- Combination of tube size, nozzle size and vapor content of LP side liquid can be used to minimize surge pressure

- Table below summarizes this potential for a system with HP side vapor, pressure of 110 barg and LP side liquid pressure of 10 barg.

						Surge Pressure above LP side operating pressure, bar		
Variable	LP liquid side Vapor (Vol frac)	Shell nozz.	tube	from	to	from	to	
Tube size	0	6"	-	¾" tube	½" tube	36	22	
Nozzle	0	-	¾"	6" nozz	10" nozz	36	20	
Nozz, Tube	0	10"	½"	-	-	-	-	12
Nozz, Tube, Vapor Frac.	0.01	10"	½"	-	-	-	-	7

References

[1] American Petroleum Institute, 2008, API 521 "Pressure-relieving and Depressuring Systems"

[2] The Institute of Petroleum, 2000, "Guidelines for the Design and Safe Operation of Shell and Tube Heat Exchangers to Withstand the Impact of Tube Failure"

[3] Energy Institute, 2011, "Bursting Disks for Shell and Tube Heat Exchanger Overpressure Protection", retrieved from http://www.energyinst.org/technical/safety/shell-and-tube-heat-exchangers, last accessed on March 14th 2014

[4] American Petroleum Institute, 2008, API 581 "Risk Based Inspection Technology"

[5] Tijsseling A.S. & Anderson A., 2006, "The Joukowsky Equation for Fluids and Solids", retrieved from http://www.win.tue.nl/analysis/reports/rana06-08.pdf, last accessed on March 14th 2014

[6] Wylie E.B. & Streeter V. L., 1993 by Prentice-Hall, Inc., "Fluid Transients in Systems"

Determination of Structural Composite Thermal Properties Using an Oscillating Boundary Temperature

William M. Chirdon, Ph.D.
wchirdon@louisiana.edu
Department of Chemical Engineering

March 31, 2014

UNIVERSITY OF LOUISIANA Lafayette

Motivation

Thermal properties are essential for thermal modeling

(garbage in = garbage out)

Thermal properties of structural composites are important for

- calculating heat transfer/loss through structures
- modeling thermal gradients in structures which cause cracking
- modeling thermal stresses which cause strength loss

Thermal cracking is a very important and expensive problem with $18-21 billion spent on concrete maintenance and repair in US annualy

(from "Vision 2020," Strategic Development Council (SDC), 6/16/06)

- solar heating (daily and seasonal cycles)
- curing (exothermic hydration reaction)
- fire

Oscillating Boundary Temperature (OBT) method may be applied to other materials

- biological tissues
- polymers and composites
- other reactive solid and semi-solid systems

Research for a reason.

UNIVERSITY OF LOUISIANA Lafayette

Motivation

Thermal properties in reactive systems are complex

- Transient (dependent on time or reaction coordinate)
- Temperature dependent
- Reaction rates depend on temperature

Different techniques lead to different thermal properties and trends

Source	Diffusivity Change
RILEM Commission, *Mater. Construct.* 1970	Constant
Brown, TD, et al. *Mater. Construct.* 1982	-15%
Hansen, PF, et al. *Proc. Rilem Int. Conf. on Concrete at Early Ages.* 1982	+12%
Marechal, J. *Essais Mes.* 1973.	Maximum @ 50hr

Research for a reason.

UNIVERSITY of LOUISIANA Lafayette

Issues with Steady-State Methods

- Requires large sample sized and long time to reach steady-state
- Prolonged heating can affect or damage specimens
- Difficult or impossible to track transient thermal properties
- Yields steady-state thermal properties which may differ from transient properties

Issues with Other Transient Methods

- Some requires up to 8 hours to reach thermal equilibrium between runs
- Often requires a large heat pulse
- No mechanism for temperature control
- Yields thermal properties which may differ from steady-state properties

Research for a reason.

UNIVERSITY of LOUISIANA Lafayette

Problem of Heat Generation in Transient Hot-Wire Methods

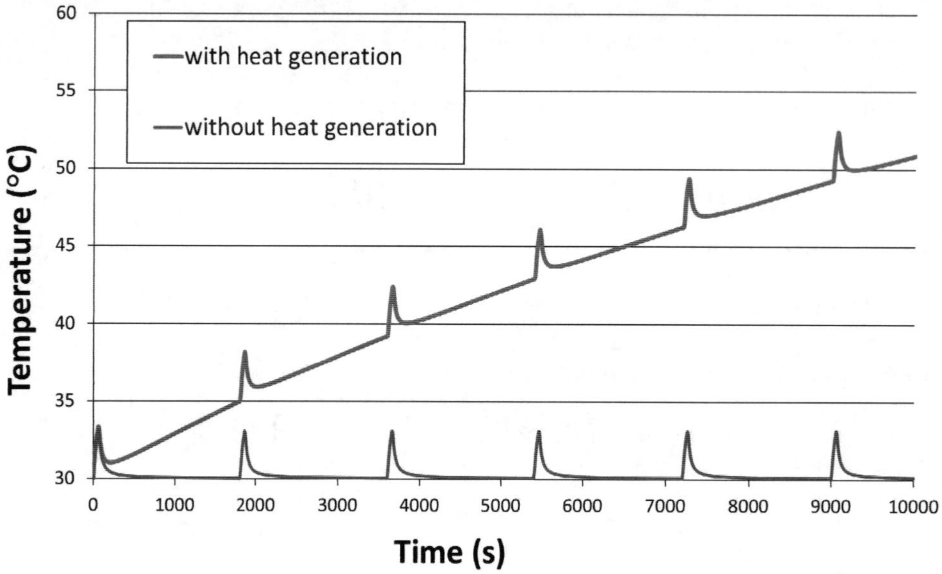

Research for a reason.

UNIVERSITY OF LOUISIANA Lafayette

Problem of Heat Generation in Transient Hot-Wire Methods

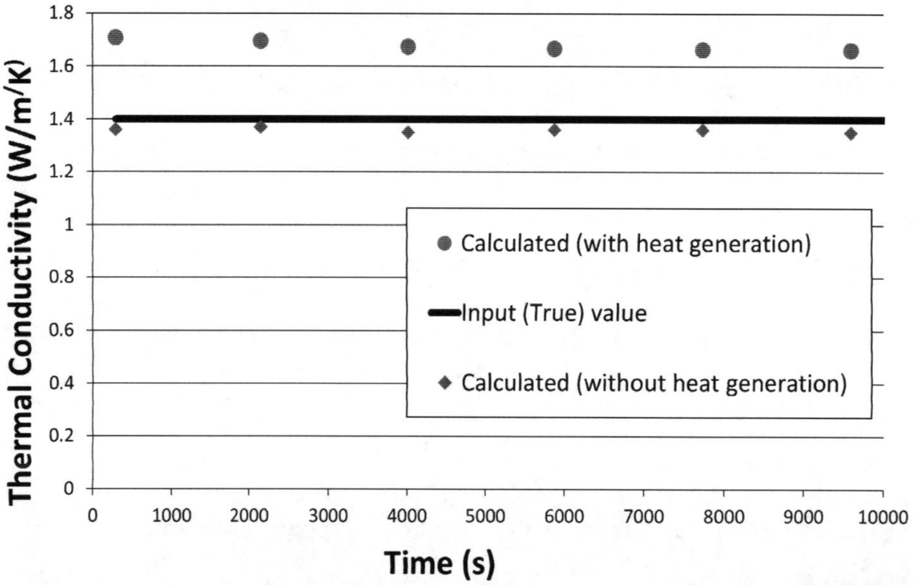

Research for a reason.

UNIVERSITY OF LOUISIANA Lafayette

Oscillating Boundary Temperature (OBT) Method

- Uses a small amplitude temperature oscillation and internal temperatures to get:
  - Thermal Diffusivity
  - Thermal Conductivity
  - Volumetric Heat Capacity
  - Intrinsic Heat Generation
- Controls the <u>temperature</u> of the boundary, not the heat flux/input
- Provides temperature control in the sample
- Works by decoupling oscillatory heat transfer from heat generation

Oscillating Boundary Temperature (OBT) Method

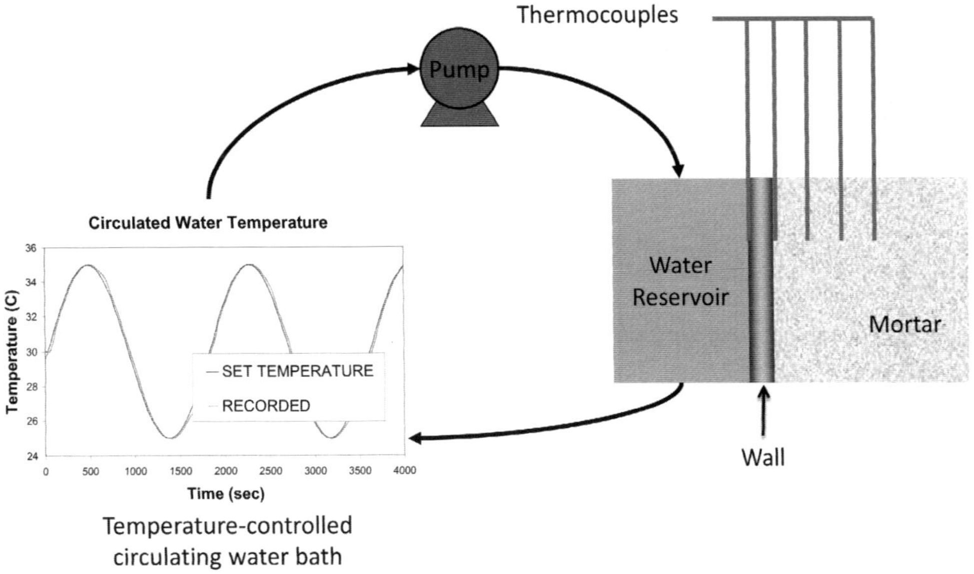

Decoupling Heat Generation from Thermal Oscillation

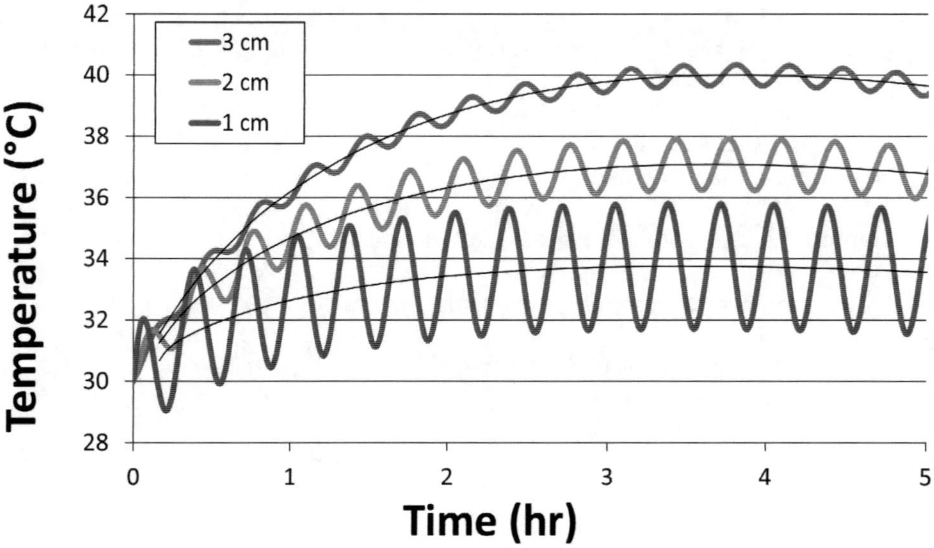

Research for a reason.

UNIVERSITY OF LOUISIANA Lafayette

Decoupling Heat Generation from Thermal Oscillation

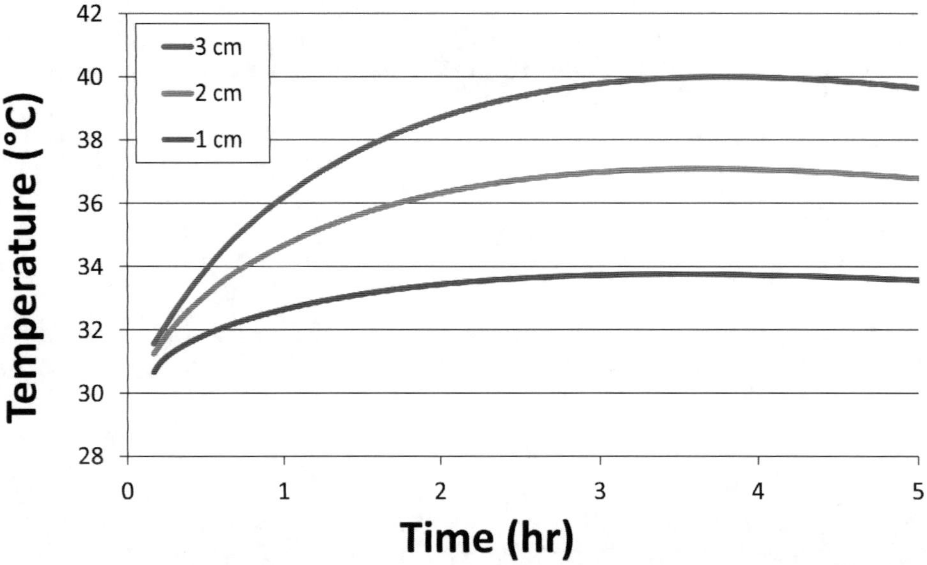

Research for a reason.

UNIVERSITY OF LOUISIANA Lafayette

Decoupling Heat Generation from Thermal Oscillation

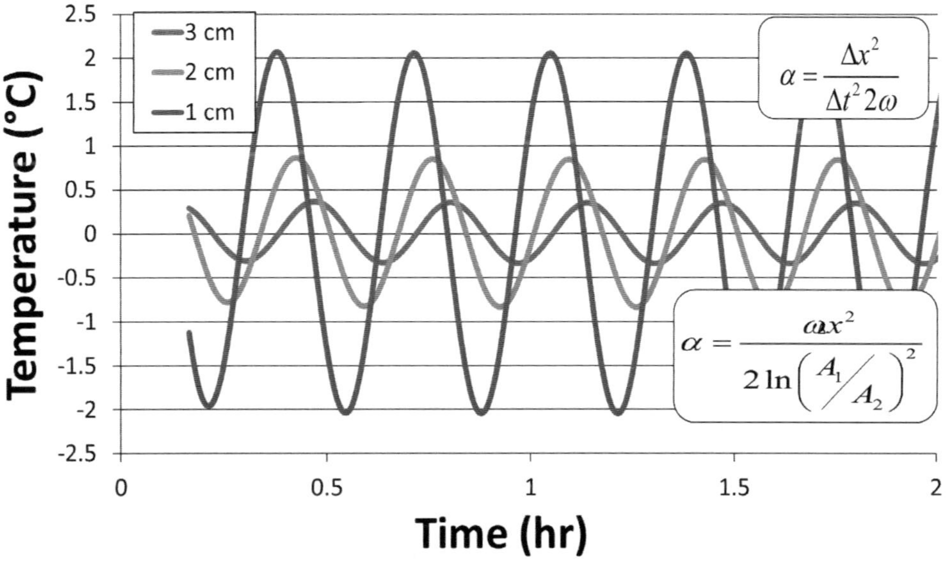

Research for a reason.

Thermal Diffusivity Determination

	Literature Value x 10^7 *	Experimental Value x 10^7	Standard Deviation x 10^7
Wet sand	7.43	7.46	0.16
Moist sand	8.40	8.39	0.41
Dry sand	2.33	1.70	0.04

*W.R.V.Wijk, Physics of plant environment. 1963, Amsterdam, New York: North-Holland Pub. Co.; Interscience Publishers.

Research for a reason.

Thermal Diffusivity:

Simulation Results

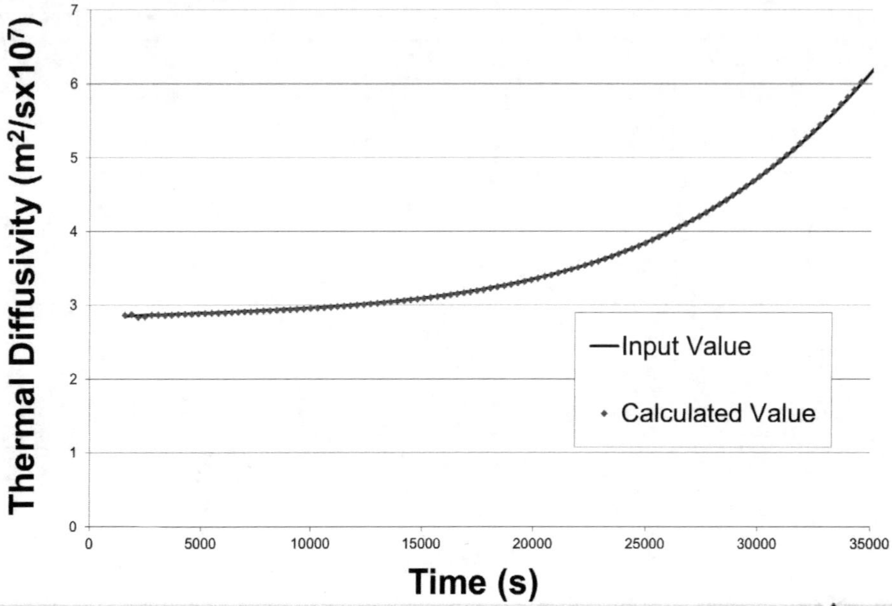

Thermal Conductivity Determination

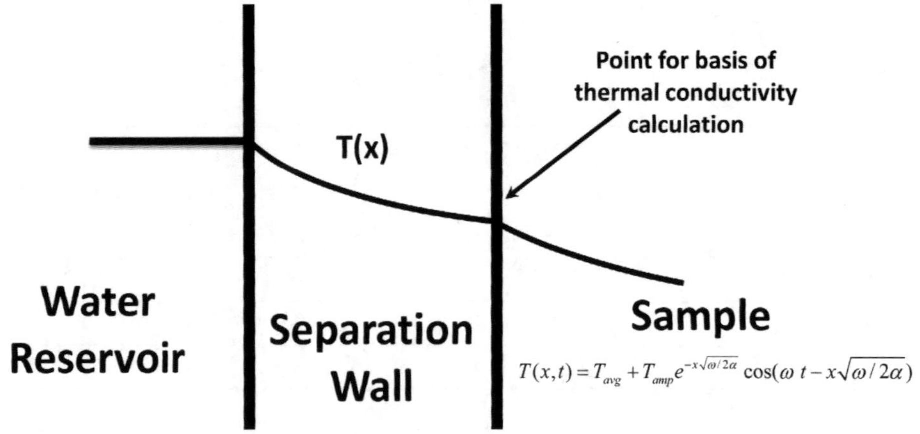

$$\frac{q}{A} = -k_{\text{wall}}\left(\left.\frac{\partial T}{\partial x}\right|_{\text{LHS, interface}}\right) = -k_{\text{sample}}\left(\left.\frac{\partial T}{\partial x}\right|_{\text{RHS, interface}}\right)$$

$\left.\frac{\partial T}{\partial x}\right|_{\text{RHS, interface}} = T_{\text{amp}}\sqrt{\omega / 2\alpha}$ at max T(t)

Research for a reason.

UNIVERSITY of LOUISIANA Lafayette

Thermal Conductivity:

Simulation Results

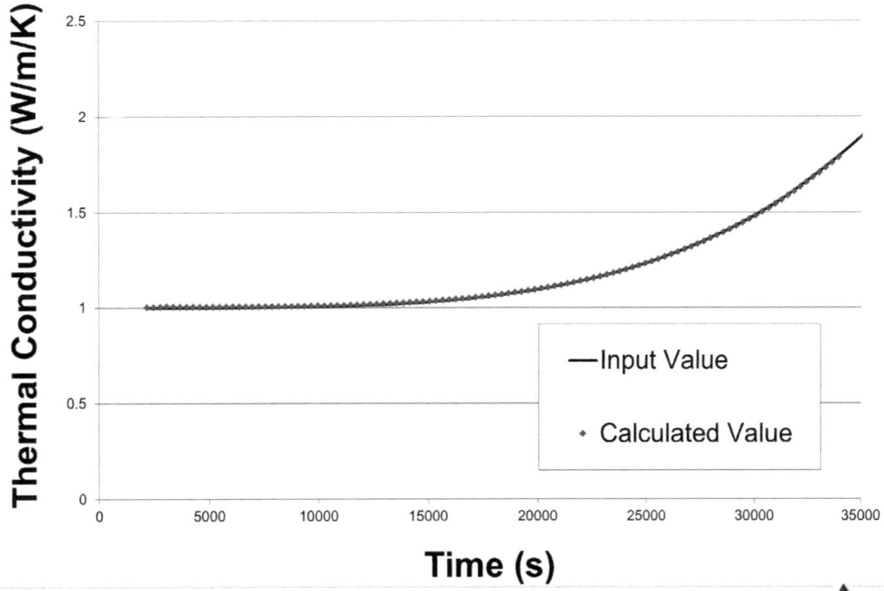

Volumetric Heat Capacity

- Trivial calculation = k / α
- Requires density to convert to mass-basis

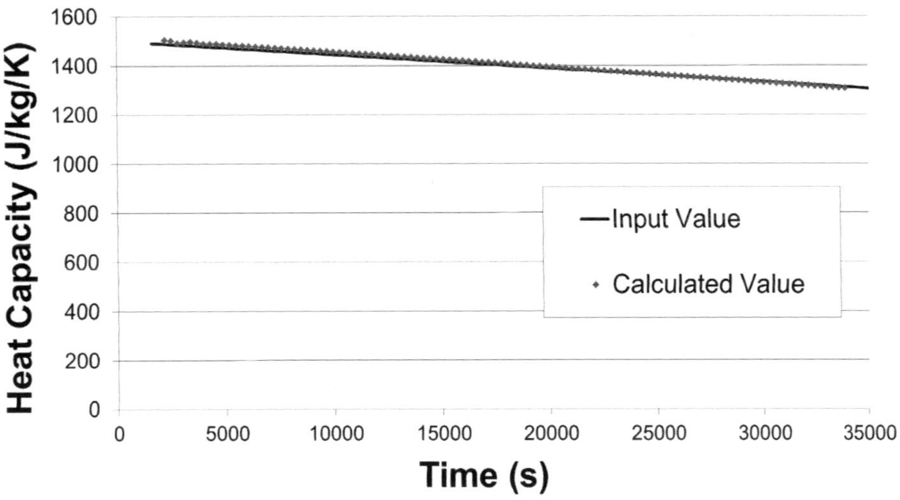

Intrinsic Heat Generation

Calculated from temperature dependence on *x* and *t* from 1-D conduction equation:

$$\frac{\partial T}{\partial t}=\alpha(t)\frac{\partial^2 T}{\partial x^2}+g(t),\quad x\in(0,\infty),t\in[0,t_{\max}]$$

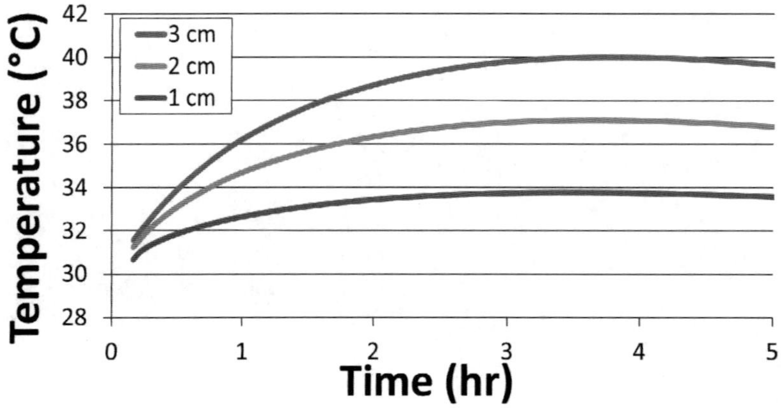

Research for a reason.

Intrinsic Heat Generation

Computer Models

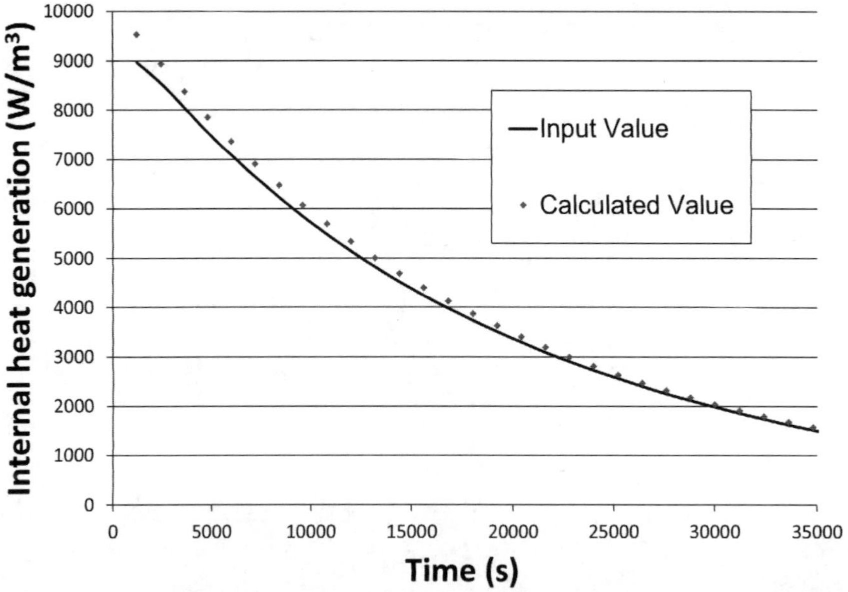

Research for a reason.

Intrinsic Heat Generation

Experimental Results – Mortar Specimen

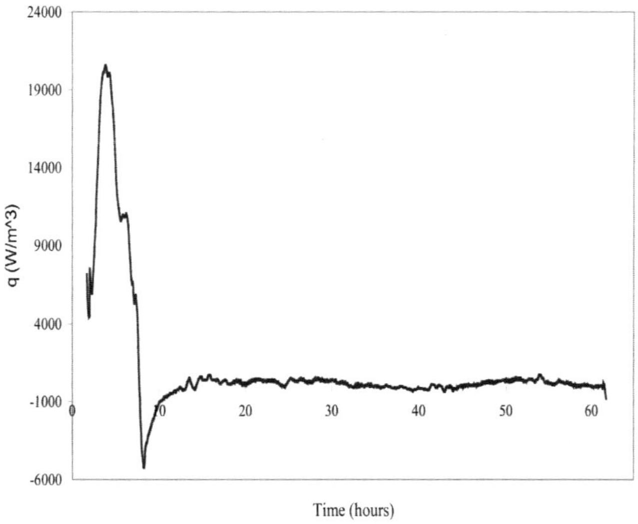

Intrinsic Heat Generation

Experimental Results

Mortar Specimen with 0.1°C error in first thermocouple

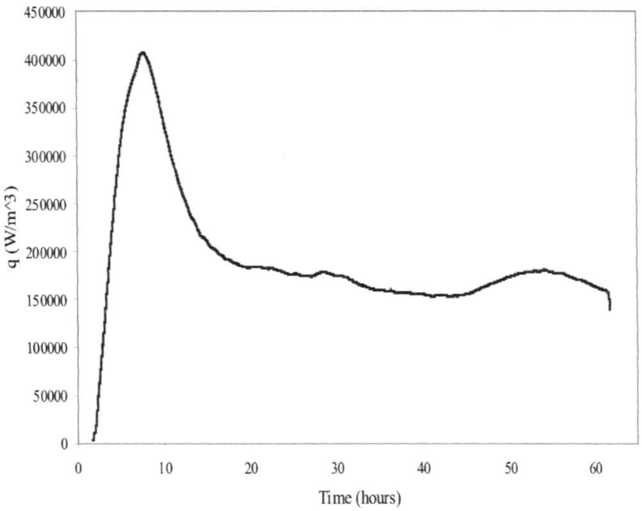

Research for a reason.

Intrinsic Heat Generation

Experimental Results

$$\frac{\partial T}{\partial t} = \alpha(t)\frac{\partial^2 T}{\partial x^2} + g(t), \quad x \in (0,\infty), t \in [0, t_{\max}]$$

- Heat generation effects can be easily decoupled
- Works well in simulations without experimental error
- The proposed analysis was impractical due to sensitivity to errors
- Different analytical technique required
  - Inverse problem solving techniques
  - Fit $g(t)$ to $T(x,t)$

Advantages of OBT Method

- Can control the temperature over a narrow range, important for
  - Materials with highly temperature-dependent properties
  - Tracking material properties in reactive systems
  - Biological systems
- Provides some temperature control
  - Experiments are near-isothermal
  - Most transient methods are near-adiabatic
- May be useful for specimens that cannot be heated significantly
  - Biological tissues
  - Reactive solids/semi-solids
- Result in many data points for statistical analysis
- Can track transient properties
- Allow for separation of heat generation and heat transfer effects
- Allows for simultaneous determination of:
  - Thermal diffusivity
  - Thermal conductivity
  - Volumetric heat capacity
  - Intrinsic heat generation rate
- Allows for study of time-scale

Time-Scale Dependence:

Portland Cement Mortar – Thermal Diffusivity

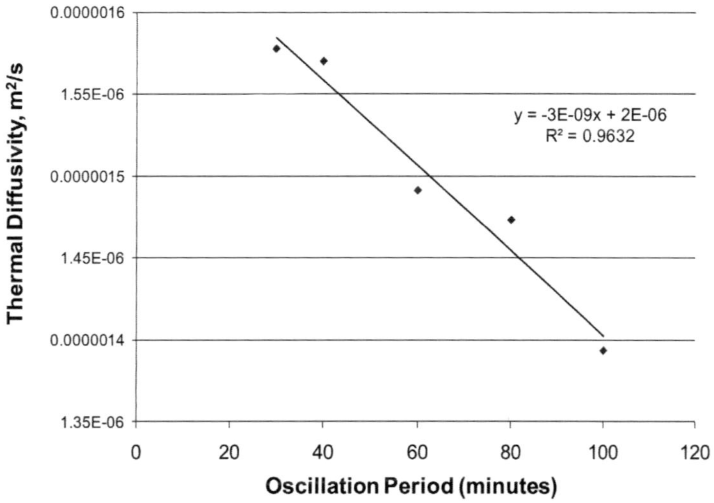

LTRC Project: 09-1TIRE

Research for a reason.

Disadvantages of OBT Method

- Standard equipment does not exist, you will need to build your own
- Results in an ill-posed problem if the system parameters are poorly chosen, especially:
  - Separation wall thickness
  - Thermal amplitude and oscillation frequency
  - Thermocouple choice and placement
- Hence, modeling for optimal experimental parameters is recommended
- Individual data points may have more error

Research for a reason.

Design Considerations

- Highly recommend optimizing experiment using thermal models first
- Oscillation should occur on a faster time scale than thermal property changes
- Higher thermal amplitudes may yield better signal-to-noise ratio, but:
  - May require more powerful heating and cooling techniques
  - Will acquire the thermal properties over a wider range
- Thermocouples far from the wall have amplitudes difficult to measure
- The relative error in distance increases as the thermocouples become closer
- Thermocouples should be placed across a representative sample of composite
- Thermocouple distances can be determined after the experiment by sectioning
- Wall thickness and material should be carefully chosen to achieve an amplitude drop across the wall that can be measured accurately
- Alternatively, the wall calculation may be replaced by a heat-flux meter
- Design to minimize thermal contact resistance
  - Wall-sample contact resistance will cause large errors in conductivity results
  - Thermocouple-sample contact resistance may also cause minor errors
  - Thermal contact may be difficult to maintain when shrinkage occurs
  - Problem for nearly any thermal property detection method (not unique to OBT)
  - Solutions highly dependent on application and material

Research for a reason.

UNIVERSITY of LOUISIANA Lafayette

Thank you for your attention

wchirdon@louisiana.edu

UNIVERSITY of LOUISIANA Lafayette

PROSYS™ Maximize Your Plant.

Operator Errors and What Can be Done to Minimize

Global Congress on Process Safety

March 31, 2014

10TH GLOBAL CONGRESS ON PROCESS SAFETY

PROSYS™ Maximize Your Plant.™

10TH GLOBAL CONGRESS ON PROCESS SAFETY

Presenter

Tom Nolan

- **Graduated from Ohio University with B.S. in Chemical Engineering**
- **24 years experience in chemical process industry in a variety of roles**
- **MAIC and DFSS Six Sigma Black Belt**

PROSYS Maximize Your Plant™

The Cost of Errors

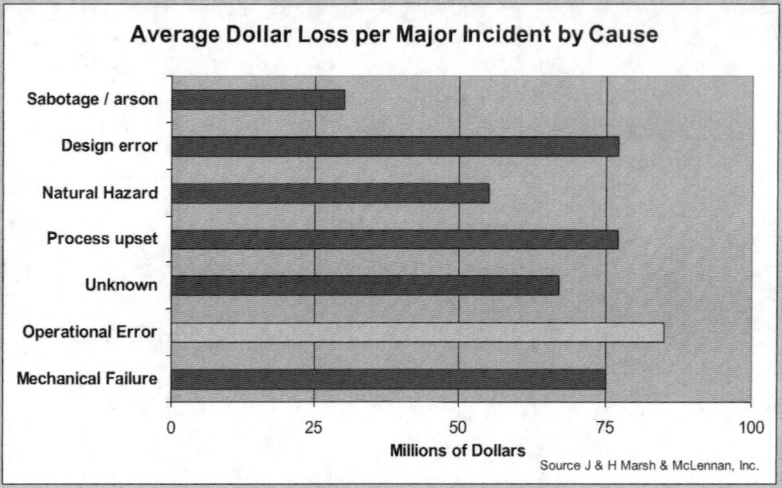

PROSYS Maximize Your Plant™

The Cost of Operator Errors

- **ASM estimates total loss due to operator error is $8B per year**
- **Chemical Safety Topical Committee - average of one chemical incident per day - cost of over $2 million per incident to <u>comply with requirements</u>**
- **Errors cause 42% of unscheduled shutdowns**
- **70% of process incidents occur during start-up or shutdown**

PROSYS Maximize Your Plant™

10th Global Congress on Process Safety

Reasons for Errors

- Lack of Skill
- Lack of Knowledge
- Carelessness
- System Design
  - Operator set up to make errors by inappropriate design – built in errors
  - Errors that are predictable are preventable by better design

PROSYS Maximize Your Plant™

10th Global Congress on Process Safety

Focus of this Discussion

- System Design
  - **Alarm Management**
  - **Operator Graphics**
  - Operator set up to make errors by inappropriate design – built in errors
  - Errors that are predictable are preventable by better design

PROSYS Maximize Your Plant.™

10th GLOBAL CONGRESS on PROCESS SAFETY

Alarm Floods

In a number of industrial incidents, alarm floods were identified as a significant contributing cause to the incident...

As found by EEMUA in 1999 and CSB

Alarm Flood defined by ISA 18.2 as -
"10 or more annunciated alarms in any 10 minute period per operator"

PROSYS Maximize Your Plant.™

10th GLOBAL CONGRESS on PROCESS SAFETY

Why do Alarm Floods Occur?

One reason is not providing dynamic alarm management

- **Alarms need to indicate abnormal situations that require operator action**
- **Processes do not operate in one state**
- **What is normal vs abnormal changes with state**
- **Alarms are typically configured for run therefore many alarms are triggered upon a Change of State: Run to Shutdown**
- **Many of the alarms are not applicable or actionable for the new state impleading the operators ability to act quickly on what is important**

PROSYS Maximize Your Plant.™

10th GLOBAL CONGRESS ON PROCESS SAFETY

What is Impacted by Alarm Floods?

- **Product quality**
- **Operability or profitability of the process**
- **Loss of equipment**
- **Loss of containment – environmental releases**
- **Injury and loss of life in plant or community**

PROSYS Maximize Your Plant.™

10th GLOBAL CONGRESS ON PROCESS SAFETY

What Makes Alarm Floods so Dangerous?

Can be a problem for three reasons:

- **A deluge of alarms can cause critical alarms to be missed**
- **Floods can be a significant distraction when dealing with process upsets**
- **Can be an indicator of larger systemic safety issues**

PROSYS Maximize Your Plant.™

10th GLOBAL CONGRESS on PROCESS SAFETY

Impacting Alarm Management Design

- **Alarm rationalization is not a process to eliminate alarms - its about quality**
- **Good rationalization will add alarms when appropriate**
- **Typical (Static) rationalization is only for run mode**
- **Dynamic rationalization considers all plant modes**

PROSYS Maximize Your Plant.™

10th GLOBAL CONGRESS on PROCESS SAFETY

Dynamic Alarm Management

- **Dynamic rationalization does everything a static rationalization does plus asks "when" for each alarm**
- **Requires dynamic software to make changes based upon operating mode of the plant**
- **Eliminates redundant alarms and lowers operator loading during transitions**
- **Only alarms what is abnormal and actionable for the given state**

PROSYS Maximize Your Plant.™

10th GLOBAL CONGRESS ON PROCESS SAFETY

Actual Performance Metrics vs ISA 18.2

Type	Avg Alarm Rate/hr	Low/Hi of Avg Rate/hr	Peak Alarm Rate/hr	Low/Hi of Peak Rate/hr	% Time in Flood	Standing Alarms >24 hrs
Before Rationalization	30	Low=6.2 Hi=61	638	Low=152 Hi=2402	17.2%	9
ISA 18.2 Target Metrics	6	---	<60	---	<1%	<5
After Dynamic Alarm Management	2	Low=0.09 Hi=2.5	25	Low=13 Hi=42	0.25%	4

PROSYS Maximize Your Plant.™

10th GLOBAL CONGRESS ON PROCESS SAFETY

Results After Dynamic Alarm Management

Annunciated Alarms Per Hour (Monthly) **Area: G1-G5-G6** **June 2013**

Average Counts | Peak Counts | Total Counts

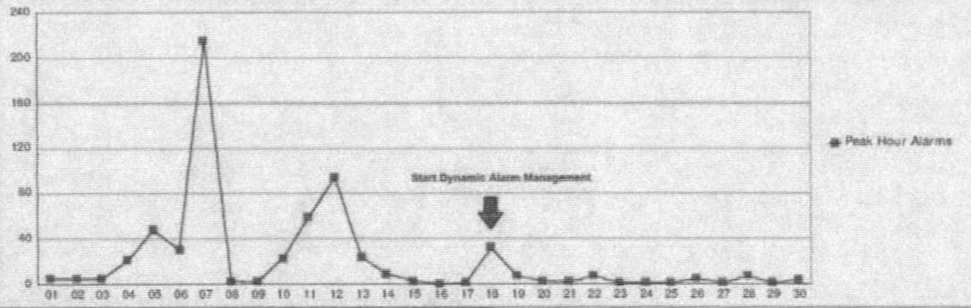

Day	Min Hour Alarms	Avg Hour Alarms	Peak Hour Alarms	Total Alarms
01-Jun-2013	0	0.3	5	6
02-Jun-2013	0	0.4	5	9
03-Jun-2013	0	0.5	6	13
04-Jun-2013	0	1.1	21	27
05-Jun-2013	0	13.3	48	319
06-Jun-2013	0	3.6	30	86
07-Jun-2013	0	11.9	215	285
08-Jun-2013	0	0.4	3	10
09-Jun-2013	0	0.3	2	7
10-Jun-2013	0	1.6	23	35
11-Jun-2013	0	4.7	59	112
12-Jun-2013	0	11.9	94	285
13-Jun-2013	0	3.8	24	90
14-Jun-2013	0	1.3	9	32
15-Jun-2013	0	0.3	3	8

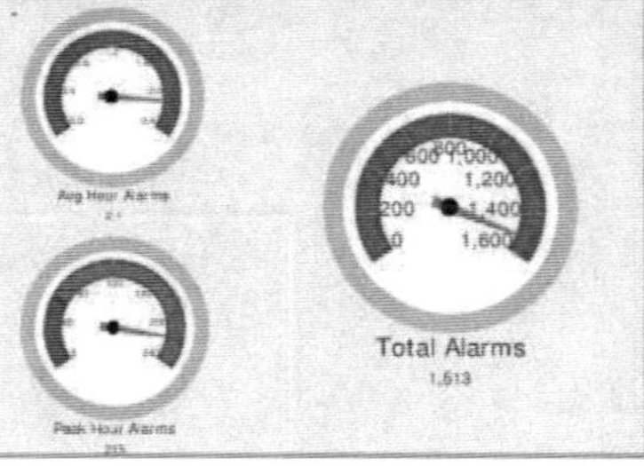

PROSYS Maximize Your Plant™

10th GLOBAL CONGRESS ON PROCESS SAFETY

Operator Graphics – The Risk of Changing Focus

- **Distraction of changing focus from process graphic to faceplate window**
- **Added workload from managing multiple open windows**
- **Increased probability of errors when changing values for un-intended tag with multiple faceplates open**
- **All of above become multiplied and more complex when process is transitioning from one state to another**

PROSYS Maximize Your Plant™

10th GLOBAL CONGRESS ON PROCESS SAFETY

Faceplate Operation

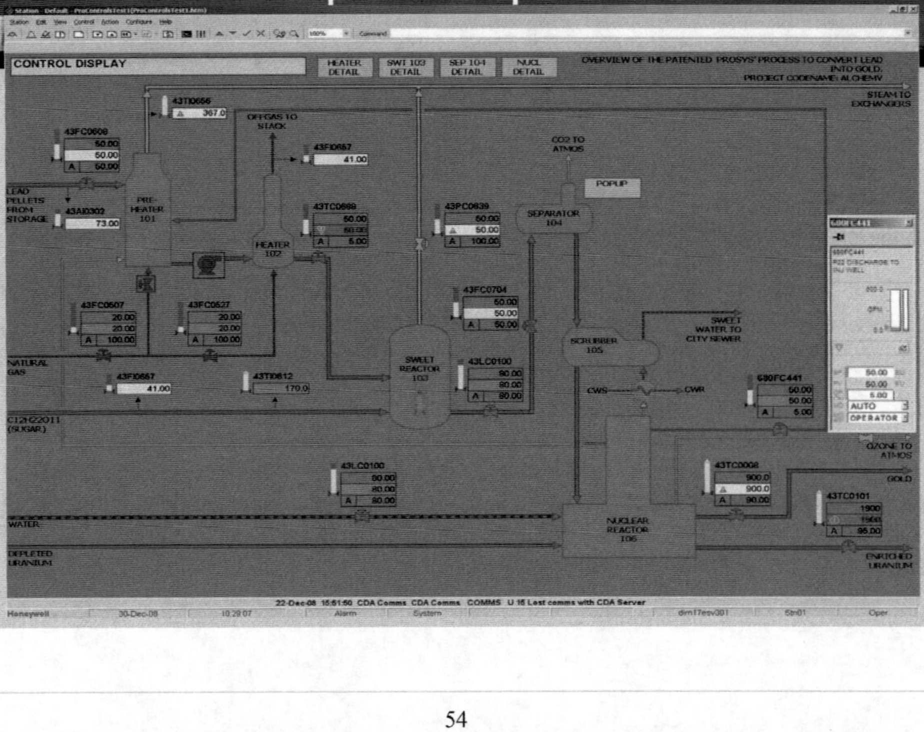

PROSYS™ Maximize Your Plant™

10th GLOBAL CONGRESS ON PROCESS SAFETY

Potential for Errors

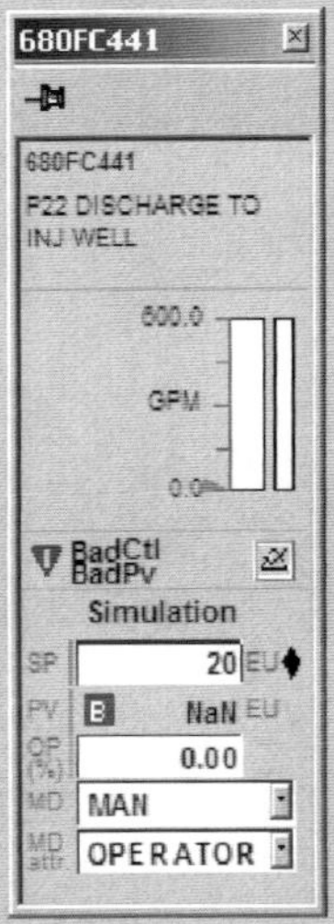

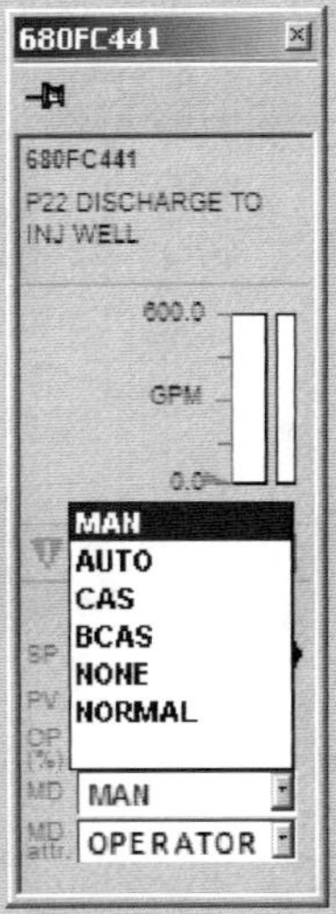

Faceplate Design -can introduce additional errors

- Mode drop down list covers SP, PV and OP values
- Operator may select wrong Mode if SP, PV and OP values are hidden from view
- Mode list offers more options than necessary

PROSYS™ Maximize Your Plant™

10th GLOBAL CONGRESS ON PROCESS SAFETY

Direct Entry Operation

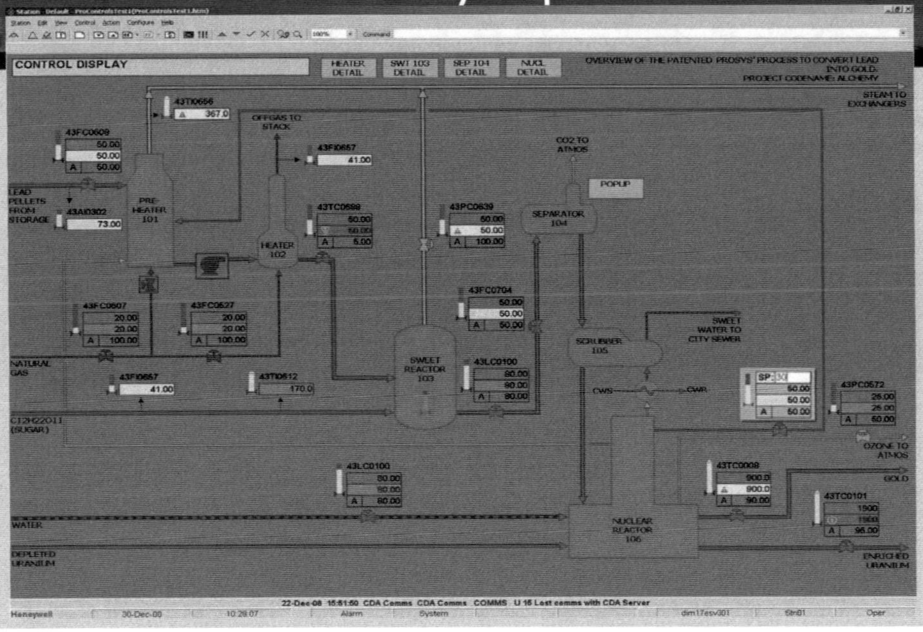

PROSYS Maximize Your Plant™

Direct Entry Fields

Values can be entered by:

- Select point
- Typing value via keypad
- Pressing enter

Key Factors:

- By maintaining focus of points after value change, it improves safety for input errors, multiple clicking functions and is more time efficient
- Easily recognizable operator enterable fields

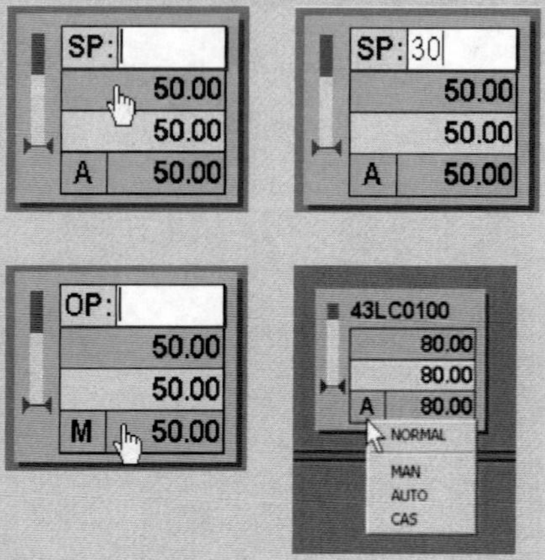

PROSYS Maximize Your Plant™

Evaluation of Methods

Comparison of Faceplate to Direct Entry

- **Keystroke Level Analysis**
- **Operator Loading Analysis**
- **Risk Analysis**

PROSYS Maximize Your Plant™

Keystroke Level Model

Symbol	Time (s)	Description
K	0.28	Keystroke
P	1.1	Point to object
BB	0.2	Click on object
H	0.4	Home hands on keyboard or mouse
M	1.2+	Mental act or routine thinking

PROSYS Maximize Your Plant™

Estimate Execution Time Analysis

- **Evaluates execution time by an operator comparing faceplates to direct entry for SP/OP and Mode Changes**

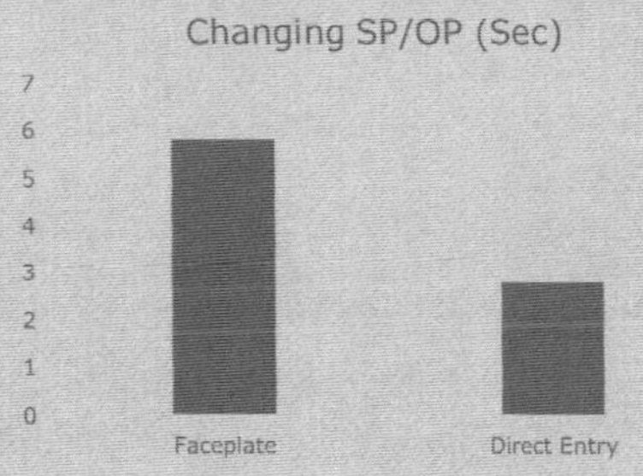

- **50 % Reduction**

PROSYS Maximize Your Plant™

10th GLOBAL CONGRESS ON PROCESS SAFETY

Estimated Operator Loading Analysis

- **Evaluates Mental and Physical Load on Operator**
  - Tasks that require very little thought are removed – such as (BB), (H)

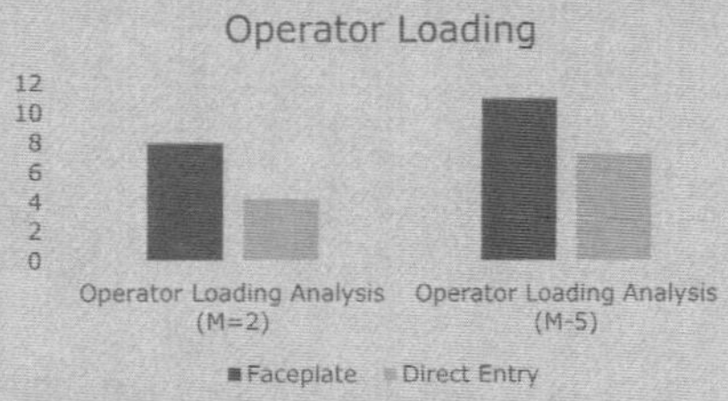

  - Mental and Physical tasks are weighted based on the amount of fatigue they cause – assigned a weight of M=2.0 to 5.0

PROSYS Maximize Your Plant™

10th GLOBAL CONGRESS ON PROCESS SAFETY

Risk Analysis

- **Identifies actions where operator entry errors can occur**

PROSYS Maximize Your Plant.™

Risk Analysis

Faceplate operation KLM for changing SP/OP	Potential Error
Move to shape (P)	Select wrong point
Click on shape (BB)	
Move to faceplate (P)	Point in faceplate is previous point
Click on SP/OP field (BB)	Change wrong parameter
Move hands to keyboard (H)	
Type in value and press enter (4K)	Mistype value and press enter
Move hands to mouse (H)	
Move to close faceplate (P)	
Click to close faceplate (BB)	
Potential error count	**4**

PROSYS Maximize Your Plant.™

Risk Analysis

Direct Entry operation KLM for changing SP/OP	Potential Error
Move to SP/OP (P)	Select wrong parameter or point
Click on SP/OP field (BB)	
Move hands to keyboard (H)	
Type in value and press enter (4K)	Mistype value and press enter
Potential error count	**2**

PROSYS Maximize Your Plant.™

10th GLOBAL CONGRESS ON PROCESS SAFETY

Direct Entry Vs Faceplate

Summary of Benefits

- **Execution Time SP/OP Changes - 51.5% less**
- **Execution Time for Mode Changes – 50% less**
- **Operator Loading – 33 % to 47% less**
- **Risk of Errors – 50% less**

PROSYS Maximize Your Plant.™

10th GLOBAL CONGRESS ON PROCESS SAFETY

Conclusion

- **Many industrial errors are a result of operators using control systems with flawed designs**
  - Poor design and performance by alarm management distracts operators and/or occludes critical alarms
  - Operator graphics with multiple open faceplates can cause a change intended for one controller to be entered into the faceplate of another
- **The two mechanisms listed above are predictable and therefore preventable through better design available today**

PROSYS™ Maximize Your Plant.

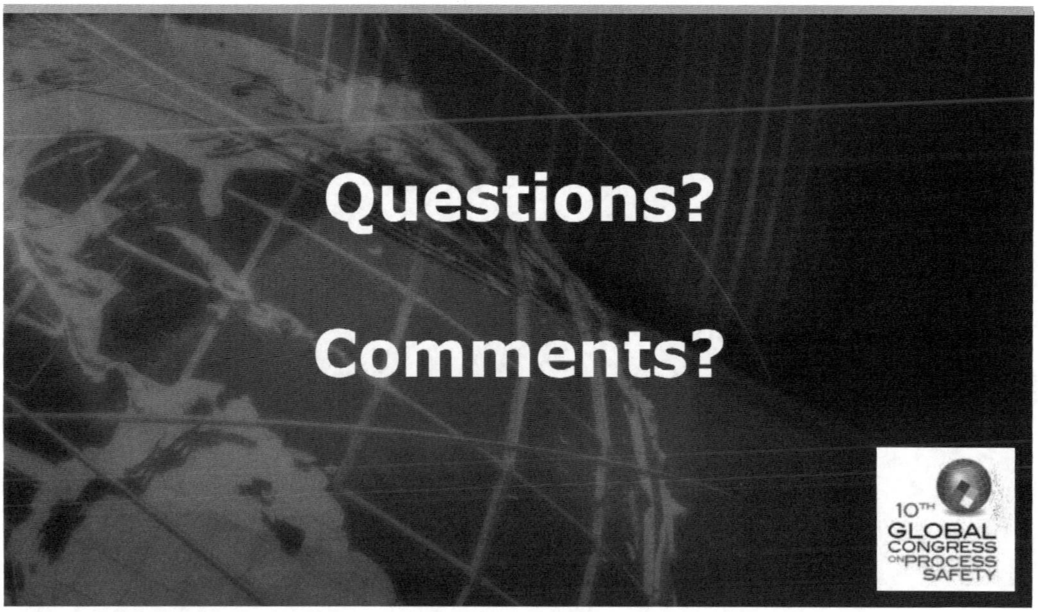

visium KMS
Process Safety Management Solution
When Sharing is not Enough… HUAA
10th Global Congress on Process Safety – New Orleans, LA
Expert Insight Delivered
© 2014 Optimized Systems and Solutions Inc. All rights reserved.
OSYS
Part of the Rolls-Royce Group

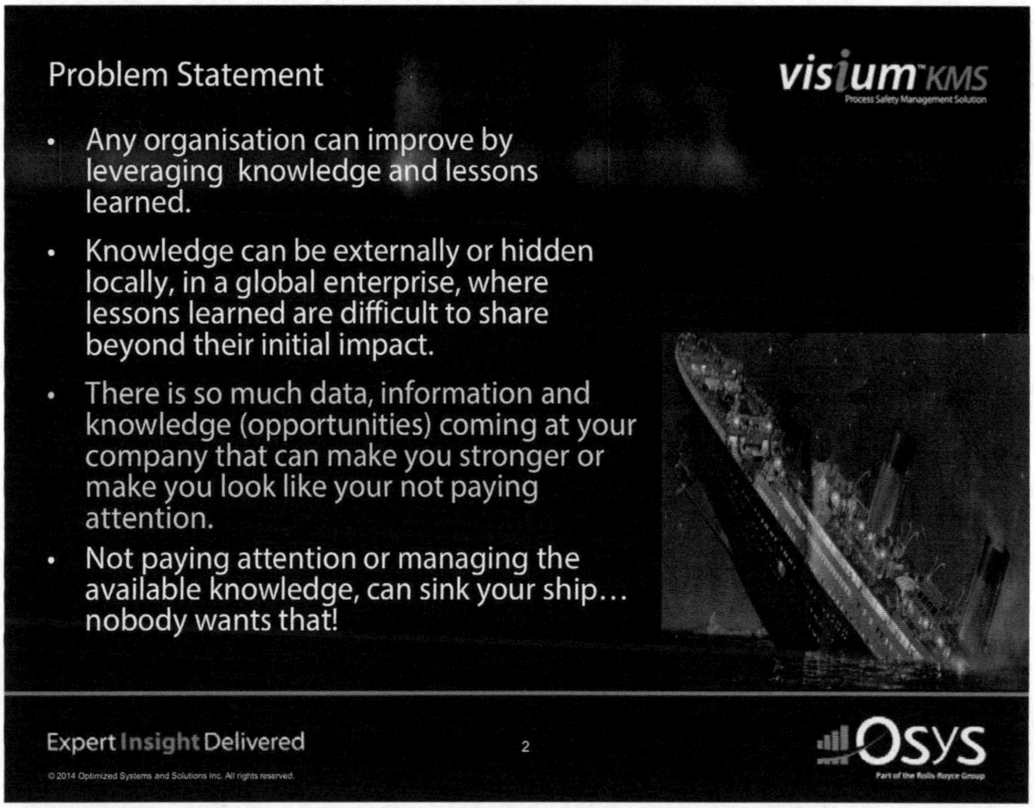
Problem Statement
visium KMS
Process Safety Management Solution
• Any organisation can improve by leveraging knowledge and lessons learned.
• Knowledge can be externally or hidden locally, in a global enterprise, where lessons learned are difficult to share beyond their initial impact.
• There is so much data, information and knowledge (opportunities) coming at your company that can make you stronger or make you look like your not paying attention.
• Not paying attention or managing the available knowledge, can sink your ship… nobody wants that!
Expert Insight Delivered
2
© 2014 Optimized Systems and Solutions Inc. All rights reserved.
OSYS
Part of the Rolls-Royce Group

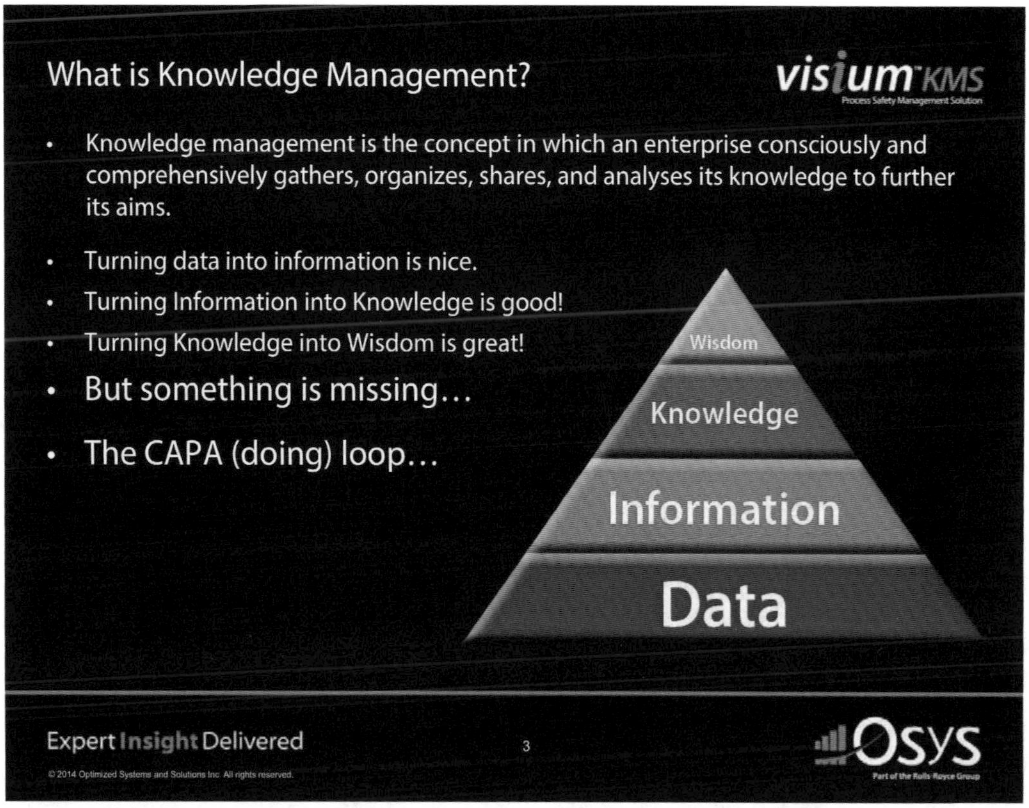
What is Knowledge Management?
visium KMS
Process Safety Management Solution
• Knowledge management is the concept in which an enterprise consciously and comprehensively gathers, organizes, shares, and analyses its knowledge to further its aims.
• Turning data into information is nice.
• Turning Information into Knowledge is good!
• Turning Knowledge into Wisdom is great!
• But something is missing…
• The CAPA (doing) loop…
Wisdom
Knowledge
Information
Data
Expert Insight Delivered
3
© 2014 Optimized Systems and Solutions Inc. All rights reserved.
OSyS
Part of the Rolls-Royce Group

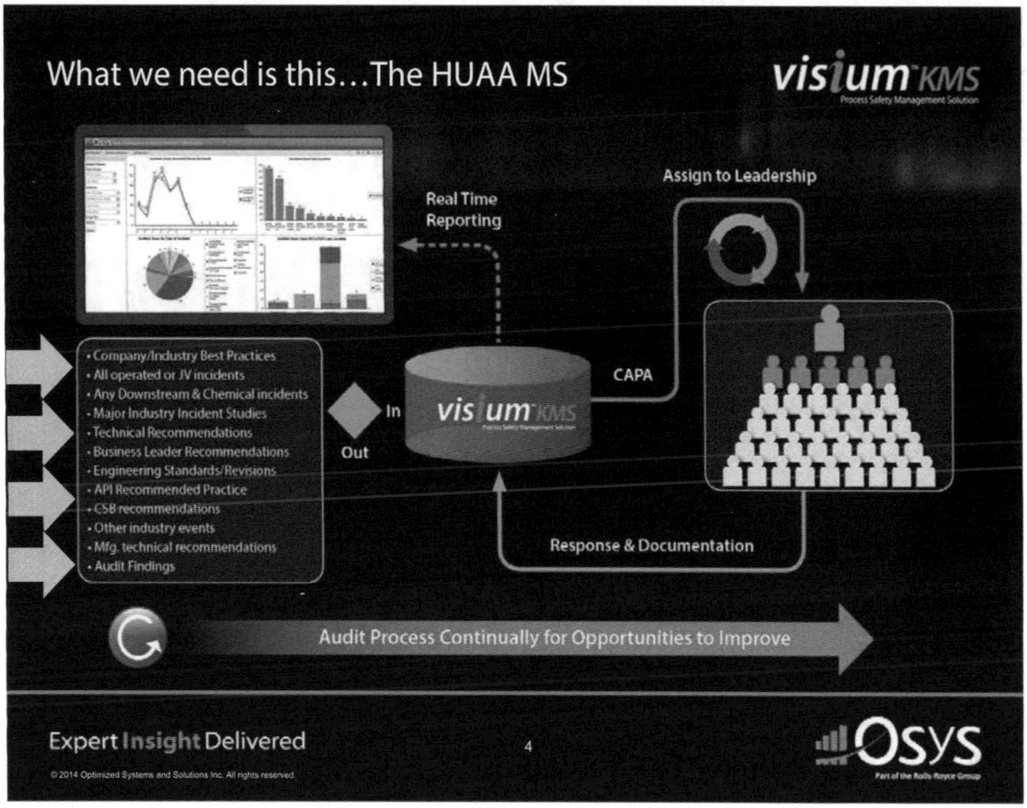
What we need is this…The HUAA MS
visium KMS
Process Safety Management Solution
Real Time Reporting
Assign to Leadership
• Company/Industry Best Practices
• All operated or JV incidents
• Any Downstream & Chemical incidents
• Major Industry Incident Studies
• Technical Recommendations
• Business Leader Recommendations
• Engineering Standards/Revisions
• API Recommended Practice
• CSB recommendations
• Other industry events
• Mfg. technical recommendations
• Audit Findings
In
Out
visium KMS
CAPA
Response & Documentation
Audit Process Continually for Opportunities to Improve
Expert Insight Delivered
4
© 2014 Optimized Systems and Solutions Inc. All rights reserved.
OSyS
Part of the Rolls-Royce Group

What is the HUAA Management System…

visium KMS
Process Safety Management Solution

HUAA – Heard, Understood, Acknowledged, Actioned.

1. Opportunities identified (heard) by your experts,
2. Captured by an electronic system,
3. Expertly Reviewed (understood),
4. Accepted or rejected (acknowledged) and
5. Assigned (actioned) to Sr. Managers for resolution and tracked to closure.

Expert Insight Delivered 5 Osys
Part of the Rolls-Royce Group

HUAA – Management Systems - Detail

visium KMS
Process Safety Management Solution

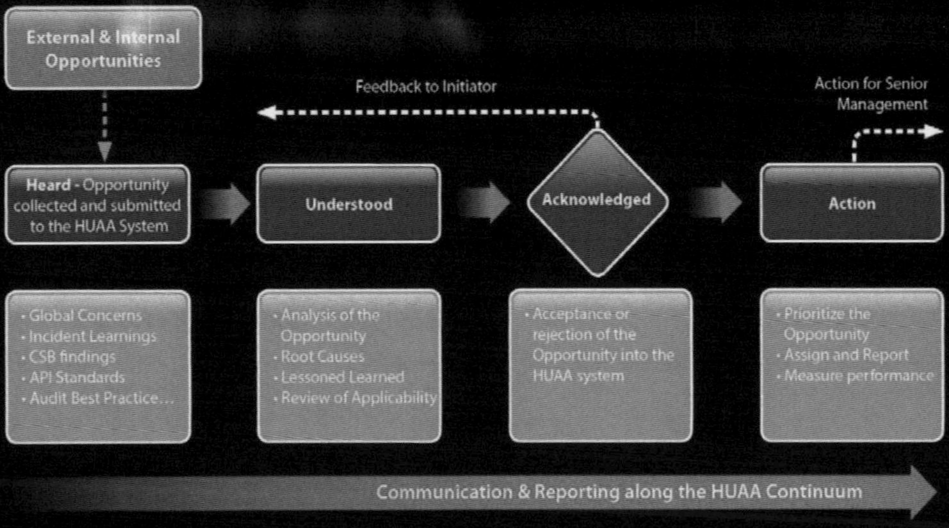

Expert Insight Delivered 6 Osys
Part of the Rolls-Royce Group

HUAA Results

visium KMS
Process Safety Management Solution

- Efficient Collection - Important Opportunities are identified and collected from many sources (internal and external), and entered into the HUAA management system.
- Expert Vetting - These opportunities are vetted, parsed and resolved for inclusion or exclusion; documenting & communicating both.
- Accountable Assignment - Included opportunities are systematically assigned to Sr. Leaders and managed across the enterprise at SVP and C levels.
- Action - Improvement will be assured through the accountable actions of these leaders.
- Results - The organization will gain competitive advantage by active listening, learning from their experience, as well as the experience of others, and by taking ACTION!

Expert Insight Delivered 7

Osys
Part of the Rolls-Royce Group

Sharing is not enough to keep your ship afloat!

visium KMS
Process Safety Management Solution

Expert Insight Delivered 8

Osys
Part of the Rolls-Royce Group

352631 Autocad - Smart Grid – Agriculture - Architecture and the United States Government

Abstract:

The computational methods, process control, process design, process safety, public policy of the smart chip for agriculture industry, and architecture industry in the United States utilizing the computational methods of AutoCAD in the structural development stages of the smart grid for device/component/equipment and geographical development and public policy for grant funding.

352747 Production, Characterization and Catalytic Studies of Biobased Carbon Materials

Abstract:

Carbon-based nano materials were produced from wood chips, lignin and bio-char. The influence of reaction conditions on the production of nano-carbons was investigated. Carbon-based nanomaterials were characterized using scanning electron microscopy (SEM), transmission electron microscopy (TEM), high resolution transmission electron microscopy (HRTEM), and X-ray diffraction (XRD). The catalytic performance of these nanomaterials as syngas cleaning adsorbents and chemical conversion catalysts for bio-fuel processes has been evaluated.

347835 Uncertainty in Sour Gas Viscosities Estimation, What Is the Effect on Your Reservoir Inflow and Tubing Performance

Abstract:

In light of the need to find an applicable, consistent and promising correlation to accurately calculating sour gases viscosity under several reservoir conditions, this study aims to compare the existing correlations and other industry standard methods, check their accuracy and range of applicability for sour/acidic gases. Moreover, investigating the impact of uncertainties in estimating sour gas viscosities on reservoir flow performance relationship (IPR), reserves and resources estimations, tubing performance, pipeline and facilities design, the design of compressors and more importantly the need to find such correlation to be used in further advancements of reservoir simulation.

354052 Chromonic Nanocarriers for Chemotherapeutics: Size Distribution and Control Release Studies

Abstract:

Conventional chemotherapy fails to release the drug in the controlled manner; rather they release the drug instantly due to which re-administration is required as therapeutic index falls. Moreover, they target both the cancer cells and non-cancer cells which make the treatment of the cancer cells highly ineffective due to excessive toxicities. Recently, the nanoparticle mediated drug delivery for chemotherapeutics has gained much attention. The present study reports the use of a novel class of materials – *Chromonics* in cancer drug delivery. Folic acid as a chromonic material has been explored to develop nanocarrier for sustained release of chemotherapeutic drugs *in-vitro*. With the help of a biocompatible polymer as a tool, folate nano-domains were formed from self-assembled liquid crystalline folate solutions. These nano-domains were crosslinked to form a stable nanoparticle system. The effect of different factors like crosslinking agent, relative polymer and folate concentration, on the size distribution was investigated in detail. Based on the observations, we have reported that not only the synthesis but also the size of the developed nanoparticles can be controlled by varying the folate: HPMC weight ratios. The present study then explored the release of drugs from these nanocarriers. We have investigated the release of folic acid as a model drug from these nanoparticles as well as release of methotrexate and cytarabine as an anticancer drug for the period of 60 days. It was observed that release of drug was nanoparticle's size-dependent, release medium dependent, pH of the release medium dependent, crosslinking agent and its concentration dependent. The significant part of the current study is the synthesis of chromonic nanoparticles, how the size can be controlled, how controlled release can be obtained as well as how the release profiles can be modified depending on the dosage forms. Therefore, we can say that folate nanoparticles can be a potent nanocarrier for the sustained release of cancer drugs.

Double Containment Piping Systems: A Fail-Safe, Leak-Free Solution

IPEX
Committed to Excellence

Darin Johnson, P.Eng.
Applications Engineer
IPEX USA LLC

AIChE – 2014 Spring Meeting
New Orleans, LA

Monday, March 31, 2014

Why Double Containment Piping Systems?

IPEX
Committed to Excellence

- December 1988 – law change within the ***Resource, Conversation, and Recovery Act*** in the United States regarding underground piping.
- U.S. Environmental Protection Agency (EPA) requirement: Code of Federal Regulations (CFR), Title 40, Part 280.
  - Petroleum Products
  - Hazardous Substances (over 3200, CFR Title 40 Part 302.4)
  - Hazardous Wastes

Why Double Containment Piping Systems?

- Buried tank - all buried piping connected to it must be double contained.
- Above ground tank – associated buried piping consists of 10% or more of the tank volume, it also must be double contained.

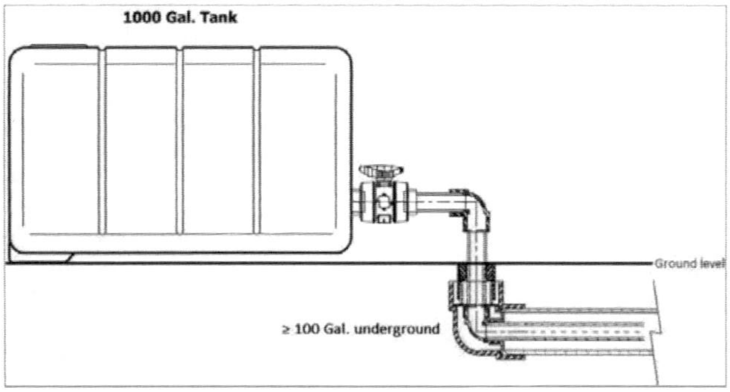

Why Double Containment Piping Systems?

- If the piping is less than 10% of the volume of the aboveground tank, and you choose not to double contain the buried piping, ***<u>responsibility for any clean-up or remediation resulting from a release will still fall on the Client.</u>***

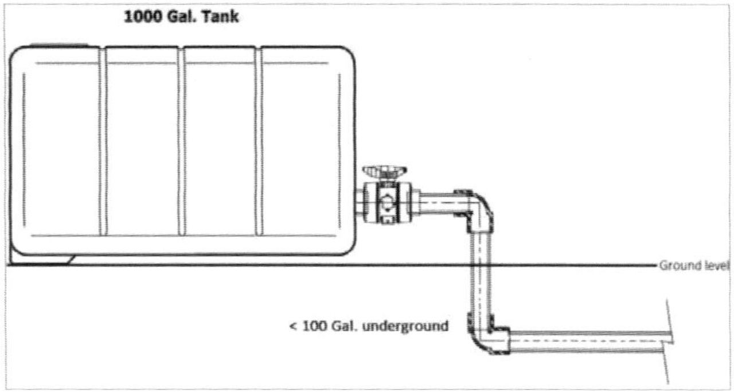

Why Double Containment Piping Systems?

- **Above Ground Piping Systems**
  - Different jurisdictions have to ability to require chemicals be double contained.

Nomenclature of Double Containment Piping

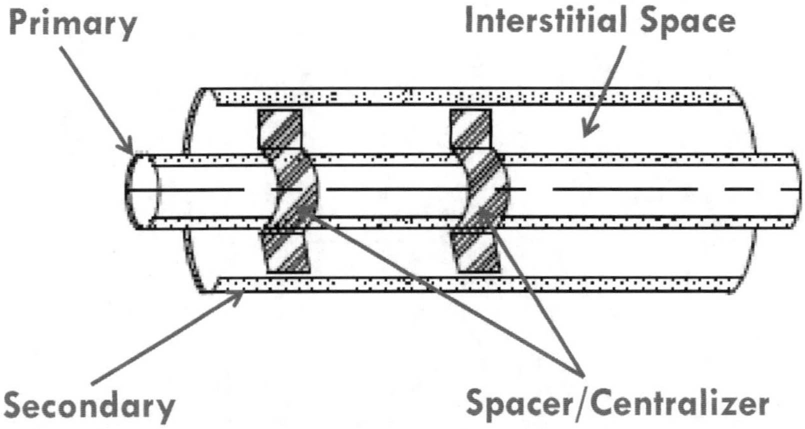

Double Containment Piping Systems: Material Combinations

SYSTEM Carrier	Containment	Carrier SIZE inches
PVC 80	PVC 80	1/2 - 12
PVC 80	PVC 40	1/2 - 12
PVC 40	PVC 40	1/2 - 12
PVC DWV	PVC DWV	1-1/2 - 12
PVC 80	CS SCH 10	1/2 - 12
CPVC 80	CPVC 80	1/2 - 8
CPVC 80	PVC 80	1/2 - 12
CPVC 80	PVC 40	1/2 - 12
CPVC 80	CS SCH 10	1/2 - 12
NAT PP 80	PVC 80	1/2 - 2
NAT PP 80	PVC 40	1/2 - 2
RED PVDF 80	CPVC 80	1/2 - 2
RED PVDF 80	PVC 80	1/2 - 2
NAT PVDF 80	PVC 80	1/2 - 2
ENFUSION PP	PVC DWV	1/2 - 6
S40 A53B CS BW	PVC 80	1/2 - 8
S40 A106B CS BW	PVC 80	1/2 - 8

SYSTEM Carrier	Containment	Carrier SIZE inches
S40 A53 BLK SW	FRP	1/2 - 2
S40 A53 BLK SW	PVC 80	1/2 - 2
S80 A53B CS SW	PVC 80	1/2 - 2
S10 T304L SS BW	PVC 80	1/2 - 8
S10 T304L SS BW	PVC 40	1/2 - 8
S10 T0304L SS BW	CPVC 80	1/2 - 8
S10 T304L SS BW	FRP	1/2 - 12
S10 T304L SS BW	S10 T304L	1/2 - 12
S40 T316L SS BW	PVC 80	1/2 - 8
S40 T316L SS BW	CPVC 80	1/2 - 8
S40 T316L SS BW	FRP	1/2 - 12
S40 T304L SS SW	PVC 80	1/2 - 2
S40 T304L SS SW	PVC 40	1/2 - 2
S40 T304L SS SW	CPVC 80	1/2 - 2
S40 T304L SS SW	FRP	1/2 - 2
S10 T316L SS SW	PVC 80	1/2 - 2
S10 T316L SS SW	CPVC 80	1/2 - 2

Double Containment Piping Systems: Design Considerations

- **Material Selection**
  - Primary Pipe
    - Chemical Compatibility
    - Pressure Capabilities
    - Thermal Properties
  - Secondary Pipe
    - Chemical Compatibility
    - Loading (Static & Live) for buried applications
    - Hanging requirements for above ground
    - Leak detection requirements
    - UV requirements

Double Containment Piping Systems: Design Considerations

- **Thermal Expansion & Contraction**
  - Two basic thermal relationships may be present:
    - When only one of the two systems experiences thermal expansion/contraction.
    - When both systems experience different magnitudes of thermal expansion/contraction.
  - The Floating System
  - The Restrained System

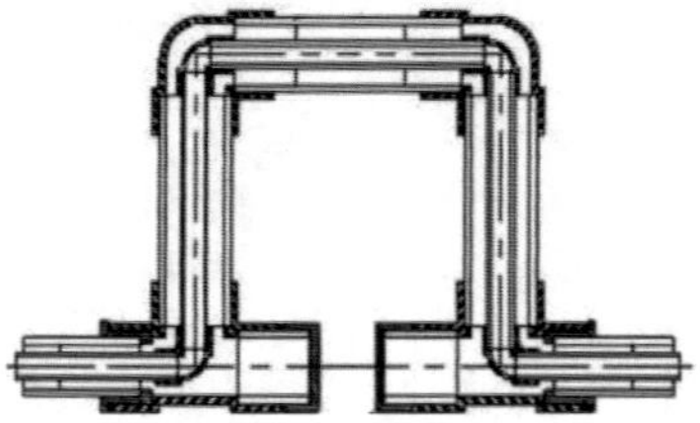

Double Containment Piping Systems: Leak Detection Requirements

- EPA requirement: (CFR), Title 40, Part 280
  - Pressurized systems require electronic leak detection that must be checked once per year.
  - Drainage systems (gravity flow) require visual leak detection with monthly manual inspection for a media leak.

Double Containment Piping Systems: Leak Detection Requirements

- The following forms of leak detection are acceptable as per regulations
  - Visual (drainage applications only)
  - Electronic low point probes
  - Continuous cable

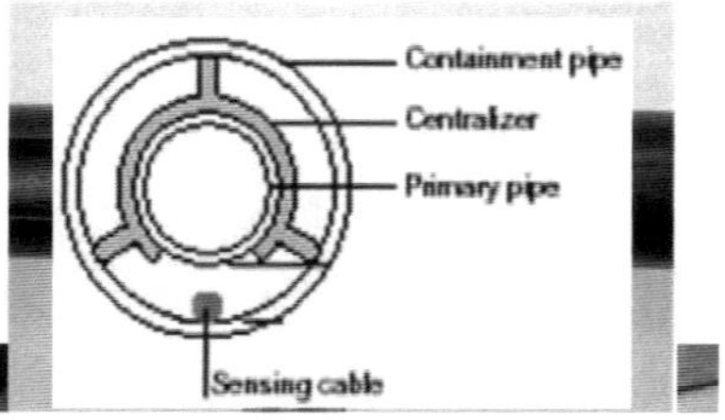

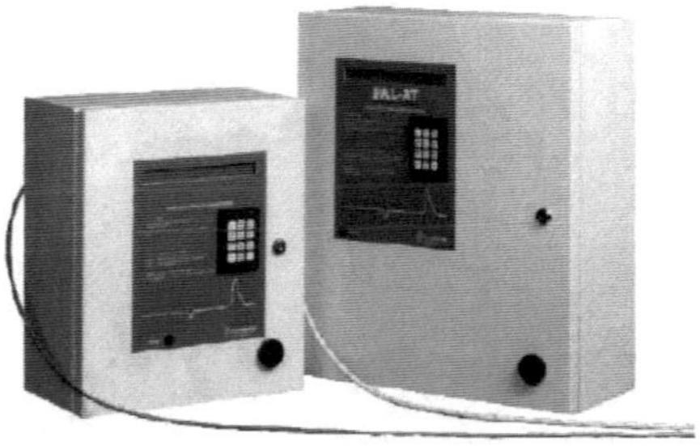

Double Containment Piping Systems: Applications

- Underground Applications
- WTP & WWTP
- Oil & Gas
- Hospitals
- Museums
- Food and Beverage
- Pharmaceuticals
- Laboratories
- Data Centers

Questions?

Thank you!

DTP Process

"Now Ready to License"

JGC Corporation
Mitsuo Morita

2014

JGC CORPORATION

MITSUBISHI CHEMICAL

What is DTP Process ?

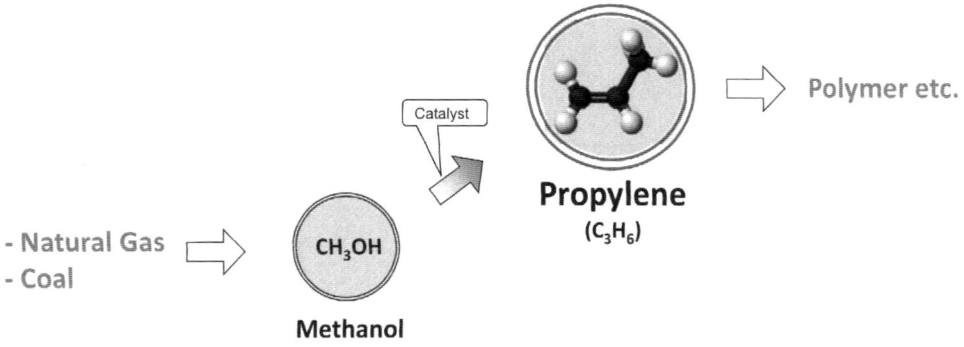

-On-Purpose Propylene Production from Methanol

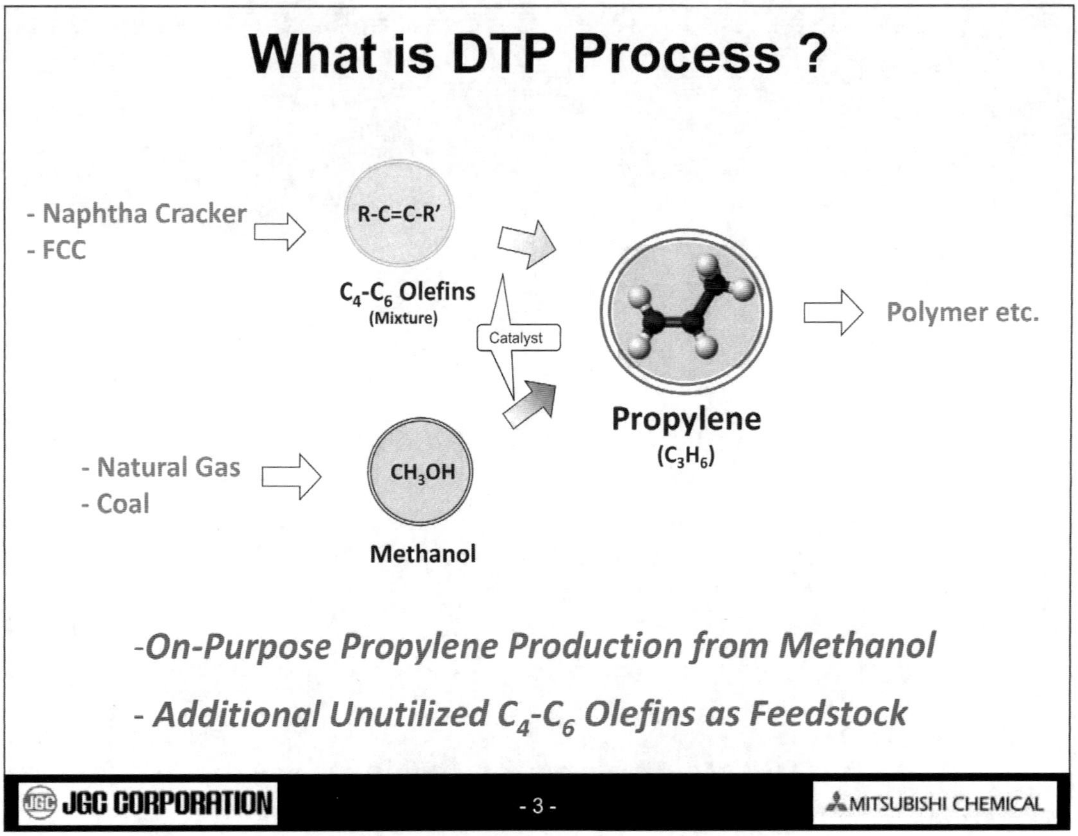
What is DTP Process ?
- Naphtha Cracker
- FCC
R-C=C-R'
C_4-C_6 Olefins
(Mixture)
Catalyst
Propylene
(C_3H_6)
Polymer etc.
- Natural Gas
- Coal
CH_3OH
Methanol
-On-Purpose Propylene Production from Methanol
- Additional Unutilized C_4-C_6 Olefins as Feedstock
JGC CORPORATION
- 3 -
MITSUBISHI CHEMICAL

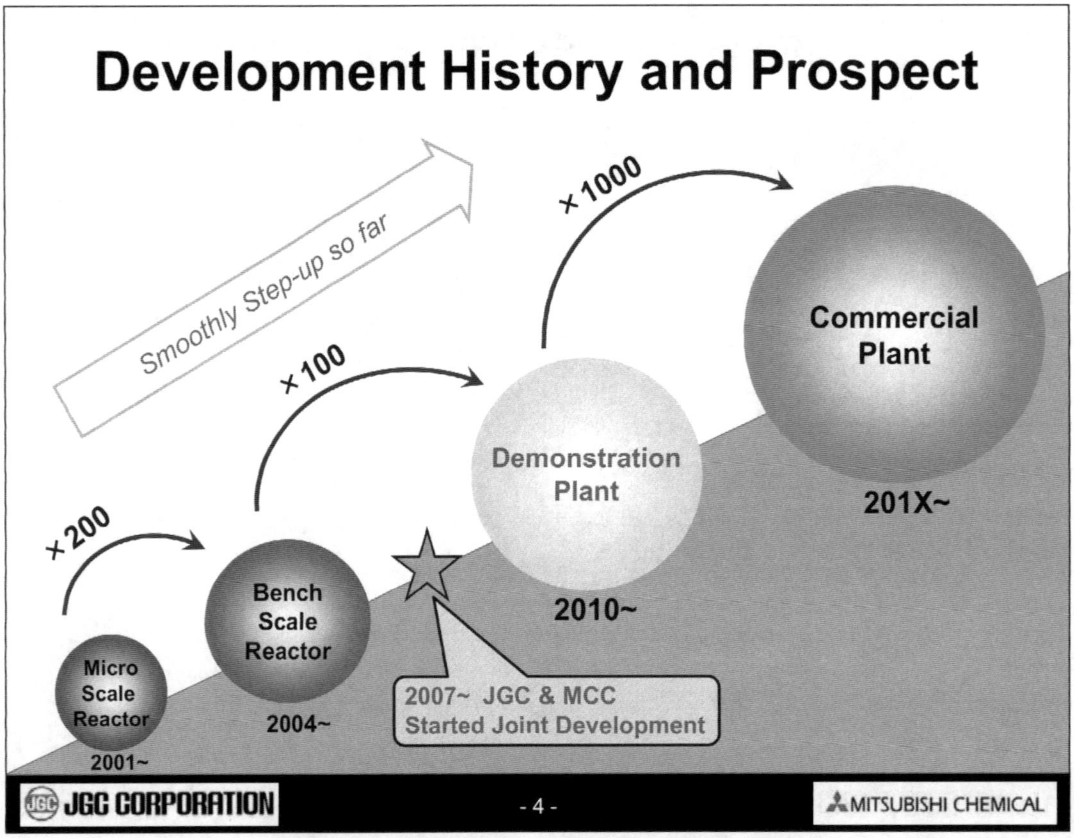
Development History and Prospect
Smoothly Step-up so far
×200
×100
×1000
Micro Scale Reactor
2001~
Bench Scale Reactor
2004~
Demonstration Plant
2010~
Commercial Plant
201X~
2007~ JGC & MCC Started Joint Development
JGC CORPORATION
- 4 -
MITSUBISHI CHEMICAL

Demonstration Plant Overview

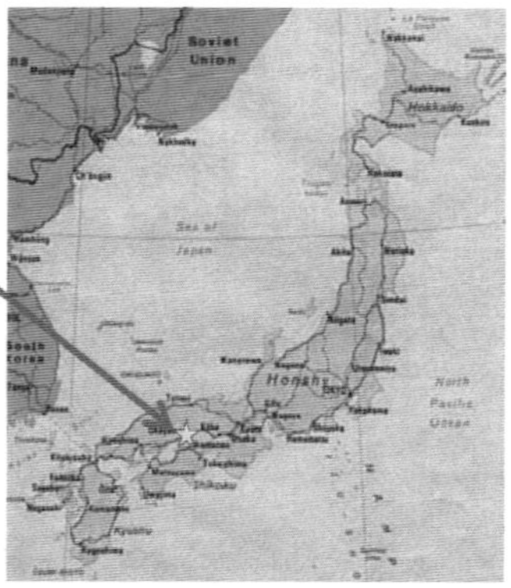

- **Located in Mitsubishi Chemical Mizushima Plant**
- **Successfully Start-up in August 2010**
- **Continuous Operation for Two Years**

JGC CORPORATION - 5 - MITSUBISHI CHEMICAL

Process Flow (Standalone Case)

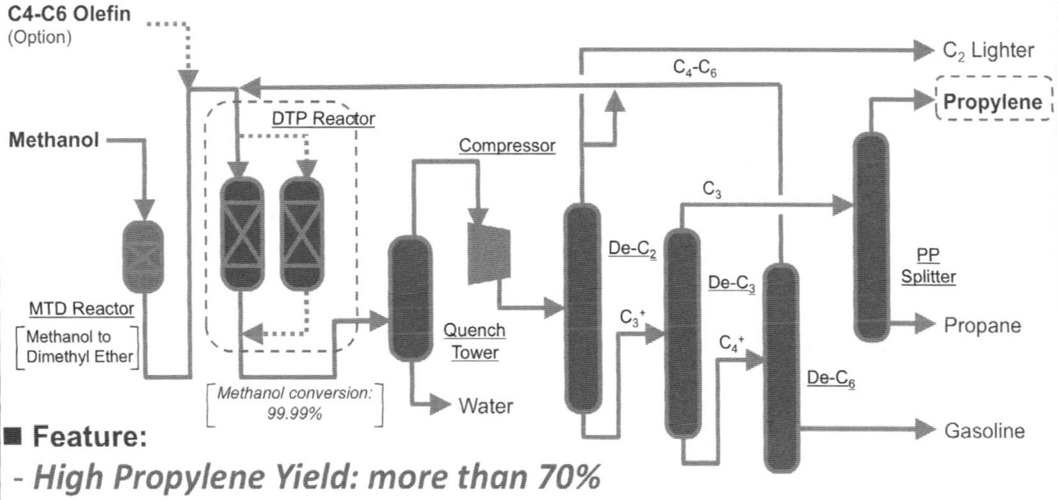

■ **Feature:**

- ***High Propylene Yield: more than 70%***
- ***Applied Existing Proven Technologies***

JGC CORPORATION - 6 - MITSUBISHI CHEMICAL

Process Flow (Integration Case)

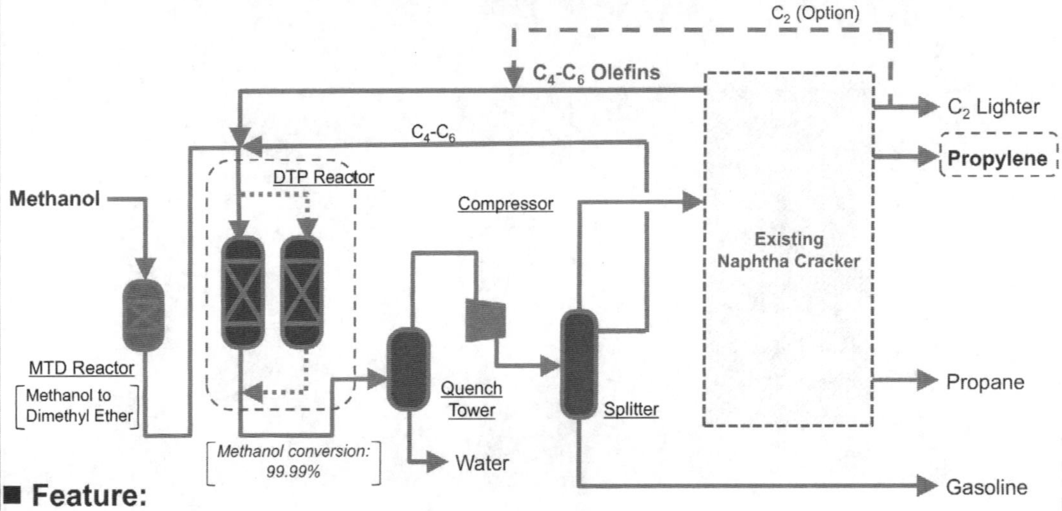

■ **Feature:**

- *High Propylene Yield: more than 70%*
- *Applied Existing Proven Technologies*
- *Low Investment Cost in case of Integration with Naphtha Cracker*

Application

(1) Naphtha Cracker Integration Case (Propylene Capacity: 20KTA – 200KTA)

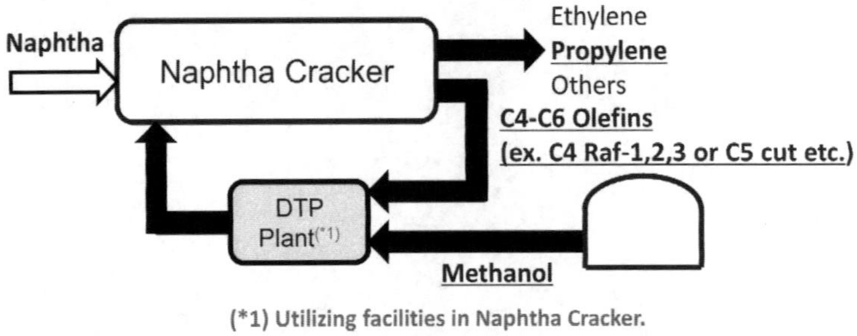

(*1) Utilizing facilities in Naphtha Cracker.

(2) Standalone Case (Propylene Capacity: 250KTA – 500KTA)

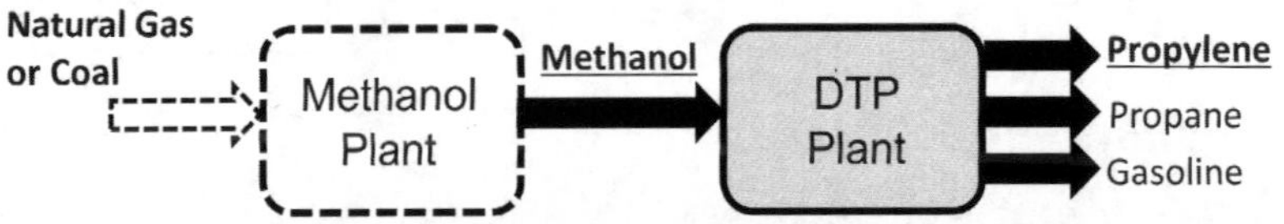

- *Two Cases available depending on Client requirements !!*

DTP is...

- *High Yield Propylene Production Process*
- *Flexible Feed Methanol to Olefin Process*
- *Process that increase P/E Ratio for Cracker*

Thank you very much !!

Contact:
Mitsuo Morita
Market Development Dept.
Global Marketing Division
2-3-1, Minato Mirai, Nishi-ku,
Yokohama 220-6001, Japan

Phone: 81-45-682-8371
FAX: 81-45-682-8183
E-Mail: morita.mitsuo@jgc.co.jp

352533 Self-Cleaning "Bernoulli" Type Filters Used in Onshore and Offshore Applications

Abstract:

Georg Schuenemann GmbH (SAB-BREMEN) has developed a new Filter Type F-480, dubbed the "Compact Giant" for applications in onshore water intake and cooling water systems, sea water desalination plants, and in offshore projects, such as FPSO's, Drillships, and Production Platforms.

This paper will describe the essential design features of the filters, based on the "Bernoulli principle" which result in a fully automated self-cleaning filter system, and choice of materials in a sea water service with high corrosion rates.

A case study will be presented in which stringent project requirements such as substantial flow rates, adaptibility to space limitations and pipe connections had to be observed. Reliability of the system, extended run times without the need for frequent replacement of filter elements, as well as ease of monitoring and operator involvement have been achieved.

352653 Effects of Support on Sulfur Tolerance and Regeneration of Pt Catalysts Measured By Ethylene Hydrogenation and EXAFS

Abstract:

We investigated the effect of support on sulfur tolerance and regenerability under reducing environments by measuring the rate of ethylene hydrogenation, combined with hydrogen chemisorption, and extended x-ray absorption fine structure (EXAFS). Alumina and titania (P25) supports were used to prepare Pt catalysts at 1 wt.%. The catalysts were tested after treatment in H_2S/H_2 at 250 °C followed by regeneration treatments in H_2 at 250 °C, 350 °C and 450 °C. The sulfided catalysts exhibited a 20-27 times decrease in the rate of ethylene hydrogenation and a 4-6 times drop in the Pt surface area. Regenerations up to 450 °C were unable to remove all the sulfur, as shown by the presence of Pt-S bonds by EXAFS. However, a partial recovery of the hydrogenation rate per mole of Pt was observed on sulfided Pt/Al_2O_3 after reduction at 450 °C, while the induction of strong metal support interactions (SMSI) at reduction temperature above 350 °C was observed on Pt/P25, regardless of the presence of sulfur. For Pt/P25, the reversal of the SMSI state together with sulfur removal by mild oxidation suggests that sequential reduction/oxidation treatments may be more effective in restoring the S-free state of TiO_2-supported catalysts.

352799 Improved Operational Efficiency and Reliability through Insulation Materials Selection

Abstract:

This paper will explore the implications of insulation materials selection on the energy efficiency and reliability for Ethylene production units. Considerations will include insulation properties, structural support, installation methods, installation and maintenance costs.

BOILER MACT COMPLIANCE: MERCURY IN FUEL GAS

PATRICK LAINE, WES YONGE, & VIC VICKERY
PEI, HOUSTON, TX 77043

Patrick Laine, Ph.D.
Chief Scientist
Houston, TX
850-866-6686
plaine@pei-tx.com

Wes Yonge, PE
Chief Engineer
Houston, TX
713-806-8396
wyonge@pei-tx.com

James 'Vic' Vickery
Chief Technology Officer
Houston, TX
281-536-0899
vvickery@pei-tx.com

1 Introduction & Objectives

For most ethylene producers the primary objective of this sampling and analysis effort will be to verify and "establish a mercury fuel specification through which gas-fired boilers that use a fuel other than natural gas or refinery gas may be considered Gas 1 unit" as identified in 40 CFR Part 63.7521.(f).

Measurement and Monitoring Requirements: Decision Rules

❑ If Hg concentration ≤ 50% of Gas 1 specification* then no further testing required

❑ If Hg concentration > 50% and ≤ 75% of the specification* then semi-annual sampling required

❑ If Hg concentration > 75% of the specification* then monthly sampling required

*Specification is **40 µg/Sm³**

The goal will be to execute a sampling and analysis campaign that will produce compliance quality data. The key to producing and validating compliance quality data is a well-designed and executed QA/QC protocol. At a minimum, a performance based QA/QC protocol is recommended that includes the following components:

- Sorbent Trap Breakthrough Bed:
  - ❑ Sorbent traps consisting of a minimum of two sorbent beds in series so that the trailing bed can be analyzed to verify that no breakthrough occurred from the leading bed.
- Duplicate Samples:
  - ❑ The sampling system should be configured with two trains to accommodate the collection of simultaneous duplicate sorbent trap samples, in parallel, providing a means of verifying precision in the sample collection process.
- NIST (National Institute of Standards and Technology) Traceable Spike:
  - ❑ Sorbent traps with a third bed (downstream) that is spiked with a NIST traceable spiked mass of mercury serves as a matrix spike and provides a means of verifying that components in the matrix (the fuel gas) are not affecting the trapping efficiency of the sorbent trap.

2 Methods

Sorbent Trap Mercury Sampling Methods Comparison

Method	Description	Method Type: S = Sampling A = Analysis	Evaluation	Sample Collection Phase QA/QC
ISO 6978	Hg in Natural Gas by Thermal Desorption (CVAFS)	S & A	Method calls for low sample volumes can lead to non representative results.	No
EPA Method 30B	Determination of Gaseous Hg Emissions from Coal Fired Combustion Sources using Carbon Sorbent Traps	S	Requires modification (Modified M30B)	Yes
ASTM D6350	Hg S&A in Natural Gas (AFS)	S & A	Method calls for short term samples (i.e., low sample volume) which leads to poor accuracy.	No
ASTM D5954	Hg S&A in Natural Gas (AAS)	S & A	AAS is less sensitive than AFS.	No
EPA Method 1631	Measurement of Hg in Water	A	No	Yes

3 EPA Method 30B QA/QC Performance Criteria

QA/QC Test or Specification	Acceptance Criteria	Frequency
Analyzer calibration	Correlation Coefficient: 0.999	Daily
Transfer efficiency of analyzer	%Transfer Efficiency: 90%–110%	Once per sorbent trap
Initial calibration blank	< 0.5 ng	Once before each calibration
Sorbent Trap Section B breakthrough	≤ 10% of Section A Hg mass, for Hg concentrations > 1µg/dscm ≤ 20% of Section A Hg mass, for Hg concentrations ≤ 1µg/dscm	At each sample point
Spike Recovery Test	Average recovery between 75% and 125 % for Hg_0	Once per sample point
Paired Sorbent Traps Agreement	≤ 10% RD for Hg concentration > 1µg/dscm ≤ 20% RD or 0.2 µg/dscm absolute difference for Hg concentrations ≤ 1µg/dscm	At each sample point

4 Equipment

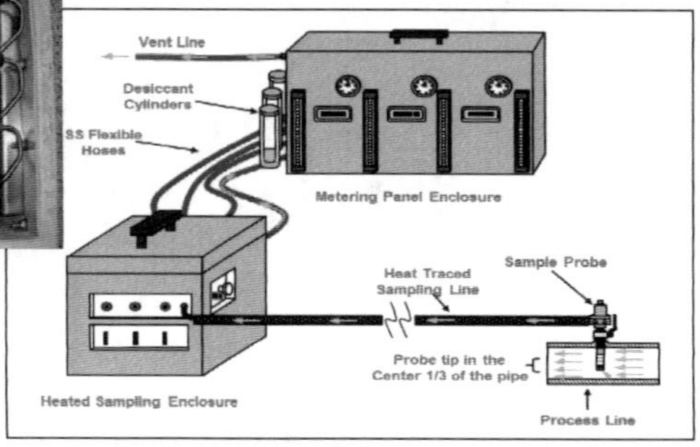

Flow

	Primary Bed	Breakthrough Bed	Spiked Bed

QA/QC Acceptance Criteria:	99% of Total Collected Hg	≤ 10% of Primary Mass	Recovery between 75% and 125%

Gold Trap:

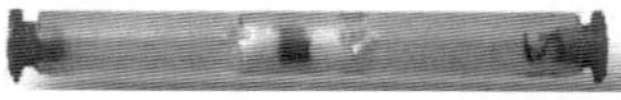

5 Conclusions/PEI Recomendations

❑ There are five methods accepted by the Boiler MACT rule. PEI has extensive experience and expertise employing **all five** of these methods.

❑ Operators can achieve compliance without submitting a Fuel Analysis Plan by using ASTM D6350, however, we have noticed variable results with this method.

❑ The **PEI recommendation** for obtaining the most accurate compliance quality data is to use Modified EPA Method 30B in conjunction with EPA 1631 where traps should be analyzed in a certified ISO 17025 laboratory. The recommended pathway includes a rigorous, performance based QA/QC protocol that validates accuracy and precision of all sample collection procedures.

351376 Young Professional Simulation Tutorial

Abstract:

18C00 Young Professional Simulation Tutorial

Authors: Naomi Hua and Mike Donahue, Invensys

Process simulation is a complex skill to master. Even with years of experience, a typical user is constantly facing new problems that need to be addressed. There are many issues that can lead to unexpected and unnecessary errors with the use of process simulation tools. Regardless of what simulation program you are using, these issues are common from one program to another. In this presentation, targeted for the young professional, we will discuss the "Most Common Pitfalls in Process Simulation". For each topic, we will discuss various techniques and methods these troublesome problems and prevent their occurrence.

Topics to be discussed include:

1. Erroneous data entry

2. Simulation defaults

3. Bad plant data

4. Thermodynamic method selection

5. Simulation tolerances and convergence

6. Recycle identification and handling

7. Model complexity

8. Difficult or impossible specifications

9. Assay pseudo-component development

10. Assay property blending

1. Erroneous data entry is the most common mistake made when building a process simulation model. This presentation will cover methods that can help ensure proper data entry in your process simulation model, such as checking the data you entered, particularly the units of measure, using "alternate forms" of data input such as copy/paste and data import, and reviewing all warning and error messages.

2. Defaults are methods, values, and conditions that are automatically selected for you by the program when you begin a new simulation model. These defaults will affect the results calculated by your model and every simulation program has them. Most defaults within simulation programs fall under categories such as convergence criteria, thermodynamic methods, tolerance and units of measure. The problem with defaults is unintended results from not fully understanding the effects a particular default has on the model. Take the time to understand the defaults of the program and how these defaults will affect the results. If you do not agree with the defaults selected by the program, most simulation programs will allow the user to change the default.

3. Plant data changes over time. The feed information and process conditions of the unit operations may be different. Not only does plant data change over time, but measurement information is often very inaccurate. This could be due to faulty or poorly calibrated measurement equipment, leaks in lines, sensor failures, and other issues. Because of these changes and inaccuracies over time, you always have to validate the information that is entered in your model to ensure that you are obtaining the proper results. To do so, you can utilize programs that can help with data validation and reconciliation like SimSci's ROMeo online optimization program. With it, you can connect your model to your data historian to constantly import current data measurements into your model and then reconcile these measurements to determine when problems arise within your plant and where those problems are occurring. You will also need to use your engineering judgment to determine if the data is valid. More details will be provided during the presentation.

4. The proper selection of thermodynamic methods for your simulation is the most important decision that you make when modeling a process. The method you select could have a drastic effect upon your results. Often, any number of thermodynamic methods is suitable for a given application. It is up to the user to determine the best method to use based upon knowledge, personal experience, and best judgment. Some important factors that need to be considered are the components being used and the operating conditions.

Oftentimes, the selected thermodynamic method still needs to have additional inputs such as binary interaction parameters to simulate the process accurately.

5. Many of the calculations in process simulation are iterative. When these calculations iterate, the program compares the results calculated from one iteration to the results calculated in the next iteration in order to determine

whether it is converged to a solution or needs to continue calculating. This method of comparing results is where tolerances come into play and simulation programs use tolerances to determine when a solution is reached. One of the major mistakes when modeling a process is not understanding or using the correct tolerance values. A looser tolerance may make the file easier and quicker to converge, but it also reduces the accuracy of the model. A tighter tolerance may be more accurate, but might make it impossible for the model to solve or increase the convergence time exponentially. A common practice is that when you have reached a converged solution with your model, tighten the tolerances within the model and solve it again. If you notice that your answers have changed significantly, you need to continue tightening the tolerance values until your results change little from one iteration to another. You will then know that you have reached an accurate solution for your model.

6. Recycles in process simulation occur when you return to a previously calculated unit operation to recalculate the unit due to a change made elsewhere in the model. They are very common and can often be difficult to converge. Some suggestions on troubleshooting recycle convergence include initially solving your model without the recycles to see how it behaves, providing initial estimates for recycle streams, using tighter tolerances on unit operations within recycle loops than the total recycle tolerances themselves, and ensuring that all components have a way in and out of a recycle by performing mass balances and using purge or makeup streams where necessary.

7. The general rule of thumb when modeling is to start simple and add complexity as you go. Each time you add a few streams or unit operations converge and save the model before moving on. Make sure you justify any additions that you make to the model. You can most likely model your system without including every unit from your PFD in the simulation. When modeling, think about the goal of your simulation and not about literally translating every real-world piece of equipment into the simulation model. Make the judgment call on what is most important and what the purpose is of every unit you add to your simulation. For example, if your real-life process shows a heat exchanger, a pump and a splitter, but all that is necessary is that the temperature, pressure, and separation are achieved properly, consider using a flash drum to accomplish all three.

8. Specifications within process simulation are everywhere and every stream and unit operation you place in your model contains specifications. In its simplest form, a specification is a constraint that you place on the model that the model needs to meet in order to successfully converge. For streams, these are usually temperatures, pressures, flowrates and compositions. For unit operations, they range from process conditions, like temperatures and pressures, to product conditions, like an overhead composition in a column. There are some pitfalls, however, that you can run into with specifications such as azeotropes in columns, high purity specifications, and the possibility of multiple solutions.

9. Petroleum assays are sets of distillation data that are used to characterize all of the components that reside in a given crude sample. When you enter distillation data into a process simulator, the program converts it into components that it can use when modeling your process. The components that are generated are called pseudo-components, and each pseudo-component represents all of the unknown components that would be within a given boiling point range in the crude. Some of the things you will want to ensure when properly developing your assay components are did you use the correct distillation curve basis, did you enter the data correctly, did you use enough cuts and are you using the correct characterization methods .

10. When you have more than 1 assay in a simulation model, you need to be concerned with assay blending and how it will affect the results of your file. Assay blending is the mixing of 2 or more groups of components (i.e. ¨C sets of cuts/pseudo-components) in order to calculate averaged component properties for cuts that are in the same boiling point range as one another. Once these new averaged component properties are determined, the program will use these averages from this point on in the simulation model any time it needs to access component

properties for a given pseudo-component. All component properties are blended, including normal boiling points, specific gravity, molecular weight and refinery inspection properties, and usually the blended component properties are calculated using a weighted average

These topics are based on experience and feedback from end-users to technical support representatives for process simulators. They often represent the most commonly asked questions or areas that most users find to be problematic. While the specific examples highlighted in each topic come from SimSci software, the intent of this tutorial presentation is to be broad-based and is particularly applicable to young professionals seeking to develop their experience with simulation.

Polymeric Sulfonic Acid Composites for Fuel Cell Applications: Effects of Casting Solvent and Sonication Time on Membrane Performance

Jimoh K. Adewole[1], *Abdullah S. Sultan*[1, *], *Amir Al-Ahmed*[2], *and S. M. Javid Zaidi*[3]

[1]*Center for Petroleum & Minerals, King Fahd University of Petroleum & Minerals, Dhahran 31261, Saudi Arabia*

[2]*Center of Research Excellence in Renewable Energy, King Fahd University of Petroleum & Minerals, Dhahran 31261, Saudi Arabia*

[3]*School of Chemical Engineering, The University of Queensland, Australia*

Abstract

Polymer blends comprising sulfonated poly (ether ether ketone) (SPEEK) and 3wt% (2-acrylamido-2-methyl-1-propanesulfonic acid) (PAMPS) were prepared by solution casting method. Two solvents (N,N-Dimethylacetamide (DMAc) and N-Methyl-2-pyrrolidone (NMP)) were uesd separately. Residual solvents in the membrane samples after drying were quantified using thermo gravimetric analysis. Effects of periods of sonication (0 to 90 minutes) and types of solvent on the membrane performance were evaluated interms of morphology and water uptake. An attempt was made to correlate membrane performance with the time of sonication in order to select the most suitable solvent and membrane casting conditions for subsequent membrane preparations. For both solvents, the water uptake was observed to first decrease at 10 minutes of sonication and then increased as the sonication time was increased. The minimum water uptake recorded for DMAc samples was 19.23% at 20 minutes sonication time while that of NMP sample was 24.24% at 10 minutes. However membrane morphology was found to deteriorate with increase in sonication time. No correlation was obtained between the membrane properties and the sonication time. In general, membrane performance was better for membranes samples using DMAc than NMP solvent. Therefore, DMAc was selected as the casting solvent for membrane preparation on which further studies will be carried out.

Keywords: morphology, water uptake, residual solvent, thermogravimetric analysis

Introduction

Water soluble polymers have received academic and industrial interests due to their potential applications in fuel cells and other electrochemical devices (Ye et al. 2012). One of the water soluble polymers that have attracted publication interest is poly (2-acrylamido-2-methyl-1-propanesulfonic acid) (PAMPS) (Florjańczyk et al. 2003). Generally, some of the major performance criteria for fuel cell materials include proton conductivity, water and methanol uptake, good mechanical properties, chemical stability, thermal stability and methanol permeability (Gil et al. 2004; Swier et al. 2005; Ye et al. 2012). PAMPS was reported to posses higher proton conductivity than partially hydrated Nafion. In addition, it has been found to be more tolerant to water fluctuation in high temperature operations. Some of the major drawbacks in the use of PAMPS are its poor mechanical properties, thermal stability, water and methanol uptake and solubility in many solvents (Florjańczyk et al. 2003). One of the approaches that have been proposed to solve these problems is development of polymer composites.

This research paper investigates the synergistic effect of combining water soluble PAMPS with sulfonated poly (ether ether ketone) (SPEEK) which is non - water soluble polymer to prepare composite membranes for application in fuel cells. Specifically, preliminary results on the performnance of the composite were presented. Performance were evaluated based on the composite morphology and water uptake. Effects of sonication time and types of casting solvent on membrane properties were also evaluated.

Material and Method

Material

Sulfonated poly (ether ether ketone) (SPEEK) purcahsed from FuMA-Tech, Germany, and poly (2-acrylamido-2-methyl-1-propanesulfonic acid) (PAMPS) purchased from Aldrich Chemical Co., are the two major polymers used in this research. The SPEEK has a DS of 0.533 and solubility of 99% in DMAc. SPEEK is derived by sulfonation of Poly (ether ether ketone) (PEEK). PEEK is a commercially successful polymer due to its excellent thermal stability and solvent resistance (Mohr et al. 1991). For sometimes, further application of PEEK have been stalled due this solubility issues. By sulfonation, PEEK can be made soluble in most commercial solvents. Importantly, the solubility of SPEEK also depends on its degree of sulfonation (DS). Since, this work involves the use of two polymers, a solvent which will disolve both the PAMPS and SPEEK must be selected. Literature review showed that PAMPS dissolves in N,N-Dimethylformamide (DMF), N,N-Dimethylacetamide (DMAc), propylene carbonate (PC), and water (Florjańczyk et al. 2003). Initial solubility studies that was carried out in our laboratory was done for N,N-Dimethylacetamide (DMAc) and N-Methyl-2-pyrrolidone (NMP). The choice of NMP was due its unique solvent properties and its structtural distiction from DMF and DMAc (Figure 1). Moreover, no literature was found where NMP was used as casting solvent for PAMPS. Thus, DMAc and NMP were used as solvents in this work. All materials were used as received from the manufacturers.

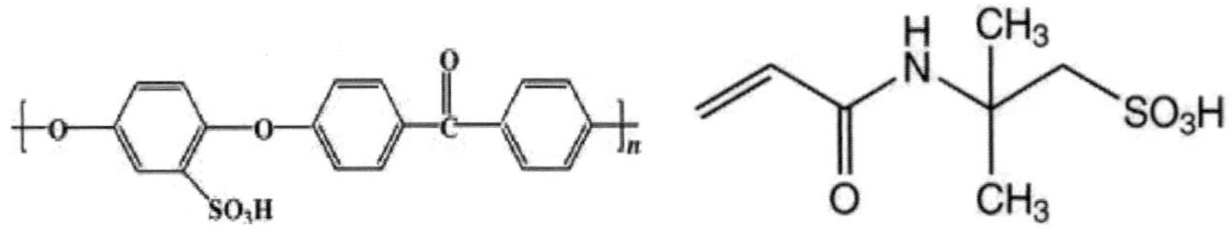

Figure 1 Structure of SPEEK and PAMPS

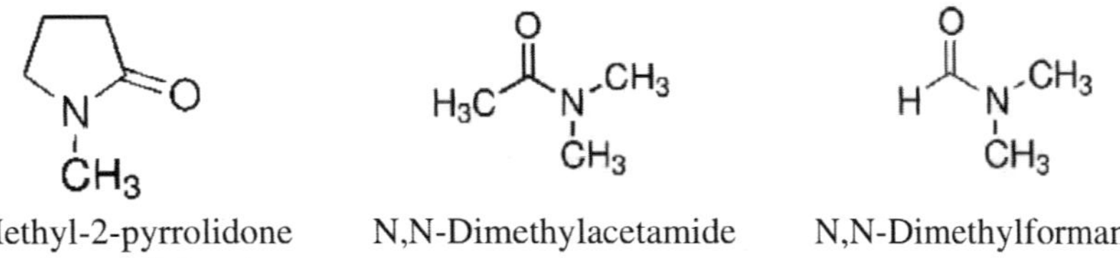

N-Methyl-2-pyrrolidone N,N-Dimethylacetamide N,N-Dimethylformamide

Figure 2 Structure of Solvents

Experimental Procedure

Membrane Preparation

SPEEK was first dissolved in solvents (DMAc and NMP) and stirred for 2hours at 70°C. PAMPS was then added and the blend was stirred for 24hours. The solution was sonicated for various periods of time ranging from 10 to 90minutes using an Ultrasonic bath. In untralsonic bath, the final temperature achieved for a given time of sonication depends on the tempearutre of the laboratory (summer or winter). It is therefoe genrally advisable that maximum temperature of the bath is determined and maintains (as the so - called equilibrium tempearture), when operating continuously under ambient conditions (Santos et al. 2009). During the sonication, equilibrium temperature was maitained by simply filling the bath with water that was preheated to sample temperature.

Membrane Characterization

Water uptake and morphological tests were performed. Water uptake was carried out by first vacuum drying all membrane samples at 100°C until constant weights were obtained. Dried samples were then immersed in deionized water for 24h at room temperature (25°C). Samples were taken out of the water after 24h, wipped to remove surface water and then weighed immediately. Water uptake was calculated using Eq. 1.

$$water - uptake = \frac{W_{wet} - W_{dry}}{W_{dry}} \quad \text{Eq. 1}$$

Surface morphology of the membrane was done using scanning electron microscopy (SEM). Thermogravimetric analysis (TGA) was done from room temperature to 350°C using TA Q600 analyzer at heating rate of 20°C/min under nitrogen atmosphere with flow rate of 20mL/min using ~ 8 -10mg membranes samples.

Results and Discussion

Membrane Morphology by Scanning Electron Microspcopy (SEM)

Results of surface morphological analysis performed on membrane were shown in Figure 3 and 4. The overall pictures of this figure depicts a dense membrane with some surface defects observed for most of the sonicated samples. For example, Figure 3A showed some kind of layered arrangement with no visible pores at the surface. Also, the PAMPS seems to be well distributed throughout the SPEEK. The micrograph in Figure 3B can be observed to be rough and possesses some uneven ramdom distribution of PAMPS agglomerates indicating that the distribution of PAMPS have been disturbed by the 10minutes sonication. An increase in sonication time to 20 minutes was observed to improve the distribution of PAMPS. This was however accompanied by some cracks believed to be due to the slight increase in the viscousity of the blend solution. The vicousity increase was brought about by the excape of some fraction of the solvent during sonication. As the sonication time was increased, PAMPS was observed to become more evenly distributed and the cracks become more visible and larger in size (Figure 3D to F). SEM images of samples at sonication time of 10 and 30 minutes were reproduced in higher magnification of 10μm (Figure 4) in order to further revealed the effect of sonication on the morphology of the somposites. Similar morphological behaviour has been reported in literature, though for inorganic/organic clusters of nanoclay/epoxy composites membranes (Lam et al. 2005).

Water Uptake

Water uptake have been reported to be affected by the degree of sulfonation, membrane pretreatment, hydration, ambient relative humidity and water temperature (Xing et al. 2004; Wu et al. 2006). Figure 5 showed the effect of sonication time and solvent type on water uptake that was measured at room temperature (25°C). For both solvents, water uptake is same at zero sonication time. There was however a further decrease in the water uptake for NMP at 20 minutes sonication time before it beging to increase. Thus, the best sonication time for both solvent lied between 10 and 20 minutes with respect to water uptake. Interestingly, this result correspond to that reported by Lam (2005) on the effect of ultrasonication on microhardness of composites.

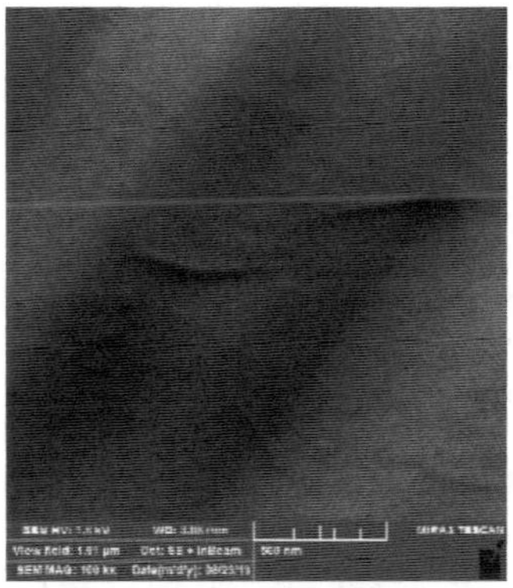

A) No sonication

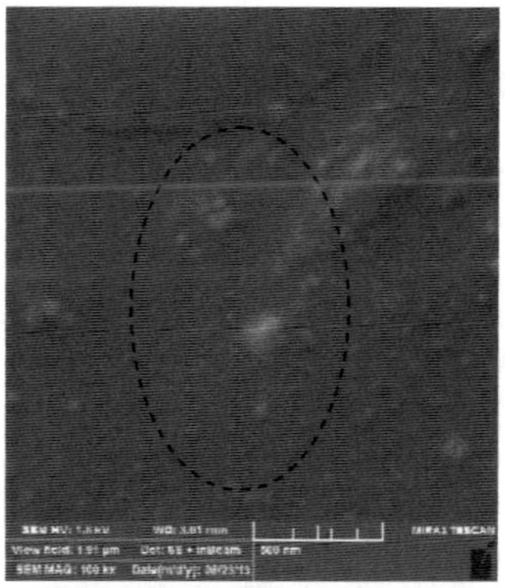

B) 10 minutes sonication

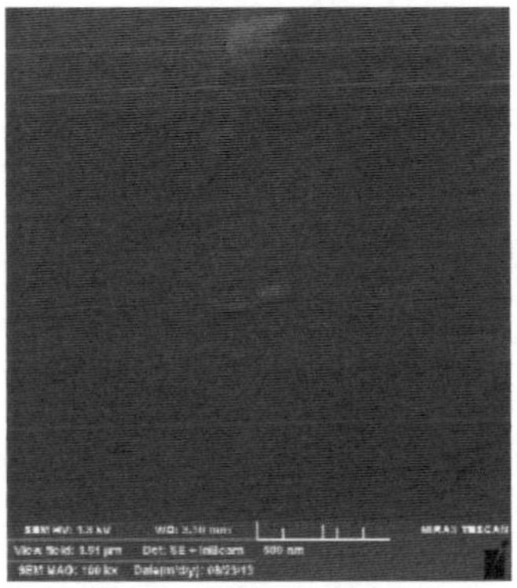

C) 20 minutes sonication

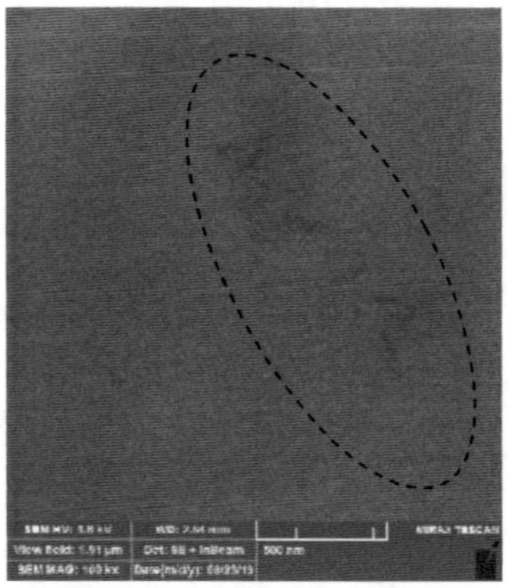

D) 30 minutes sonication

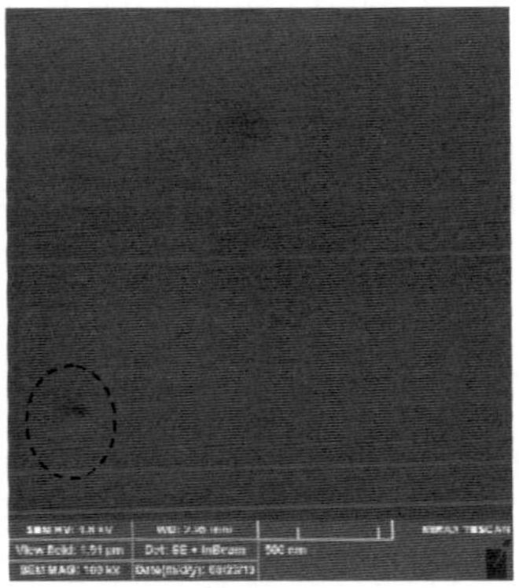

E) 60 minutes sonication

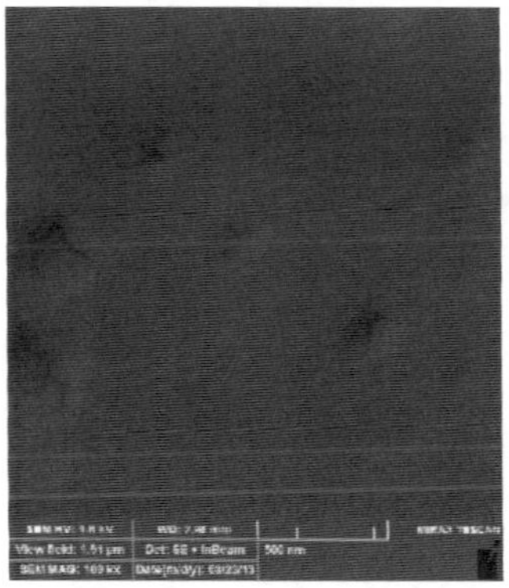

F) 90 minutes sonication

Figure 3 SEM micrograph of SPEEK-PAMS composite with various times of sonication with magnification of 500nm

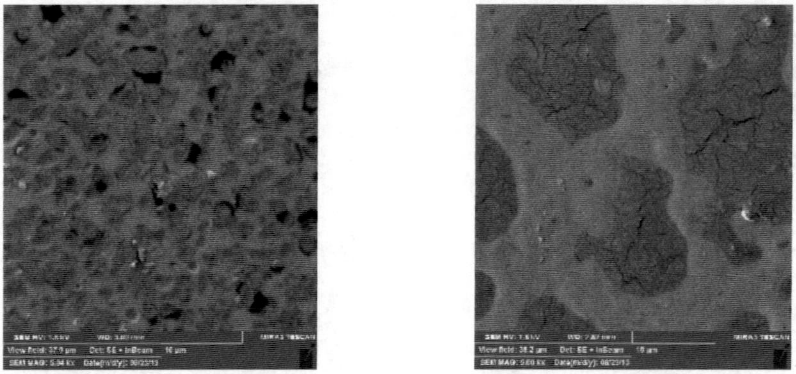

A) 10 minutes sonication B) 30 minutes sonication

Figure 4 SEM micrograph of SPEEK-PAMS composite with various times of sonication with magnification of 10μm

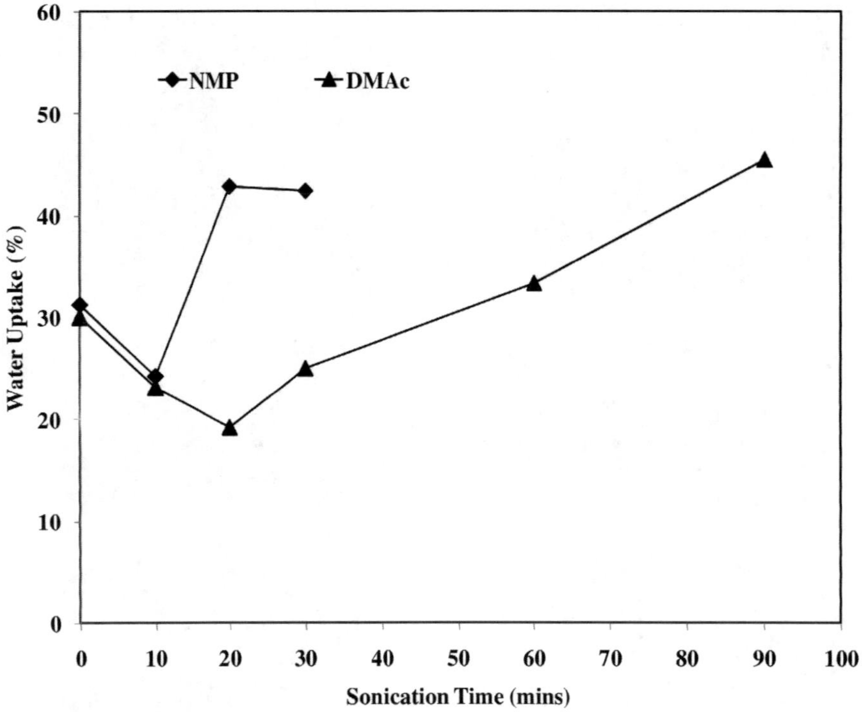

Figure 5 Water uptake of SPEEK-PAMS (3wt%) composite with variuos sonication times

Thermogravimetric analysis (TGA)

The difference in membrane performance resulting from the use different casting solvents have been attributed to the presence of residual solvents in the membrane samples after evaporation (Joly et al.

1999). TGA was thus used to quantify the residual solvent and evaluate its effect on the membrane performance as has been previously demonstrated in lieterature (Recio et al. 2007; Iqbal et al. 2008). Figure 6 shows the TGA curve for samples prepared from NMP and DMAc with different post vacuum heat treatment conditions. From this figure, weight loss can be attributed to evaporation of variety of substances, including solvents, from the samples (Rich et al. 1997). Table 2 contains the solvent and its boiling points, and weigth loss at the boiling point range that were extracted from TGA results. For sample heat treated for 24hrs, the overall percentage weight loss up to 204°C was about 8wt% of membrane sample for NMP while it is about 6.3wt% for DMAc. These will definitely include compnenents such as water, solvents and other unidentified substances. To quantify residual solvents, weight loss was estimated between 192 -214°C for SPEEK-PAMPS-NMP and between 154 and 176 for SPEEK-PAMPS-DMAc. The results obtained showed that the amount of the residual solvent in NMP is larger than that of DMAc by 1.25wt% (Table 1). To further investigate if this difference is responsible for higher value of water uptake for the NMP sample, membrane samples made from both solvents were further heat treated under vacuum for 72hrs. Residual solvent after this treatment was slightly higher in DMAc than NMP. Despite this, the value of water uptake of the NMP membrane is still higher than that of DMAc by approximately the same value as before the heat treatment (1.71wt%).

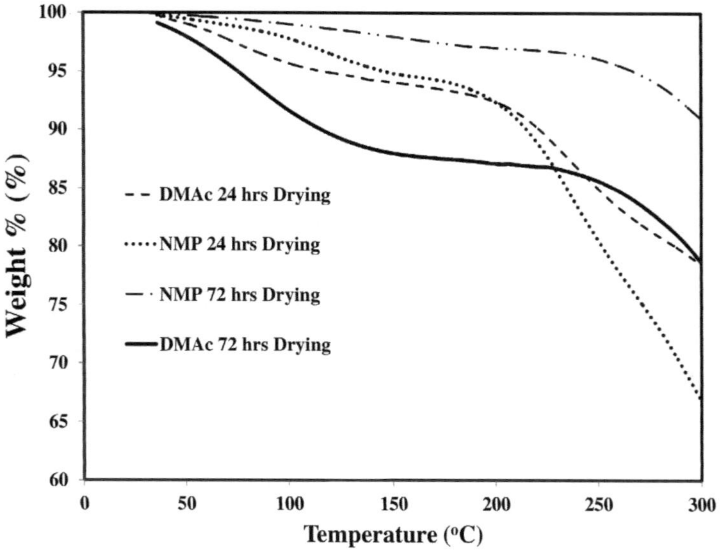

Figure 6 TGA of SPEEK-PAMPS-NMP and SPEEK-PAMPS-DMAc

Table 1 Residual solvent in SPEEK-PAMPS composites membrane samples

Membrane Sample	Solvent	B. Pt. (°C)*	Residual Solvent (wt%)	Water Uptake
DMAc 24 hrs Drying	DMAc	164.5 – 166	0.57	30.00
NMP 24 hrs Drying	NMP	202 – 204	2.85	31.25
DMAc 72 hrs Drying	DMAc	164.5 – 166	0.367	21.21
NMP 72 hrs Drying	NMP	202 – 204	0.217	22.92

*Boiling Points of Solvents

On a general note, results from Figire 1 – 5 showed that the best perfomance can be obtained for variety of sonication time depending the property of preference. Also, performance of DMAc solvent was observed to be better than that of NMP on the basis of water uptake. For this reason, DMAc was selected as the casting solvent for membrane preparation on which further studies will be carried out. Water uptake of DMAc membrane, prepared without sonication, was comapred with those reported in previuos literature. Table 2 illustrates the comparison between SPEEK-PAMPS and other relevant membranes previuosly reported in literature. In addition, this table serve to provide insight into the possible potentials of the composites. From this table, it was observed that the water uptake for SPEEK-PAMPS is within the range of values of other membranes on the table. Its water uptake is less than that of Nafion 112 which is a commercial membrane. The differences observed between the water uptake values of SPEEK – PAMPS and other membranes from the table can be explained based on three important factors - SO_3H group, the degree of sulfonation and experimental temperature (Gil et al. 2004; Shan et al. 2006; Wu et al. 2006). Water uptake of SPEEK – PAMPS is higher due to the presence of SO_3H group in the PAMPS molecules (Figure 2). SO_3H group have been reported to enhance membrane acidity and hydrophilic properties (Shan et al. 2006). Infact, the higher the SO_3H group the larger the water uptake. The degree of sulfonation has been reported to affect the water uptake of membranes. The higher the degree of sulfonation, the larger the value of water uptake (Gil et al. 2004). Thus, water uptake of SPEEK – PAMPS is lower than SPEEK -3 due their different degree of sulfonation, 53.3 and 72% respectively and the experimental temperature 24 and 60°C, respectively. The experimental temperature of SPEEK – 2 is close to that of the composite. However, the results cannot be compared in this case since the DS of SPEEK – 2 was not known. Above explanation is also true for the values of water uptake for SPEEK – 1.

Table 2 Comparison of water uptake values obtained with those from Literature (Xing et al. 2005)

Membrane	Water Uptake (wt%)	Experiment Temperature (°C)	**D S (%)	Reference
Nafion 112	22.85	N/A	-	(Xing et al. 2005)
SPEEK – 1*	16.00	30	45	(Ren et al. 2005)
SPEEK – 2*	44.90	25	N/A	(Zhong et al. 2008)
SPEEK – 3*	85.00	60	72	(Wu et al. 2006)
SPEEK– 4*	15.00	N/A	N/A	(Handayani and Dewi 2011)
SPEEK/PAMPS (3wt%)	21.21			This work

*The numbering have been included for explanation purposes.

**DS means Degree of Sulfonation

Conclusion and Recommendations

Blend of SPEEK and PAMPS was prepared by solvent casting method and its performance was evaluated with respect to its morphology and water uptake. Effect of sonication time (10 – 90minutes) and solvent types (DMAc and NMP) were also investigated. Performance of membrane prepared using DMAc was observed to be better than that of NMP. There was no correlation between the sonication time and membrane properties, but it was observed that the sonication deteriorate the membrane morphology. Membrane prepared without sonication shows better surface morphology, whereas the best water uptake was observed between 10 and 20 minutes. Thus, other properties need to be investigated in order decide the best sonication time. Therefore, further investigation such as thermal stability, methanol permeability, ion exchange capacity and gas (O_2 and H_2) permeability of the membrane are recommended for a complete evaluation of the membrane performance for direct methanol fuel cell (DMFC) application. All these investigations are in progress in our laboratory.

Acknowledgement

Authors would like to thank the National Science, Technology and Innovation Program (NSTIP) of KACST for funding this research under project number 10-ENE1374-04. Also, authors extend their appreciation to Center of Petroleum & Minearls, Center of Excellence in Renewable Energy, Mr. Dan Jerry D.Cortes of the Nanotechnology Laboratory, Chemical Engineering Department, King Fahd University of Petroeleum & Minerals, Saudi Arabis for their supports.

REFERENCE

Florjańczyk, Z., et al. (2003). "Proton conducting electrolytes based on poly(2-acrylamido-2-methyl-1-propanesulfonic acid)." Electrochimica Acta **48**(14–16): 2201-2206.

Gil, M., et al. (2004). "Direct synthesis of sulfonated aromatic poly(ether ether ketone) proton exchange membranes for fuel cell applications." Journal of Membrane Science **234**(1–2): 75-81.

Handayani, S. and E. L. Dewi (2011). "Influence of Silica/Sulfonated Polyether-Ether Ketone as Polymer Electrolyte Membrane for Hydrogen Fueled Proton Exchange Membrane Fuel Cells." Internat. J. of Sci. and Eng. **2**(2): 26-30.

Iqbal, M., et al. (2008). "Solvent effect on morphology and CO2/CH4 separation performance of asymmetric polycarbonate membranes." J. Membr. Sci. **318**(1–2): 167-175.

Joly, C., et al. (1999). "Residual solvent effect on the permeation properties of fluorinated polyimide films." Separation and Purification Technology **16**(1): 47-54.

Lam, C.-k., et al. (2005). "Effect of ultrasound sonication in nanoclay clusters of nanoclay/epoxy composites." Materials Letters **59**(11): 1369-1372.

Mohr, J. M., et al. (1991). "Gas transport properties of a series of poly(ether ketone) polymers." Polymer **32, 13**: 2387-2394

Recio, R., et al. (2007). "Gas separation of 6FDA–6FpDA membranes: Effect of the solvent on polymer surfaces and permselectivity." J. Membr. Sci. **293**(1–2): 22-28.

Ren, S., et al. (2005). "Surface modification of sulfonated poly(ether ether ketone) membranes using Nafion solution for direct methanol fuel cells." Journal of Membrane Science **247**(1–2): 59-63.

Rich, D. C., et al. (1997). "Effect of cure conditions on Probimide 32 polyamide-imide." Journal of Applied Polymer Science **63**(9): 1113-1126.

Santos, H. M., et al. (2009). The Power of Ultrasound. Ultrasound in Chemistry: Analytical Applications. J.-L. Capelo-Martinez. Weinheim, Wiley-VCH.

Shan, J., et al. (2006). "Sulfonated polyether ether ketone (PEEK-WC)/phosphotungstic acid composite: Preparation and characterization of the fuel cell membranes." Pure Appl. Chem **78**(9): 1781 - 1791.

Swier, S., et al. (2005). "Sulfonated poly(ether ketone ketone) ionomers as proton exchange membranes." Polymer Engineering & Science **45**(8): 1081-1091.

Wu, H.-L., et al. (2006). "Preparation and characterization of poly(ether sulfone)/sulfonated poly(ether ether ketone) blend membranes." European Polymer Journal **42**(7): 1688-1695.

Xing, D. M., et al. (2005). "Characterization of Sulfonated Poly(Ether Ether Ketone)/Polytetrafluoroethylene Composite Membranes for Fuel Cell Applications." Fuel Cells **5**(3): 406-411.

Xing, P., et al. (2004). "Synthesis and characterization of sulfonated poly(ether ether ketone) for proton exchange membranes." Journal of Membrane Science **229**(1–2): 95-106.

Ye, Y.-S., et al. (2012). "Water Soluble Polymers as Proton Exchange Membranes for Fuel Cells." Polymers **4**(2): 913-963.

Zhong, S., et al. (2008). "Modification of sulfonated poly(ether ether ketone) proton exchange membrane for reducing methanol crossover." Journal of Power Sources **180**(1): 23-28.

Fixed Plant Facility Siting – A Mega Project Lifecycle Framework

M.F. Haider
mech_haider@yahoo.com
mhaider@suncor.com

Prepared for Presentation at
American Institute of Chemical Engineers
2014 Spring Meeting
10th Global Congress on Process Safety
New Orleans, LA
March 30 – April 2, 2014

Fixed Plant Facility Siting – A Mega Project Lifecycle Framework

M.F. Haider
mhaider@suncor.com
mech_haider@yahoo.com

Keywords: Facility Siting, Project Management, Mega Project, Consequence Analysis (CA), Quantitative Risk Assessment (QRA), Adaptive, Change driven, Agile methodology

Abstract

Facility Siting based on tools such as Consequence Analysis (CA) and Quantitative Risk Assessment (QRA) is often times successfully applied for Brownfield operating unit projects, projects with relatively smaller footprints and for projects lasting relatively short time-periods. Challenge remains to apply and realize the full potential and value of Facility Siting for Mega-Projects. This is mostly due to project complexity and length – this paper will introduce a Management Framework to successfully apply CA and QRA as part of Facility Siting for Mega-Projects. The paper will also outline key strategies to ensure successful (i.e. value adding) implementation of facility siting, suggest a list of activities, associated sequence and pitfalls to avoid. Project Management team (e.g. Project Sponsor, Project Engineers, Managers) of both owner/operator and contractor/vendor may find this paper more useful as it provides a broad overview (i.e. when to initiate, how to plan, execute and close such assignment to maximize value) including real life examples of facility siting process over the project lifecycle.

1. Introduction

The driving force for process safety has been primarily based on catastrophic events [1]. With an increasing number of incidents, the process industry and governments started to take measures such as Facility Siting initiatives with hope to minimize loss of life and property. In the 1970s and 1980s, some of the world's most tragic industrial accidents took place. The Flixborough explosion in 1974 was reinforced by that of the Seveso tragedy in 1976 [2]. The Macondo blowout in the Gulf of Mexico on April 20, 2010, killed 11 employees and led to an uncontrolled oil spill lasting 87 days. This blowout was the most significant offshore incident in the US to-date. Finally, the Fukushima Daiichi nuclear plant incident in Japan in March 2011 drew the attention of the global process and power industries and encouraged them to incorporate natural disaster risks in a hazard analysis study [3]. As part of inherently safer design, many companies now-a-days apply CA and QRA as part of facility siting for their new projects. Due to the scarcity of information at the beginning, the studies in many cases start later in the project lifecycle for them to be effective and value adding. This paper will focus more on how to manage such initiatives for new Mega-Projects over the technical intricacies of the CA/QRA processes. The intention of the paper is to enhance the Body of Knowledge publicly available for management team in as much non technical languaga as possible to make their greenfiled Mega-Project initatives a success i.e. safer for people, environment and property.

So, what is different for industrial Mega-Projects?

- Industrial Megaprojects, as the name implies, are very large. To provide a simple and simply applied definition, we are defining a megaproject as any project with a total capital cost of more than $1 billion (U.S. dollars) as measured on January 1, 2003 [4].
- Most often engineering and construction activities run in parallel and not in sequence
- Multiple sub-projects under the 'Mega-Project umbrella' run in parallel
- Complex – design is almost 'never' complete in early stages – goes through many iterations
- Progressive elaboration is evident
- Scarcity of critical workforce with mega-project experience
- Scarcity of information – requires project team to make assumptions for CA/QRA
- Timeline: 2-5 years

2. Key Strategies

2.1 Facility Siting Charter

It is not only important or good to do but critical that the stakeholders need to identify, understand and agree on the purpose and value to conduct facility siting for their projects. The Author's personal experience over the years has been in order to maximize value of Facility Siting the project team has to start 'early' to develop project Charter and start with the end in mind. As can be seen from the following generalized Figure 1 that as the company starts to appraise opportunities, it is

Figure 1: Stages and Milestones to apply CA/QRA in a Mega-Project

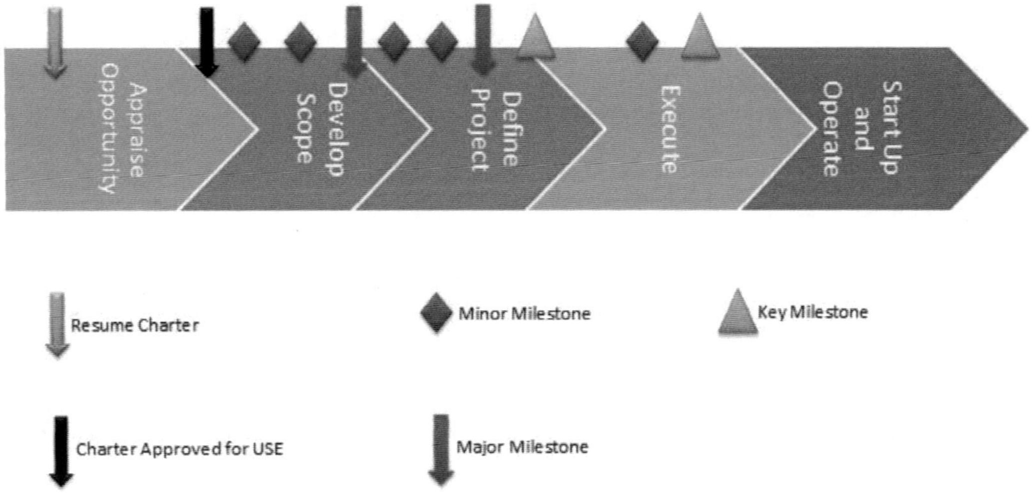

prudent to resume developing the Facility Siting Charter by bringing in the stakeholders together. Figure 1 shows a generalized project lifecycle for a typical Petro-Chemical Mega-Project. The

Figure also shows that the Key Deliverable is developed through a successive series of repeated cycles over the project life-cycle. The Milestones will be defined later in the paper. Part of starting with the end in mind is to clearly outlining ultimate goal/s, e.g. vision is to design a safer facility and utilize Mitigated QRA Report to formally document and communicate that facility is within the Risk Threshold or justification for accepting risk profile as is by Senior Management. The Vision in the Charter also provides opportunity to strike a balance between business goals and climbing the Risk Mountain – at each stage of increasing safety (decreasing risk). As part of striking a balance, the Project Engineer has to Identify options up-front if and/or when it becomes apparent that aggregate Residual Risk will not be within the Risk Threshold or acceptable so that team can put more effort in developing justification for either accepting or rejecting the risk profile by Management

Figure 2: Strike a balance between Business goal and Climbing the *expensive* Risk Mountain

2.2 Management Strategy

In order to successfully manage Facility Siting, CA and QRA for a Mega-Project, the stakeholders have to align on a management strategy. There are two primary management approaches for the purpose of managing such work:

1. Predictive or Plan driven
2. Adaptive or Change driven

The Author's past experience has been if the organization undertook very 'similar' projects in the near past (<5 years), critical experienced workforce is retained, Lessons Learned is readily available, and most importantly, the past project is still considered to be a success, it is advisable to use Plan driven approach. On the other hand, if the organization is introducing new technology, has limited past experience with similar project, involves relatively high number of stakeholders, capital cost > $1B, project life spans >2 years, it is recommended to utilize Change driven approach to maximize value. It is to be noted that it is not essential to meet all the suggested criteria mentioned above to pick one management strategy over the other.

2.2.1 Benefits of Adaptive or Change Driven Approach

In Adaptive/Change Driven Planning, project team intentionally repeats one or more project activities as the project team's understanding of the product increases ....Also known as ...agile methods .....intended to respond to high levels of change and ongoing stakeholder involvement [5]. The heart of Facility Siting involves making timely but informed decision. Some of the characteristics of this framework:

1. To add small incremental improvements that will deliver value to stakeholders
2. Milestones develop the product (e.g. Mitigated QRA Report) through a series of repeated cycles, while increments successively add to the functionality of the product
3. For each Milestone, a deliverable (read 'model') or set of deliverables will be completed. Future iterations may enhance those deliverables or create new ones. Each iteration incrementally builds the deliverables until the exit criteria for the phase are met, allowing the project team to incorporate feedback.

The biggest challenge with Mega-Project is scarcity of key information needed as inputs to develop scenarios and models. Some of the critical benefits, a project may gain by applying Change Driven approach:

1. The approach acts as Leading Indicator on Process Safety for Plant Layout - allows for multiple opportunities to make critical decisions on plant layout over the project life-cycle capitalizing on what is known at a given time. This lets the project team to use "what they know", to ensure "what they don't know" falls into the right place
2. Opportunity to make decisions over the project lifecycle not only works to save life but also helps with striking a balance between business needs and climbing the risk mountain
3. Aides design-development with progressive elaboration while producing final CA/QRA deliverable for the project

2.3 Execution Strategy

The main idea of Change Driven process starts with decomposing the known preliminary layout and process information into a set of requirements, work to be performed and their priority sequence, sometimes referred to as a *Product Backlog*.

It is advisable that the owner/operator forms a multi-use agreement (contract) with the chosen vendor. This way the plant areas under the Mega-Project umbrella can get their specific work done while the CA/QRA Project Team provides overall oversight and guidance.

At the beginning of an iteration, the team will work to determine how many of the highest priority items on the backlog list can be delivered within the next iteration. The time-length of each iteration should be less than 4 weeks. As shown in Figure 1, during "Develop Scope" and "Define Project", the length of each iteration could be even less than 2 weeks. This iterations (otherwise known as "Minor Milestone" as shown on Figure 1) will develop scenarios, assumptions and preliminary CA/QRA models. Assumptions are required due to lack of available quality information in the early stages of a project.

At the end of each iteration, the Minor Milestone/deliverable should be ready for review by the customers/stakeholders. This does not mean that the customer is required to accept delivery, just that the product should not include unfinished and/or incomplete features.

The sponsor and customer representatives should be continuously engaged with the other project stakeholders to provide feedback to Contractor on deliverables as they are created and to ensure that the *Product Backlog* reflects Sponsor/stakeholders' current needs. Below, you will find two examples using two different management strategies. For confidentiality purpose, the specific name/location is kept generic and few items were modified to suit.

2.3.1 Project X Facility Siting Using Change Driven Approach

A CA/QRA team was given the assignment to support design development during "Develop Scope" phase (Figure 1). The Sponsors already selected the site for the project. But the fixed plants were yet to be located on the site. The Vision for the initiative was for the QRA result to come under the organization Risk Threshold. Multiple EPC's were given to engineer different plant areas. The CA/QRA team selected a vendor to support them with defining the *Product Backlog* based on the preliminary Process Design Basis, intended product the facility is expected to sell, the feedstock necessary etc. The CA/QRA team had prioritized the Product Backlog based on hazard level of feedstock, preliminary PHA schedule and project schedule. This also involved obtaining feedback from different Process Engineering and Operations teams to collect necessary inputs to put together the assumptions and scenarios to be modeled within the 1st iteration.

Out of 25+ scenarios, only 7 perceived to be the most hazardous were modeled for CA that gave ample input (e.g. explosion, toxic and fire contours) for the CA/QRA team to work with the stakeholders to recommend location for specific process units on the given site. Note that, CA/QRA team integrated the potential consequence to understand local incident and global consequence events. At this time QRA was not performed. This was used as input for minor modification of preliminary plant layout. It is to be noted that the site plan was still preliminary after the first iteration but assisted different EPC accountable for their specific area to recognize where to locate the specific plants i.e. finalized key plans for the overall site plan.

With further design-development, more refined information became available from each EPC. This new information led to 130+ scenarios and was used for further CA/QRA iterations that created either new model and/or embellished the past ones. Each EPC was accountable for facility siting within their boundary while the CA/QRA team provided technical oversight and integration support for the overall project. The Sponsor/stakeholders were engaged on a regular basis to ensure that the Product Backlog reflects future modeling requirements to properly site facilities e.g. Fire Water Pump house, Tank Farms, site Medical Centre etc. Early engagement of CA/QRA team and higher management willingness to incorporate their feedback with the progressive elaboration of the project led to QRA result meeting the organization risk threshold and ultimately safer facility. Changes suggested based on CA/QRA work in earlier stage (i.e. Develop Scope, Define Project) were easier to incorporate into the design.

2.3.2 Project Y Facility Siting Using Plan Driven Approach

The sponsors for project Y selected the site for the project and decided to hold on to not perform CA/QRA work with the progressive elaboration of the project activities. This was mostly decided based on the fact that not all information was available to properly conduct the 'studies'. Fundamentally, the sponsors looked CA/QRA as deliverables requiring a more 'complete' plan. Once the Develop Scope (as referred in Figure 1) is almost complete, the sponsors decided to engage CA/QRA team and a vendor. By this time, the project had significantly high number of stakeholders. Many of the stakeholders were not very familiar with CA/QRA and the value it could bring to them. There was not much time to schedule and train the stakeholders as well to bring them up to speed. As a result, the stakeholders were unable to align on a Vision. The Vendor was asked to provide explosion, toxic and fire contour maps for specific scenarios and locations that were almost 'fixed' on the site plan at that point in time. The resultant contour maps showed significant consequence on nearby facility with over 700+ people. It was also noted that the plant area consuming the most hazardous substance would impact the nearby facilities significantly from consequence perspective. It should be noted that the result was only based on a selected few scenarios. By the time the project reached the "Execution" phase, more scenarios were added as part of the scope. But it was already too late to incorporate most of the recommended changes resulting from CA/QRA work. The project went through several area by area risk reduction workshops only to accept (by the company executives per organization risk matrix) the aggregate residual risk level which was also found to be not within the organization risk threshold.

2.3.3 Importance of Starting Early

Figure 3: Impact of variable based on project time [5]

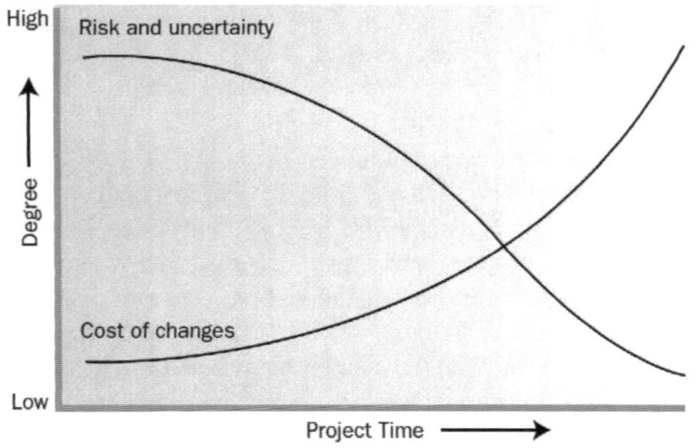

The above two examples highlight the importance of starting CA/QRA work early enough to realize and maximize the value CA/QRA brings to a Mega-Project. The high-level view provides common frame of reference for comparing projects for CA/QRA initiative, even if the projects were dissimilar. Change driven methodology allows for applying the known information at early stage in a way that supports laying out plant facilities to reduce risk and thus maximize potential value. While these characteristics remain present to some extent in almost all project life cycles, they are not always present to the same degree [5]. Change Driven framework, in particular, keeps stakeholder influences higher and the costs of changes lower especially in the early stages of the life cycle than in predictive/plan life cycles. The Figure 3 highlights a well-known fact of project management that even though risk and uncertainty remains high in the early stages of a project, cost of incorporating changes remains much lower. This translates into a key opportunity for project stakeholders to align on laying out the key areas on a site plan intelligently to reduce risk and maximize value by applying tools such as CA and QRA.

2.4 Key Decision Points

Facility Siting, from a management point of view, is all about bringing the stakeholders closer and find the 'right' spot for the facilities. Finding the 'right' spots requires making timely decisions. The challenge has always been to make timely decisions as the stakeholders more often find themselves forced to make those decisions based on information that are frequently of poor quality and at times uncertain, especially in the early stages of a project. At times, due to the scarcity of information, stakeholders rely on their past experience to make those decisions on plant layout. As a result, depending on the individual experience, two different teams for a project area may come up with two very different plant layout. CA/QRA work, if started early enough, will also depend on input that most often requires embellishment through assumptions to create needed scenarios and models. Even with this limitation, CA/QRA provides a key 'input' in allocating real estate for key facilities on a plant site at early stage of a project lifecycle. With each iteration and deliverable 'Change driven' methodology allows for multiple opportunities to make critical decisions on plant layout over the project life-cycle. It allows the stakeholders to decide when to 'sharpen the pencil' (e.g. PHAST to Computational Fluid Dynamics model), relocate facility or accept the consequence and/or risk, 'pull the plug' on further modeling/refinement and start developing justification, if required, to recommend management to accept/reject the aggregate risk profile.

2.5 Alignment with Process Hazard Analysis (PHA) Planning

CA/QRA results are a useful input to PHA (e.g. HAZOP, LOPA) activities. This is particularly true as projects move forward after establishing the plant layout on a high level and starts to conduct "Define Project" (Figure 1). PHA activities may also result in recommendations (i.e. new scenarios) that goes into the *Product Backlog* and requires project team to develop new models or refine old ones. Mega-projects have multiple sub-projects running in parallel most of the time, having an Integrated Schedule for all the reviews is a good strategy so that the right stakeholders are booked for the sessions. The integrated PHA schedule also supports prioritization of *Product Backlog* for CA/QRA activities over the lifecycle of the project.

2.6 Integration Strategy

It is critical for a Mega-Project involving multiple plant areas/silos to understand the BIG picture so that site risk can be dealt with on an aggregate basis. There is strong possibility of one area to impact other plants in the vicinity. It is recommended to review integration aspect and aggregated results on a regular basis. The following Figure 4 shows that the origin of an incident may reside in one particular plant area, while it may impact other areas in the vicinity.

Figure 4: Consequence of an incident in one area impacting multiple areas

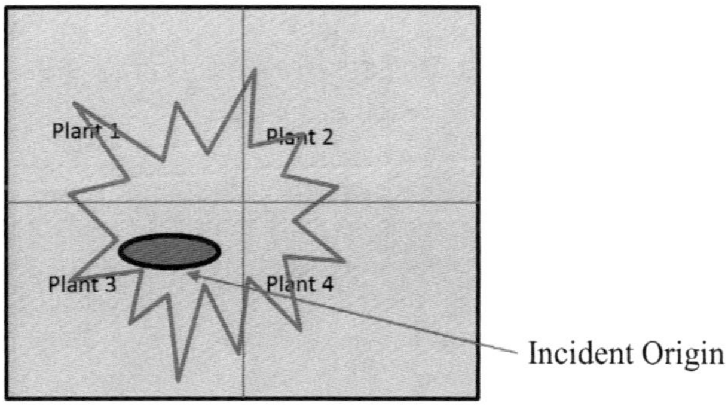

Recommended Strategy:

- Setup Strong Center - Task Force Single Point of Contact (SPOC)
- Collaborate – SPOC will promote Collaboration among Plants/Areas

3. List of Activities (suggested)

3.1 Overall Sequence

The CA/QRA work typically should start with the approved Charter as shown on Figure 1. A project Task Force i.e. SPOC is formed comprising of Project Engineer, Operations rep, Process rep and CA/QRA Technical Expert. The SPOC is ultimately accountable to produce the CA/QRA deliverable studies but also acts to support the design development and site plan finalization efforts in the early stages of the project.

The team collects 'input' such as Preliminary site Layout and design information, works with area process and operations team to identify scenarios, develop assumptions for known-unknowns and unknown-unknowns, develop modeling 'Product backlog', prioritize backlog based on PHA schedule and produce modeling output (i.e. deliverable) for stakeholder use. The stakeholder usually has the option of accepting or rejecting consequence. If the result of an iteration is acceptable within area causing the incident – communicate to outside stakeholders to ensure there is no cross boundary impact. If the result of the iteration is NOT acceptable to outside of the area – use the result as input to PHA for the areas to mitigate. Collaboration recommended between origin area and impacted area/s to identify appropriate mitigations. There is also a possibility that the consequence resulting from the iteration is not acceptable to the area causing the incident, in that case the area will initiate a PHA to mitigate the consequence. As part of developing mitigations, consider the following:

- Develop QRA if appropriate
- Perform Building Assessments for Critical Buildings
- Apply Cost-Benefit analysis to optimize risks/opportunities

- Determine Safe Haven locations and their characteristics

If Mitigation is not possible from Business perspective – perform As Low As Reasonably Practicable (ALARP) justification for Senior Management to formally accept the posed risk

- To apply mitigations on a site wide basis, review Milestone deliverables by taking on a BIG picture perspective
- Future iterations may enhance those deliverables or create new ones as new information is available. Each iteration incrementally builds the deliverables to produce the Mitigated CA/QRA report to support Facility Siting for the stage

The overall effort is expected to produce a maximum of two single reports for the overall site:

1. Mitigated version of QRA Report
2. and Consolidated CA Report

It is recommended to develop as-built, pre-start up CA/QRA reports prior to start-up. Table 1 and 2 outlines typical activities performed by SPOC and plant areas respectively once key process areas are established/finalized on Site Plan. Please note that the activities are project dependent and may warrant modifications to suit specific situation.

Table 1. SPOC Sequence of Activities to conduct Facility Siting

Initiative	Key Steps
Setup Facility Siting Team	Sponsor to support to setup a team (i.e. Project Engineer, Technical Expert, Operation, Process representative) to act as SPOC
Integration	Provide functional (technical and management) guidance to Areas
	Assess cross boundary impact/opportunity
	Communicate incremental outputs after each Area PHA/Risk Reduction
IFA versions (prior to "Define Project" stage)	Consolidated CA Report
	Mitigated and Integrated QRA Report
IFD CA/QRA (prior to "Execution" stage)	Re-validate IFA CA/QRA post PHA and vendor PHA
As-built versions	Post-oil version

3.2 Typical Area Activities

Table 2. Area (Typical) Sequence of Activities to conduct Facility Siting

Initiative	Key Steps
Plant 1 (Typical)	Perform Modeling and Calculations
	Conduct Risk Reduction Workshop (if appropriate) – Apply mitigations
	Incremental output for use based on reduced risk level
	Assess Buildings/facilities
	Identify Safe Haven/shelter Locations (if required)
	Estimate Area Project Scope impact
	Perform area scope updates/changes e.g. passive fire protection requirement
	Carry out QRA (if appropriate)
	Update Product Backlog
	Develop/update contract to support area specific activities

4. Pitfalls to Avoid

The following observations were noted by the Author during his time to manage Facility Siting scope:

1. Facility Siting does not replace good and common sense engineering practice e.g. locating a plant that uses high volume of hazardous material will significantly increase risks to other plants/areas (e.g. camp) within the vicinity
2. Evaluate effectiveness of 'Cost Reduction' philosophies as opposed to added risk e.g. different modularization techniques may decrease the plant footprint by reducing pipe lengths but makes the plant more compact and likely to increase severity of explosion consequence
3. Avoid developing models with hope to realize risk reduction
4. Sharpening the pencil for models most likely may not reduce risk
5. Avoid jumping into CFD or other similar methods without evaluating cost-benefit for undertaking such effort
6. Utilization of QRA results is much more controversial than the methodology. Watson [6] has suggested that QRA should be considered as an argument, rather than a declaration of truth. In this view – focus on accuracy over precision
7. CA/QRA Task Force should be part of Management of Change (MOC) process
8. Evaluate carefully before locating facilities e.g. tank farms close to main entrance/egress as part of Facility Siting

9. Keep the focus on VISION for the overall effort to avoid scope creep
10. Note the difference between Facility Siting as a Deliverable and tool for design-development activities

5. Conclusion

This paper presented Adaptive/Change driven framework to manage initiatives like facility siting/CA/QRA as part of Mega-Projects. Change driven framework does not mean "lack of planning or no planning". As demonstrated in the paper, this methodology starts with a vision and then develops the ultimate product both iteratively and incrementally when it is possible to define small incremental improvements that will deliver value to stakeholders. This allows making key decisions on laying out the facilities in early stages of a project rather than waiting for design to develop to a significant extent. Often times, project team wait too long for design to mature enough before applying CA/QRA to validate consequence impact and/or risk level. Most often, especially for Mega Projects, this leads to delays in executing the CA/QRA work properly at the right time to influence site layout. This delay in execution leads to scenarios when scope changes cannot be incorporated anymore. This translates into value erosion as project was unable to reap the full benefit of Facility Siting. The management framework recommended in this paper i.e. Change driven/adaptive framework allows for the stakeholders to start early in the project lifecycle to apply CA/QRA in laying out the plant facilities. The key is to start EARLY but START with the END in mind. Alignment among stakeholder on a VISION up-front is also important as this is expected to save life while striking a balance between business needs and climbing the Risk Mountain. Vision also allows avoiding scope creep while applying Adaptive/Change driven approach. It is highly recommended that stakeholders focus more on accuracy over precision. In the early stages of project lifecycle, quality information is scarce. Change driven/Adaptive framework gives you as the project stakeholder the opportunity to use "what you know" intelligently to ensure "what you don't know" falls into the right place.

6. References

[1] M. Mannan, A. Chowdhury and O. Reyes-Valdes, "A portrait of process safety: From its start to present day," College Station, 2012.

[2] M. Mannan, Lees' Loss Prevention in the Process Industries, Volumes 1–3 (3rd Edition), , 2005, Elsevier, 2005.

[3] K. McAndrews, "Consequences of Macondo: A Summary of Recently Proposed and Enacted Changes to US Offshore Drilling Safety and Environmental Regulation," in *Society of Petroleum Engineers, Americas E&P Health, Safety, Security and Environmental Conference*, 2012.

[4] E. W. Merrow, Industrial Megaprojects Concepts, Strategies and Practices for Success, New Jersey.: John Wiley & Sons, Inc., 2011.

[5] A Guide to the Project Management Body of Knowledge (PMBOK® Guide) – Fifth Edition, Newtown Square: Project Management Institute, Inc., 2013.

[6] S. Watson, "The Meaning of Probability in Probabilistic Risk Assessment," *Reliability Engineering and System Safety,* vol. 45, pp. 261-269, 1994.

348480 Global Energy and Transportation and Mobil Oil

Abstract:

Biofuels, Computational Methods, Engineering Economics, Polymers, Thermodynamics. Polymers-plastic, Thermodynamics- oil, Engineering Economics – the money it takes to finish the method of Global Engineering Research, Computational Methods – numerical input of analytical data, Biofuels – the transport fluids that utilized in more than one method from industrial usage to everyday living which enters the atmosphere. Exxon has completed a $600 million dollar deal for production of algae biofuel, amongst Mobil Oil has performed major research Global Energy and the effects of oil. Reading a most recent article on Mobil website.

348748 Characterization of Iron Phthalocyanine As the Active Material for Lithium Batteries

Abstract:

Results on the determination of the lithium-ion diffusion coefficient in the iron phthalocyanine (FePc) active material as a function of intercalation level, using a modified electrochemical impedance spectroscopy technique, will be presented. Also, developed thermodynamic functions such as the Gibbs free energy, entropy, and enthalpy changes for the lithium insertion reaction with FePc, derived from cell open-circuit voltage measurements as a function of intercalation level, are presented. Also presented are estimates of the maximum amount of lithium that can be intercalated into the FePc based on constant volume and temperature conditions.

349086 Design of a Free-Fall Reactor for Fast Pyrolysis of Waste Plastics

Abstract:

Thermal pyrolysis has earned significant reputation in the recovery of valuable chemicals, especially petrochemical based fuels from various materials, including biomass and plastics. Of many available technologies, *fast* pyrolysis was preferred in his study owing to its minimal heat and mass transfer limitations and better control on process conditions. The focus of this study is towards efficient conversion of waste plastics into gaseous products consisting primarily of alkanes, alkenes and aromatics. Lab scale experiments were performed using a micro reactor to determine favorable conditions, including reaction temperature and particle size that would result in maximum yield of ethylene. Using these results, a simple industrial size radiatively heated free-fall reactor was designed for the purpose of pyrolyzing waste plastics. The goal is to design a *fast* pyrolysis reactor that improves upon inefficiencies in traditional reactors while resulting in high yield of ethylene.

349732 Biosorptive Dehydration of Ethanol/Water Azeotropes Using Compound Starch-Based Adsorbent

Abstract:

Abstract: Recently, the interest in the production of bioethanol has greatly increased as a result of the increasing fossil fuel prices and associated environmental issues. Bioethanol is made from either grains or biomass and the broth contains 6 to 12 percent ethanol. In addition, water removal represents a serious problem as ethanol form azeotrope with water at atmospheric pressure. Distillation to water-free ethanol consumes 50 to 80 percent of energy used in the fermentation ethanol manufacturing process. Various techniques have been developed to break the azeotropic point, such as salting-out method, azeotropic distillation, extractive distillation, reactive distillation, adsorptive distillation and pervaporation. Among the methods, the adsorptive option is the most attractive, especially the biosorption, which has distinct advantages over conventional methods as it is environmentally friendly, highly selective, easily available, easy to operate, cost effective and reusable in repeated cycles in the treatment of aqueous organics.

In this work, a specially formulated compound starch-based adsorbent (CSA), which consists of corn, sweet potato and foaming agent, was developed. As an exploratory study, this research was focused on the investigation of CSA for biosorptive dehydration of aqueous ethanol and optimization of conditions for its maximum adsorption under various conditions of vapor feed flow rate, bed temperature and bed height through an orthogonal design method. Furthermore, to investigate the adsorption mechanism, dynamic saturated adsorbance of CSA was studied and characterization of the adsorbent before and after adsorption, as well as after regeneration, was performed.

In the first step, the net retention time and separation factor of ethanol and water were obtained using inverse gas chromatography (IGC), the results indicated that the net retention time of water on CSA is significantly longer than that of ethanol which attested the feasibility of this adsorptive separation purpose and low temperature could be propitious to the process which was consistent with the traditional theory of adsorption.

Then, through a batch adsorption experiment, the optimum adsorption condition was determined under vapor feed flow rate, bed temperature and bed height by orthogonal design method, which indicated that the optimization condition for ethanol was at vapor feed flow rate of 1.1g/min, bed temperature of 83 and bed height of 25cm. The packing density was 445kg/m^3. The residence time is determined by vapor feed flow rate and bed height. The residence time is more influenced than bed temperature. The longer the residence time, the more the output is. But it cannot be ignored that the adsorption selectivity may decreases. Low bed temperature is benefited to the adsorption process which agrees with the result of IGC. Five cycles experiments of adsorption-regeneration were carried out at the optimum adsorption and regeneration condition, the results showed that there is practically no change in the regeneration efficiency of CSA after the cyclic process adsorption–regeneration. It is due to the properly chosen regeneration conditions assuring complete desorption of the water adsorbed without changes in the CSA structure. This unchanged adsorption capacity after a number of cycles demonstrates reusability of the adsorbent.

To study the adsorption mechanism, experiment was conducted to study the dynamic saturated adsorbance of CSA. The water adsorption capacity of CSA has been found to be similar to some molecular sieves, which is about 0.15g/g. Besides, field emission scanning electron microscopy (FESEM) and mercury porosimetry (MP) were applied to investigate the change of surface morphology and microstructure of CSA before and after the adsorption, as well as after regeneration, as shown in Fig.1 and Fig.2. The FESEM images showed that compared with the particles before the adsorption, they became plump and the void spaces became smaller after adsorption, which indicated that the CSA was saturated with water. However, the CSA particles after regeneration became wizened like the nature particles, which demonstrated that almost all the water had been desorbed from CSA. The pore size distribution (PSD) indicated that compared with the PSD curve before adsorption, the peak at 6000nm disappeared and the peak at 17000nm became remarkably smaller after adsorption, which manifested that CSA is mainly composed of macropores which play a major role in the adsorption. However, the peaks at 6000 and 17000nm reappeared after regeneration, and the curve looked exactly the same with that before adsorption, which further demonstrated that CSA after regeneration can be totally regenerated and recovered to the original structure.

These results indicated that the starch-based adsorbent can be considered as an effective and low-cost bio-sorbent to dehydrate alcohol.

349781 Purification of 2-Amino-1-Phenylethanol Enantiomers By a New Technique Combining Distillation and Crystallization

Abstract:

Stripping crystallization (SC) is a new separation technique which combines distillation and crystallization. Basically, SC is operated at a triple-point condition, in which the liquid is simultaneously vaporized and crystallized due to the three-phase equilibrium. In essence, the process is continued until the liquid phase is completely eliminated and only the pure solid crystals remain in the feed. Thus, crystal washing is not required since only pure solid crystals remain in the feed and no impurities are adhered on the crystal surfaces at the end of SC. In the previous research, SC has been successfully applied to separate p-xylene from the mixed xylenes by Shiau et al. (2005, 2008).

Phenylethanolamine, also known as 2-amino-1-phenylethanol (2-APE), is an important basic skeleton in many adrenergic drugs and is often used as ingredients of cold medicines and anorectics. As the biological activity and pharmacokinetics of the enantiomers of a chiral drug can be different, the objective of this research is to apply SC to purify R-2-APE from the chiral 2-APE mixture. The experiments show that no crystallization occurs during SC without seeding due to subcooling of the liquid enantiomers. On the other hand, seeding can eliminate subcooling during SC, leading to the R-2-APE crystals gradually formed preferentially on the seeds. Thus, seeding with magnetic stirring can be successfully employed to purify R-2-APE from the 2-APE enantiomers.

1. Shiau, L. D., Wen, C. C. and Lin, B. S. Separation and purification of p-xylene from the mixture of m-xylene and p-xylene by distillative freezing, *Ind. Eng. Chem. Res*, **44(7)**, 2258-2265, 2005.

2. Shiau, L. D., Wen, C. C. and Lin, B. S. Separation of p-xylene from the multicomponent xylene system by stripping

350841 Fabrication of Low Cost Insulating Material from Kaolin Clay for Construction Purposes

Abstract:

A new low cost, thermal insulator, construction mineral polymeric material was synthesized from local Saudi Arabia resources using Inorganic Mineral Polymerization technique (MiP). In this technique, the mineral polymer is produced from the following components: clay mineral rich in koalinite, sand, sodium hydroxide solution, and water. Inorganic polymerization experiments are conducted by mixing known amounts of clay minerals and sand in a variable speed stainless-steel mixer for some period of times until a homogenous solid mixture is obtained. Sodium hydroxide solution is added to the mixture and the polymerization reaction takes place at different mixing speeds and times. The produced mixture is divided into a number of samples of known weights. Each sample is moulded in a stainless-steel cylinder at a medium pressure. Then the moulded samples are cured in an oven for 24 hours. The curing temperature was 80°C. The mineral polymerization conditions such as the ratio of clay to sand filler, particle size of the clay, particle size of the sand, concentration and amount of sodium hydroxide, speed and time of mixing were optimized to obtain stable and hard mineral polymer. Compressive strength for different specimens of fabricated mineral polymers was tested under dry and wet conditions. Values of compressive strength with values ranging from 11.5 MPa to 27.9 MPa were obtained.

351141 Thermodynamic Study of Binary PAH (Anthracene + Phenanthrene) Solid Mixtures

Abstract:

Polycyclic aromatic hydrocarbons (PAHs) are common components of many materials, such as petroleum and various types of tars. They are generally present in mixtures, occurring both naturally and as byproducts of fuel processing operations. It is important to understand the thermodynamic properties of such mixtures in order to better understand and predict their behavior (i.e., fate and transport) in the environment and in industrial operations. The Knudsen effusion technique was combined with gas chromatography and differential scanning calorimetry to measure the vapor pressure, composition, and phase behavior of binary anthracene and phenanthrene mixtures at near ambient temperatures. Mixtures of anthracene and phenanthrene exhibited non-ideal mixture behavior: In mixtures with less than 80% phenanthrene, this more volatile component preferentially sublimed and the vapor pressures decreased as the mixtures became enriched in anthracene. The vapor pressure data do not follow Raoult's law (that is, ideal mixture) predictions. Mixtures with high phenanthrene content (> 80%) showed azeotrope like behavior, in that anthracene and phenanthrene sublimed at nearly constant vapor pressure and molecular ratio. These results confirm that the behavior of few-component PAH mixtures does not follow ideal mixture models. Hence, as tars and petroleum (and potentially other organic compound-containing mixtures) weather to few component mixtures over time, complicated, non-ideal phase behavior might be observed.

351160 How Confident Are You That Your Major Accident Risks Are Under Control?

Abstract:

Major accidents in the process industries including those at Texas City , Buncefield UK and in the Gulf of Mexico have highlighted the critical role played by senior managers and executives. Effective leadership is essential if companies are to develop a positive safety culture that remains constantly vigilant towards the risk of a major process safety incident. In the UK the HSE has made leadership a key priority for the high hazard industries and there is a cross-industry initiative to develop and deliver training focussed towards board members responsible for process safety performance.

This paper explores the leadership failings that contributed to recent major accidents, and the cost to both the company reputation and the individuals involved. Key principles are discussed that can help improve leadership in the major hazard industries. These include; attitudes and decisions of senior manager affecting the investment strategy and the safety culture of the organisation, reinforcing the importance of safety by personal example, thorough understanding of major accident hazards and key risk control systems, investigating process safety incident and near misses to find the underlying causes, developing world class safety management systems, and identifying weaknesses in these systems using targeted performance indicators.

The author bases this paper on the development of a training standard for 'Process Safety Leadership' involving a cross-industry 'expert panel'. Training of senior executives in compliance with this standard has been carried out from late 2011 and feedback from these courses and the key requirements for effective leadership will be discussed.

351236 Sil Determination: Shortcomings with the Use of LOPA

Abstract:

Proposed Topic Area: Managing Risks

For a number of years, industry has been trying to apply the principles outlined in IEC 61508. In the process sector, this has resulted in a focus on the requirements of IEC 61511 (2003) and ANSI/ISA 84.00.01 (2004). These are risk based standards. They involve the identification of potential hazardous events, the causes that could lead to each event and the necessary risk reduction for each event. Where the risk reduction involves an instrumented function (e.g. a trip or an interlock), the required performance of the instrumented function can described in terms of the average probability of failure on demand. In the standards, the performance for such instrumented functions is assigned to one of four Safety Integrity Levels. The assessment of what performance level is needed in respect to a specific hazardous event is referred to as SIL Determination. In the field of SIL Determination, IEC 61511 provides examples of a number of different approaches – these are shown in IEC 61511 Part 3. These include Event Trees, Risk Graphs, Safety Matrices, Fault Tree Analysis and Layer of Protection Analysis (LOPA).

In the first few years, following the publication of IEC 61508 Edition 1 in 2000, industry generally embraced the use of Risk Graphs as the method of choice. However, over the subsequent few years, many companies using this method found that it was (a) giving higher safety integrity level targets than really needed and (b) was less flexible than other methods and consequently more difficult to use in many real situations. This has resulted in a general move to other methods and in particular to the use of Layer of Protection Analysis. Indeed, LOPA may now be the predominant "Method of Choice" for SIL Determination. As a method, LOPA gives the impression of simplicity and appears quite straightforward. It almost invites anyone with a minimum of understanding of the method to be a practitioner.

However, there are a number of shortcomings in the use of the LOPA method that have become evident through its use across the globe.

This paper will review the application of LOPA and provide insights into what may be seen as the proper use of the method and its core principles. It will also highlight various shortcomings in use that have been found during expert reviews carried out of a number of existing studies. The paper will be of benefit both to those who are new to the LOPA method, but also of value to those who have been using the method for some time.

Shale Oil and Gas – Wellhead to Fuel

Dr. Lawrence Kremer - Baker Hughes
AIChE - Hilton New Orleans Riverside
April 1, 2014

14 AIChE
Spring Meeting, New Orleans, LA

BAKER HUGHES

Outline Unconventional Shale Oil

- Properties of shale oils
- Problems
  - Transportation
  - Refining

14 AIChE
Spring Meeting, New Orleans, LA

BAKER HUGHES

General Characteristics of Shale Oils

- API gravity typically varies 30-55+
- Very paraffinic
- Low sulfur
- Can have hydrogen sulfide and mercaptans
- Low TAN
- Very low asphaltenes
- Light crude with very little residual material
- Many samples show great quality variation

Factor	Bakken	WCS
API	39.7	21.6
S%	0.25	3.44
Light ends	2.6%	
LSR	3.5%	3%
Light Nap	12.4	3.0
Heavy Nap	9.7	5.0
Jet	10.5	10.0
LFO	27.6	12.0
Gasoil	27.2	37.0
VacTar	6.7	29.8

14 AIChE Spring Meeting, New Orleans, LA

BAKER HUGHES

Shale Oil Property Comparison

Parameter	Eagle Ford	Bakken
API	52	40.8
TAN (g KOH/g)	< 0.05	0.09
Sulfur (wt %)	<0.2	0.304
Asphaltene (wt %)	0.1	0.41
Resin (wt %)	1.6	4.95
Filterable Solids (PTB)	225	76

14 AIChE Spring Meeting, New Orleans, LA

BAKER HUGHES

Utica Shale Crude - Three Phase

Eagle Ford Crude Samples

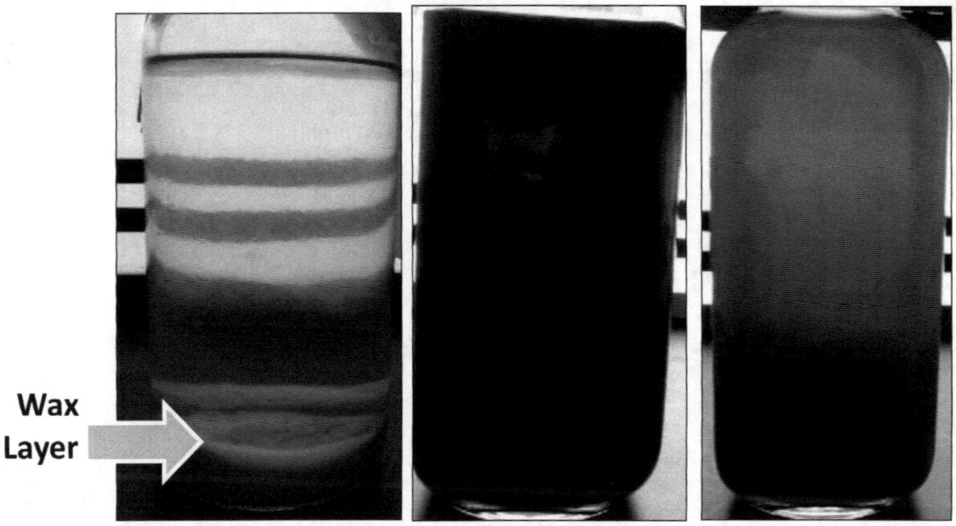

14AIChE Spring Meeting, New Orleans, LA
BAKER HUGHES

Analytical Breakdown

Value	Yellow	Red	Black
API	55.0	44.6	52.3
TAN	<0.05	0.07	<0.05
Sulfur	<0.20%	<0.2%	<0.2%
Na	1.0 ppm	1.6 ppm	1.6 ppm
K	0.3 ppm	0.4 ppm	0.5 ppm
Mg	3.4 ppm	2.9 ppm	3.0 ppm
Ca	2.6 ppm	2.8 ppm	3.8 ppm
Asphaltene	0%	0%	0.1%
Resin	0.5%	3.2%	1.6%
Filterable solids	176 PTB	295 PTB	225 PTB
P	5.3 ppm	4.6 ppm	5.0 ppm

14 AIChE Spring Meeting, New Orleans, LA

BAKER HUGHES

Micrograph of Sludge

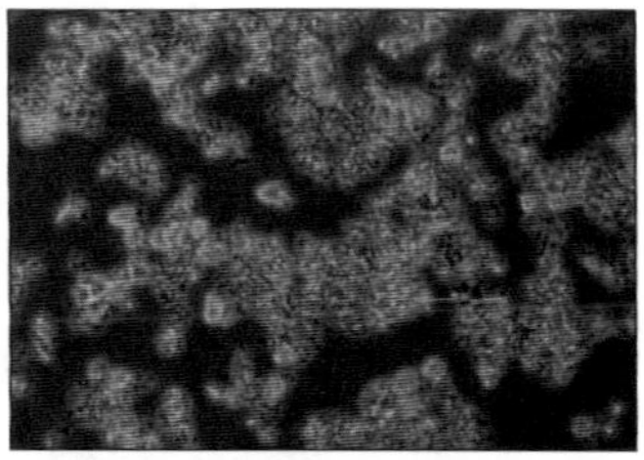

100 X

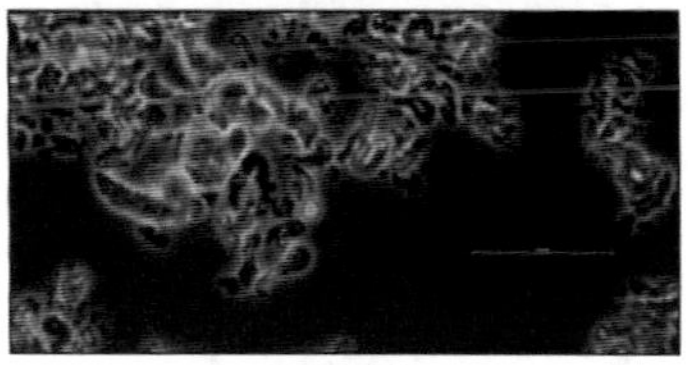

400 X

14 AIChE Spring Meeting, New Orleans, LA

BAKER HUGHES

Paraffin Analysis of Sample

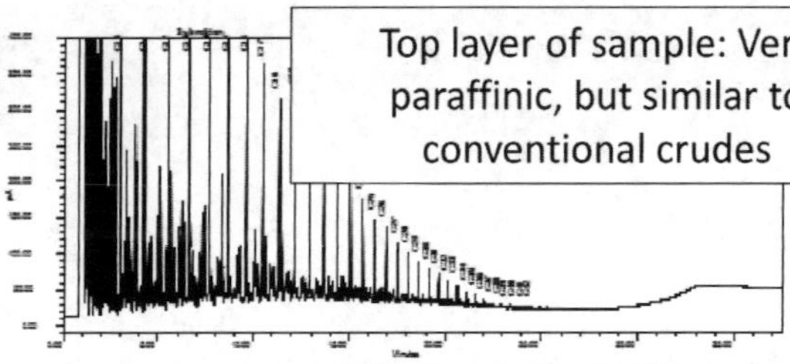

Top layer of sample: Very paraffinic, but similar to conventional crudes

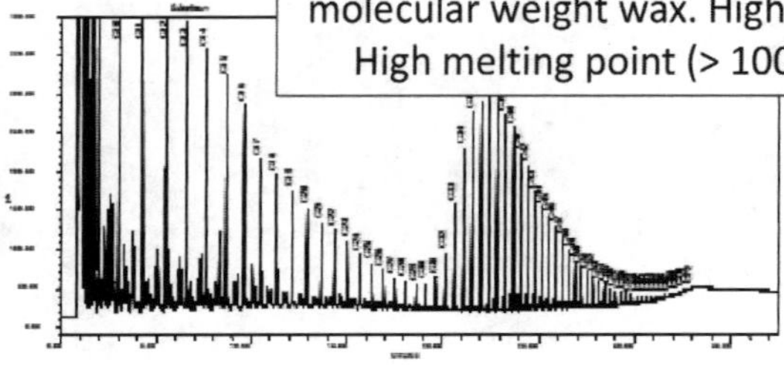

Bottom layer: Considerable high molecular weight wax. High MW = High melting point (> 100 °C)

14 AIChE
Spring Meeting, New Orleans, LA

BAKER HUGHES

Transport of Shale Production

- Infrastructure does not exist to meet needs
  - Most shale oil moved by rail or trucks
  - Limited pipelines expanding to meet transportation needs
- Unit trains to East Coast, West Coast, and Gulf Coast
- Problems:
  - Tankers and barges with heavy wax deposits ROB
  - Tight emulsions that look like oil or white condensate but high BS&W
  - Cars not uniformly treated for H_2S
  - Corrosion in rail cars

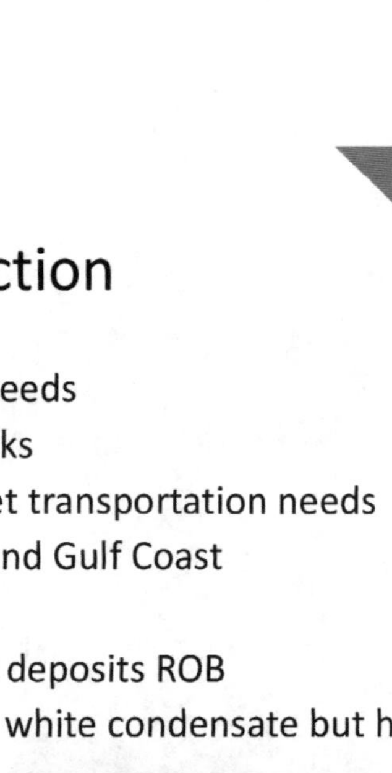

14 AIChE
Spring Meeting, New Orleans, LA

BAKER HUGHES

Viscosity of ROB Before Treatment

ROB from untreated compartment

14AIChE Spring Meeting, New Orleans, LA

BAKER HUGHES

Eagle Ford Shale Oil Wax Dispersion Testing

Several products were effective in dispersing the paraffin wax in an Eagle Ford shale oil sample at ambient temperature for over a month.

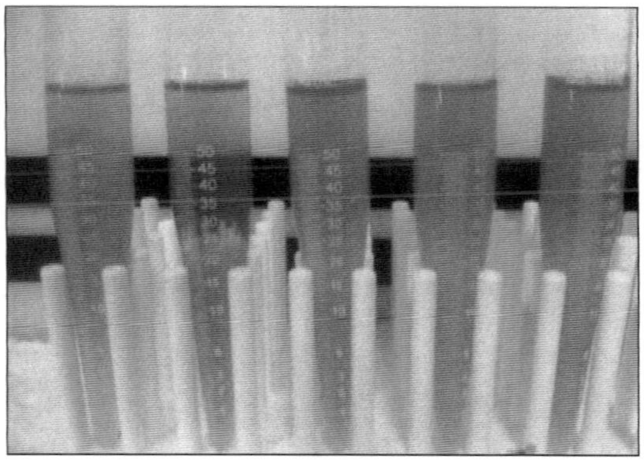

14AIChE Spring Meeting, New Orleans, LA

BAKER HUGHES

Product 'A' Effect on Wax Deposition (Utica)

Blank **200 ppm** **400 ppm**

14AIChE Spring Meeting, New Orleans, LA

BAKER HUGHES

Refinery Issues

- Crude Blending
- Fouling
- Corrosion
- Product quality

14AIChE Spring Meeting, New Orleans, LA

BAKER HUGHES

Crude Unit Fouling

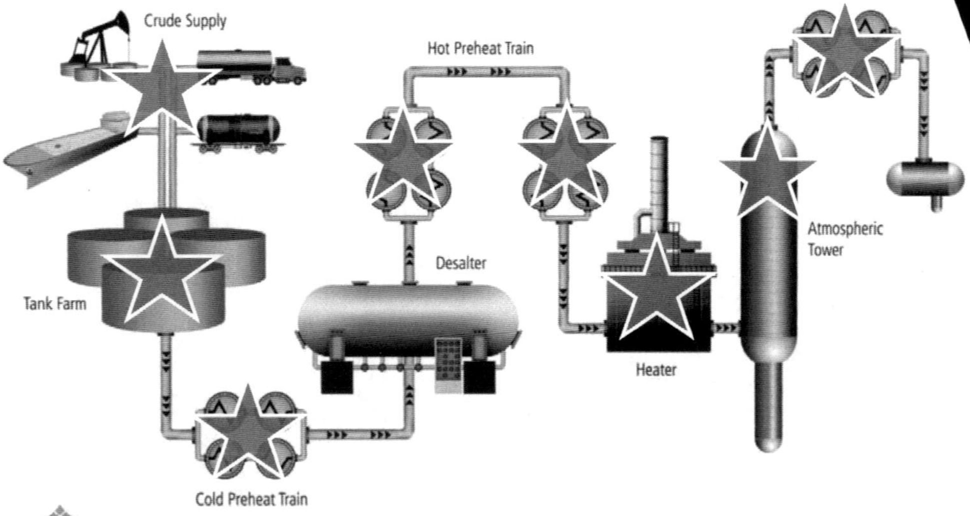

★ = Susceptible to Fouling

14 AIChE Spring Meeting, New Orleans, LA

BAKER HUGHES

Asphaltene Precipitation

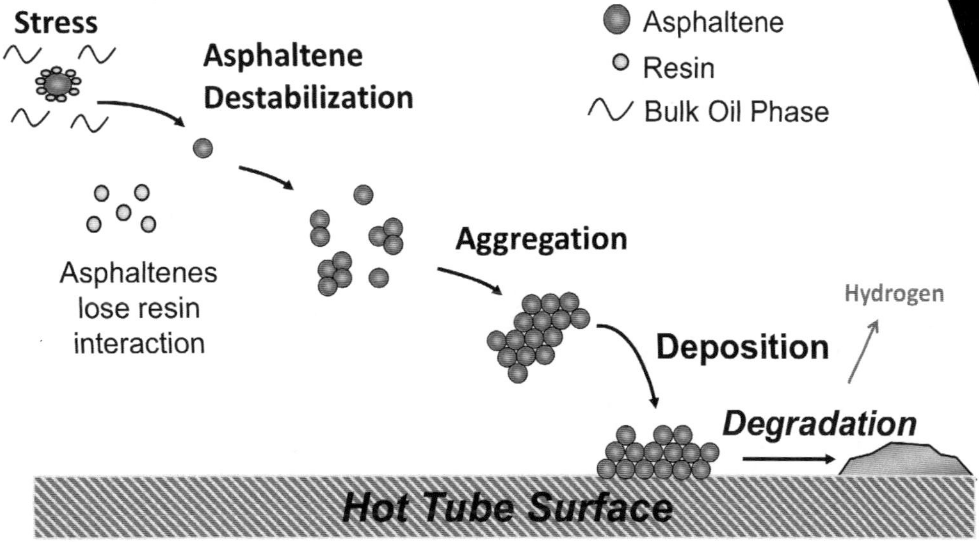

14 AIChE Spring Meeting, New Orleans, LA

BAKER HUGHES

Asphaltene Stability Index Test

Field ASIT services™ technology measures the onset of the flocculation of the asphaltenes with high accuracy by inducing the asphaltene precipitation via titration with a paraffinic solvent.

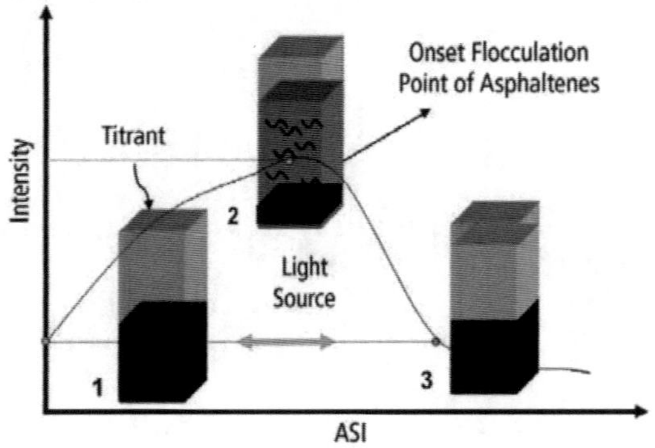

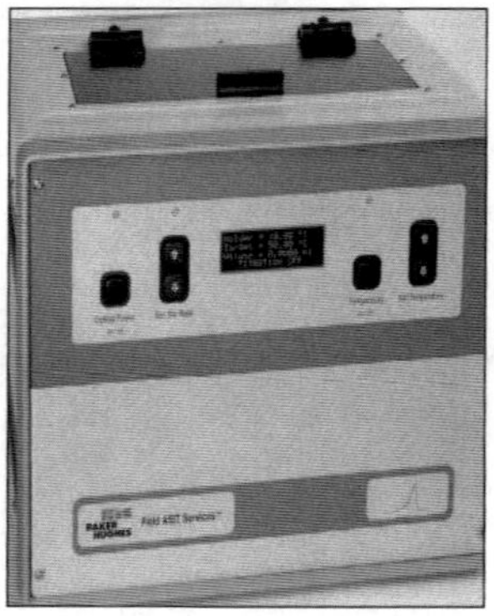

14AIChE
Spring Meeting, New Orleans, LA

Field ASIT services is a trademark of Baker Hughes Incorporated

BAKER HUGHES

ASI Shows Eagle Ford Can Destabilize Crude

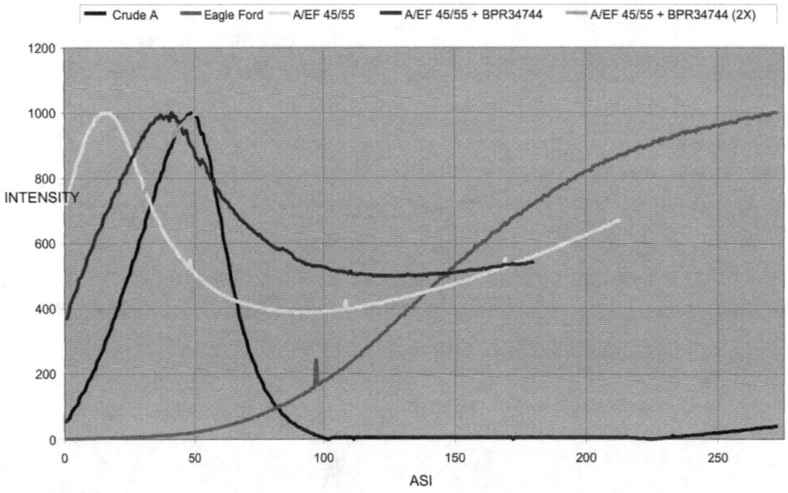

14AIChE
Spring Meeting, New Orleans, LA

BAKER HUGHES

Tower Tray Fouling - Phosphorus

- Seen by refineries running Bakken, Eagle Ford, and WTI
- Operationally manifested as increasing tower dP
  - Typically in the kero/jet zone
- Tower inspection shows deposits on trays in kero/jet zone
  - Some deposition may also show a few tray levels above and/or below kero/jet zone
  - Deposit scale typically tightly bonded to tray surface restricting flow and valve function

Above kero/jet zone

In kero/jet zone

Below kero/jet zone

14 AIChE
Spring Meeting, New Orleans, LA

BAKER HUGHES

Shale Oil Impact – MEA Contamination

Corrosion and Fouling

- Shale oils transported by truck, rail, barge.
- Many shipments are treated with H_2S scavengers:

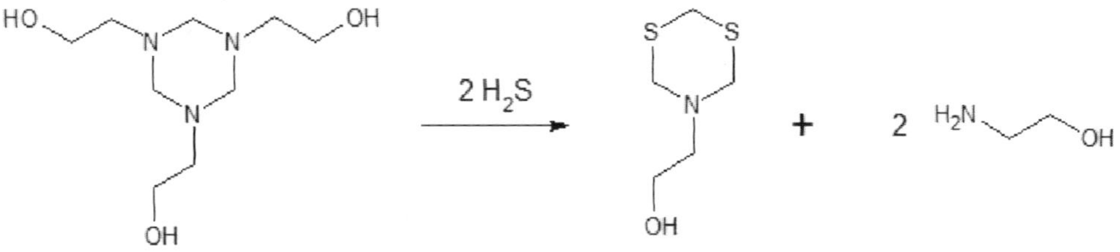

- Potential Impacts:
  - H_2S scavengers leave byproduct amine in crude oil (MEA, MA).
  - Tramp amines distill to OH system, create amine-HCl salts:
  - Corrosive to system metallurgy.

14 AIChE
Spring Meeting, New Orleans, LA

BAKER HUGHES

Impact of Tramp MEA on OH Water

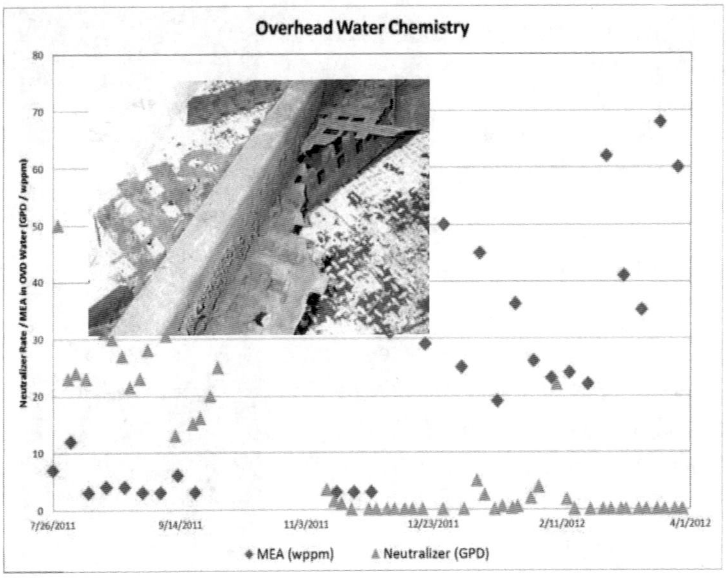

14AIChE
Spring Meeting, New Orleans, LA

BAKER HUGHES

Salt Risk w/MEA Concentration

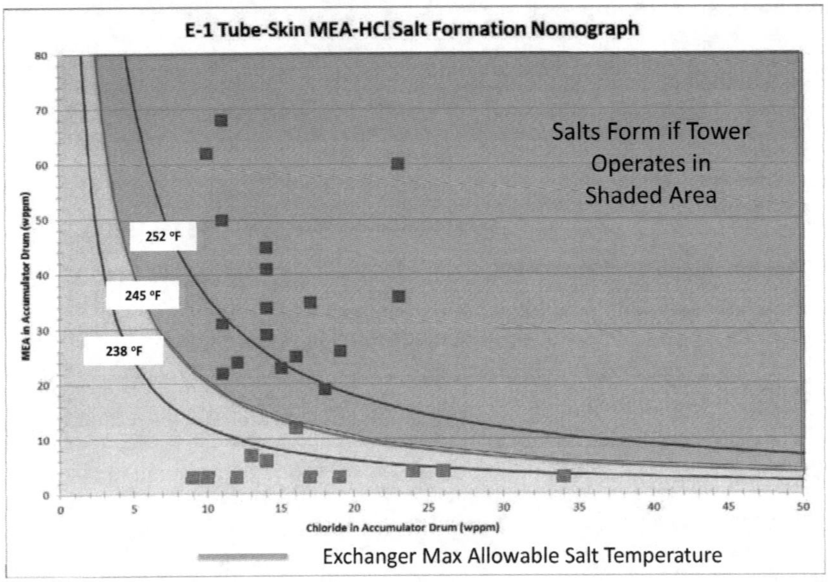

14AIChE
Spring Meeting, New Orleans, LA

BAKER HUGHES

Shale Oil – Refinery Process Concerns and Monitoring

BAKER HUGHES

CHARACTERISTIC OF SHALE OIL	CONCERN IN REFINERY	PROCESS DATA	SAMPLES AND ANALYSES		OUTPUT DATA	FREQUENCY	BHI PROGRAMS
			Samples	Analysis			
Crystalline Wax	Tank bottom sludge	• Tank Sludge Depth	• Tank bottom	• Visual sludge • GC analysis	• Tank Sludge Depth • Crystal Wax Presence	• Monthly	• XERIC™ Pretreatment
	Cold Preheat Fouling	• Desalter Temperature			• Temp vs. Time	• Update 2x/mo.	• LIFESPAN™ Wax Dispersants
	Desalter emulsions	• Emulsion layer growth	• Triline emulsions	• Visual wax • GC analysis	• Crystal Wax Presence	• After Shale Oil begins • When emulsion grows	• XERIC Demulsifiers
Asphaltene Stability in Blends	Desalter emulsions	• Emulsion layer growth	• Triline emulsions	• Precipitated Asphaltenes	• Precipitated Asphaltenes	• After Shale Oil beings • When emulsion grows	• XERIC Asphaltene Stabilizers • Field ASIT services™
	ColdPreheat Fouling	• Desalter temperature	• Crude blend	• ASIT	• Temp vs. Time • ASIT #	• Pre-Shale Oil • Update 2x/mo.	• LIFESPAN Antifoulants • Field ASIT services
	Hot Preheat Fouling	• Furnace inlet temp • Furnace tube temps • Furnace fuel usage • Preheat DP	• Crude blend	• ASIT	• Temp vs. Time • Fuel use vs. Time • ASIT #	• Pre-Shale Oil • Update 2x/mo.	• LIFESPAN Antifoulants • MILESTONE™ Furnace Fouling Control • Field ASIT services
Paraffinic	Furnace Coking	• Furnace tube temps • Furnace fuel usage	• Crude blend	• HLPS	• Coking tendency	• Pre-Shale Oil • W/Shale Oil	• MILESTONE Furnace Fouling Control
Tramp MEA	Tower/OH corrosion (Early Warning)	• OH Neutralizer rate • EXCAL usage rate	• OH water • Desalter brine	• OH pH • Brine pH	• Early Sign of Amine • Early sign of Amine	• Pre-Shale Oil • W/Shale Oil	• TOPGUARD™ Overhead Corrosion Control • EXCALIBUR™ Contaminant Removal Technology
	Tower/OH corrosion (Quantitative)	• Inputs for CRM	• OH water • Desalter ww • Desalter brine	• IC Amines	• Confirm MEA in Crude • OH MEA ppm • Salt Risk Tower • Salt Risk OH System	• Pre-Shale Oil • W/Shale Oil • Update 2x/mo.	• TOPGUARD Corrosion Control • EXCALIBUR Technology • TOPGUARD Corrosion Risk Monitor (CRM)
Phosphorus	Tower P Fouling	• Tower DP			• Tower DP Increase	• 2x/mo.	• LIFESPAN Antifoulants Dispersants
			• Crude	• ICP for P	• P in Crude	• 2x/mo.	
			• Kerosene	• Filt Solids • XRF Filter P • ICP for P	• Incr. Kero Solids • P in Solids • P in Kerosene	• 2x/mo.	
			• Kerosene • Diesel	• ASTM D86	• Incr. in Diesel IPB – Kero EP overlap	• 2x/mo.	

EXCALIBUR, FIELD ASIT SERVICES, LIFESPAN, MILESTONE, TOPGUARD and XERIC are trademarks of Baker Hughes Incorporated.

350842 Development of a Model for Mass Transfer Using Square Plate as Turbulence Promoter in Circular Tube

Abstract:

DEVELOPMENT OF A MODEL FOR MASS TRANSFER USING SQUARE PLATE AS TURBULENCE PROMOTER IN CIRCULAR TUBE

Prof.V.Sujatha*, Dr. V. Nageswara Rao*, D.S.Seetha Rama Raju*, Dr.M.Gangadhar* *

*Department of Chemical Engineering, Andhra University College of Engineering (A),

Andhra University, Visakhapatnam-530 003, Andhra Pradesh, INDIA

**Department of Chemical Engineering, G.M.R. Institute of Technology, Rajam-532127,

Andhra Pradesh, INDIA

ABSTRACT

Results are presented from experimental investigations of mass transfer in a 50mm I.D.copper tube inserted with various sizes of square plates. A similarity law approach was attempted to interpret the mass transfer results and correlate them in terms of Colburn factor J_D and Reynolds number Re. The effect of geometric parameters such as size of square plate (S_s), thickness of square plate (S_w) and spacing between square plates (h) in conduit on mass transfer coefficient are studied. The present results are compared with previously published results and a generalized correlation in terms of J_D and Re is developed which is applicable for different types of square plates. The following correlation is developed between J_D and Re

$$J_D = 195.66\ Re^{-0.9014}\ (f_1)^{0.6325}\ (f_2)^{0.1799}\ (f_3)^{-0.3484}$$

KEYWORDS: Mass transfer, Square Plate, turbulence promoter, bluff body

Extending Ethylene furnace run length and coil life through decoking procedure optimization

Abstract:

In the Ethylene manufacturing process, coke deposited in the radiant coils and Transfer Line Exchangers (TLEs) during normal operation is removed periodically by decoking the furnaces using a mixture of steam and air. There is a significant variation in operating parameters adopted by different plants for decoking. These are often either too conservative or aggressive, leading to production loss from increased off-stream time, TLE limitation and deterioration in coil life. This paper describes how the decoking procedure was modified based on furnace and TLE operating conditions and the decoking steps changed to obtain the best possible Ethylene yield, without adversely impacting coil life and TLE performance. In the particular example furnace run length was more than doubled and production significantly increased.

Introduction:

Ethylene is produced by thermal cracking of hydrocarbons (Ethane/Propane/Naphtha etc.) in the radiant coils of furnace section. The furnace effluents from radiant coils are immediately quenched in transfer line exchangers (PTLE, STLE, TTLE etc.) to arrest the chain reaction and thereby avoid secondary reactions. This cracking is accompanied by coke formation. During normal operation, coke slowly deposits on the inside walls of the radiant coils and the transfer line exchanger tubes.
This leads to

a) Reduced heat transfer to the process fluid resulting high TMT (tube metal temperature)
b) Increased pressure drop, and eventually plugged coils in extreme case
c) High PTLE out let temperature exceeding design limit of downstream piping
d) Reduction in furnace run length and feed throughput in furnaces

All the above limitations vary depending on the feedstock, severity of cracking and the type of furnace.

To burn off the coke from the coils and TLE tube walls, the furnace is periodically shut down for decoking. Typically, a combination of high temperature air and steam is used to burn out the deposited coke. In some cases only high temperature air is used to decoke the TLE. Getting the appropriate level of decoking dramatically affects the future run length of the furnaces and ongoing life of the coils.

Key operating parameters of furnace operation:

To get the desired yield from the cracking furnaces the following parameters are continuously measured and closely monitored.

1) Feed rate and composition
2) Coil pressure ratio(CPR)
3) Coil outlet temperature(COT)
4) Hydrocarbon to Dilution steam ratio(HC/DS)
5) Transfer line exchangers outlet temperature and pressure

Decoke operation:

During the pyrolysis of hydrocarbon in the furnaces, along with useful products (Ethylene, propylene etc.) a certain amount of heavy hydrocarbon and carbon is generated. This combined heavy hydrocarbon and carbon is termed as coke. The coke is partly deposited on the inner walls of radiant coils, partly with in TLE and rest is reacted with dilution steam forming CO_2 and carried away along with cracked gas. A variety of different mechanisms induces coking. This coking phenomenon inhibits the furnace performance in terms of yield and run length. Hence furnace is taken out of operation periodically to perform decoke operation.

Furnace operation is terminated and decoking done when any one of the following criteria is reached.

a) CPR reaches 0.9
b) A maximum TMT of any coil reaches the design limit of coil metallurgy(1050 ºC)
c) A maximum TLE outlet temperature limit of D/S piping is reached

Many decoke procedures are followed by the different manufacturers across the globe. Generally decoking takes 16 to 32 hours of actual decoking time.

Manufacturing units may decoke using either a rigorous (high air and steam flows and COT) procedure to burn the coke inside radiant coils and TLE exchangers or a very mild (low air and steam flows and COT) procedure for an extended period. Rigorous decoking affects the radiant coil life (coil damage) whereas very mild and extended period decoking results in significant production loss and increased utility consumption (energy loss).

It is important to adopt a holistic and scientific approach to optimize the decoke procedure for every individual Ethylene plant based on the type of furnace, feed, severity and fouling trends of TLE. Ingenero has provided this as a service to over a dozen ethylene facilities across the globe.

Historical plant operating data of various Ethylene plants shows that furnace operation and run length is limited both by high radiant coil CPR and TMT and after few months of operation TLE outlet temperatures. In both cases Ethylene yield loss occurs due to an increase in decoke cycles. In the longer run it also affects the radiant coil life and leads to coil failures (Radiant, convection and TLE). These coil failures result in down time for repair/replacement. Large production loss and decrease in plant reliability can occur.

To avoid/overcome this Ingenero developed a methodology to extend run lengths and minimize coil failures through:

1) Optimization of furnace operation (Micro Burner Management) based on various furnace operating parameters.

2) Modification and standardization of the decoke procedure.

Furnace operation optimization:

With the help an indigenously developed tool based on various furnaces operating parameters and suggested corrective actions on daily basis, coking rate in radiant coils is controlled and a fast rise in radiant coil CPR and TMT and TLE outlet temperature avoided.

Decoke procedure modification to delimit TLE out let temperature:

With normal decoke procedure cokes deposited in radiant coils gets burnt easily. But cokes deposited in TLE exchangers are not burnt or removed easily and continue accumulating. This reduces the heat removal in TLE and increases the outlet temperature and pressure drop. This starts limiting furnace operation in terms of feed throughput and severity (conversion) and results in huge losses in Ethylene production.

Ingenero modifies the decoke procedure based on furnace and TLE operating parameters to delimit the furnace operation in terms of feed rate and severity. Utmost care is taken in modified decoke procedure to avoid any impact on radiant coil, TLE tube life and connecting piping.

Key parameters considered in decoke procedure modification:

1) Coking rate in Radiant coil
2) Rate of rise in TLE outlet temperature
3) Fouling rate in TLE
4) TMT of individual Radiant coils associated with respective TLE
5) COT of individual Radiant coil

6) Excess CO_2 during decoke steps
7) Gas velocity during decoke steps
8) Dilution steam to Hydrocarbon ratio during decoke operation
9) Physical observation status of radiant coils during Furnace cracking and decoke operation

INGENERO Methodology:

- Ingenero developed a software tool, which provides deviations of individual coil CPR, TMT and TLE outlet temperatures in the furnace on daily basis
- Close monitoring of furnace parameters, Feed trims, COT trims and Burner management is done on daily basis to control CPR and TMT of coils as well as TLE out let temperatures
- Evaluating fouling trends of individual radiant coil and TLE in the furnace through proprietary software and commercially available software (HTRI/ HTFS)
- Optimizing decoke steps and modifications in decoke procedure

Following trend shows the overall control on rate of furnace radiant coils CPR through Ingenero developed micro management of burner and other furnace operating parameters.

CPR deviation trends of Furnaces

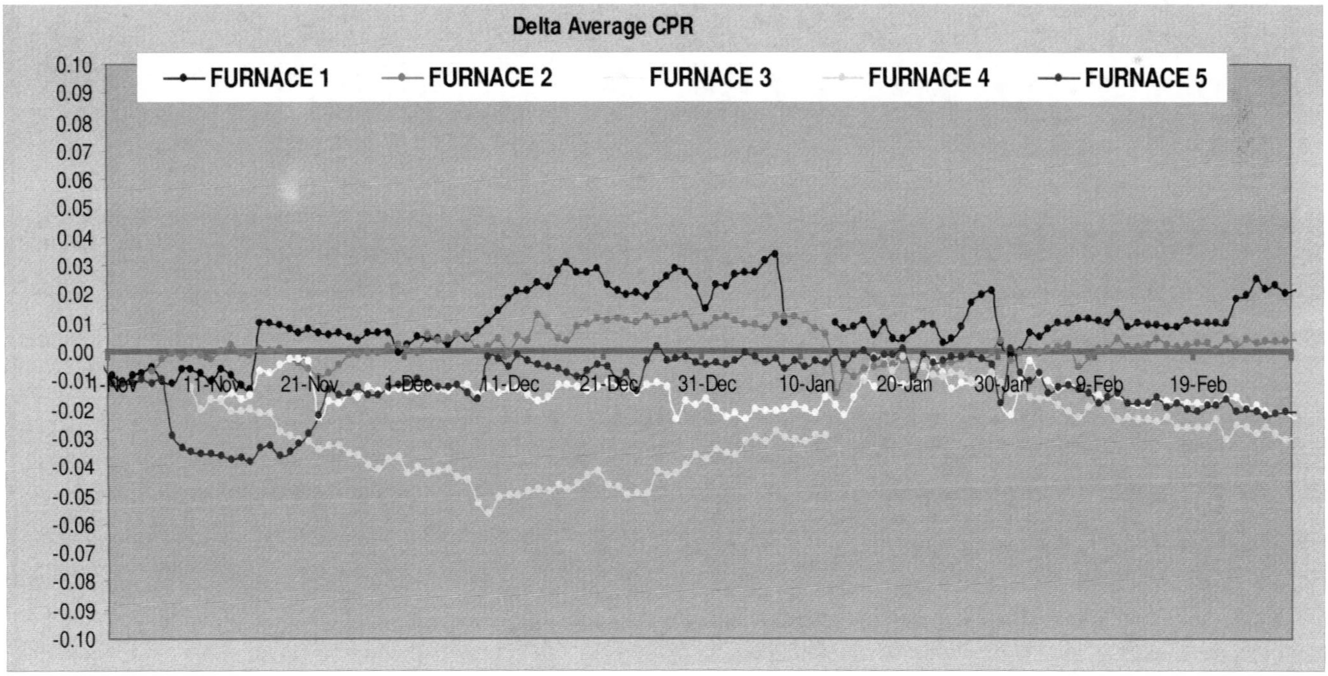

Following trend shows reduction in TLE fouling after Decoke procedure modification.

Transfer line exchanger fouling trends before and after Decoke procedure modifications

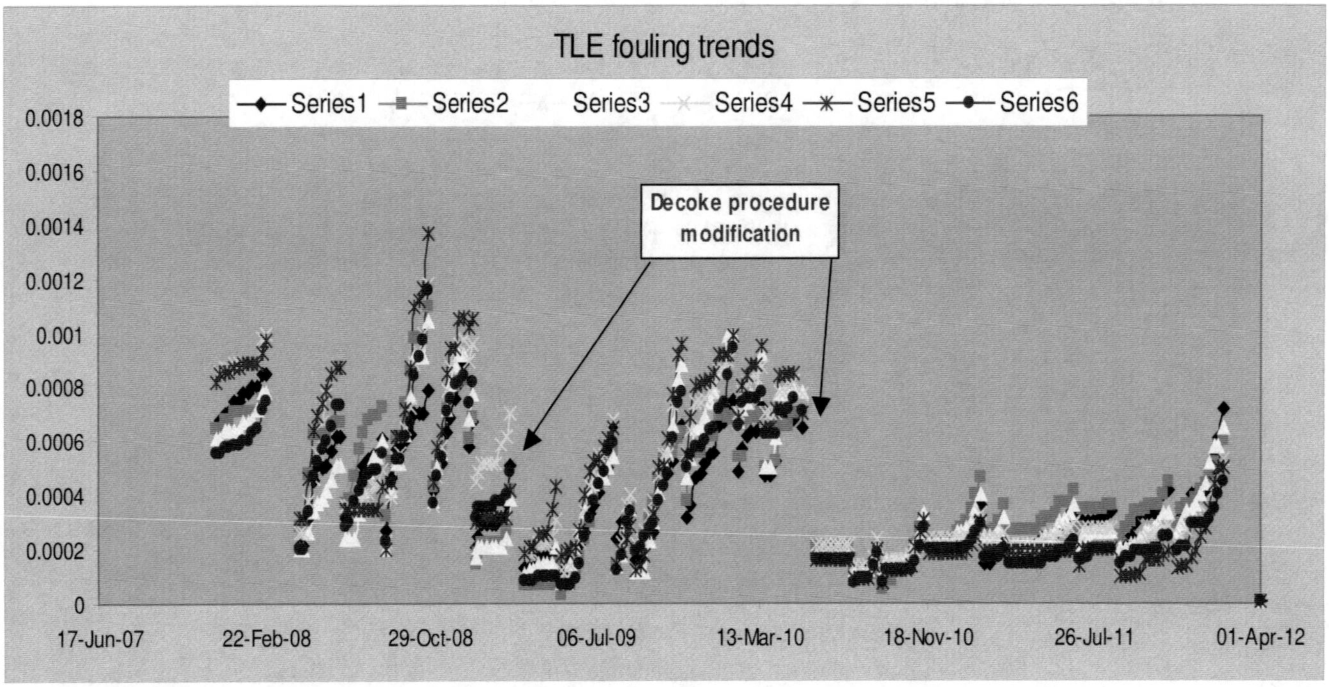

Benefits:
The gain achieved by our clients adopting this methodology across the globe:

- Furnace run length improvement (35 days to >90 days).
- Delimited TLE maximum outlet temperature
- 3 Furnace decoke operations were avoided during the year
- Equivalent to extra Ethylene production 1055 tons
- Other intangible benefits - Increased furnace coil life/ reliability due to reduction in decoke cycle
- Energy saving due to reduced decoke cycle
- Prediction and planning for mechanical cleaning of TLE

353353 Simulation Based Optimization of Stirred Reactors

Abstract:

A computational study of a continuously stirred tank operating in a turbulent regime is reported using a commercial code, STAR-CCM+ in conjunction with Optimate+ for optmization. The motion of the impeller is simulated using the moving reference frame approach and the turbulence modeling is done using a realizable k-   model. A Pareto front optimization is performed with the objective of minimizing the power consumption and maximizing the mixing. The parameters varied as part of the study include geometric dimensions such as number of impellers, number of baffles, baffle length and physical parameters such as speed of rotation of the impeller. The Pareto front obtained from the optimization study shows that there is a threshold value for power consumption, beyond which an increase in the power consumption does not lead to an increase in mixing.

351186 Conversion of Captured Carbon Dioxide to Value-Added Dimethyl Carbonate Using Reactive Distillation

Abstract:

The present design study focuses on converting captured carbon dioxide (CO2) to high-value dimethyl carbonate (DMC) that can be an intermediate for other carbonate-based products. The capture and conversion of carbon dioxide (CO2) to chemicals represents a niche opportunity of significance. However, CO2 recovery schemes may leave trace components such as Oxygen that affect the downstream use of even high purity CO2 product (>99% CO2). As a consequence, the downstream impacts of CO2 capture from reducing environments are typically much lower cost than capture from combustion environments and should prove more favorable. The source of the CO2 for this study is from a chilled methanol (Rectisol®) unit following gasification and shift as part of a proposed Petcoke gasification complex that will produce methanol and power (Oil & Gas Journal 6/6/2011).

The overall chemical pathway of the DMC production is represented by: CO2 + 2MeOH >> DMC + H2O. In the first stage 1) captured CO2 and recovered ammonia will be converted to urea and water using a commercial process; 2) the urea is then reacted with methanol to produce methyl carbamate and ammonia in a heat-integrated reactive distillation (HIRD) column with side-reactors and pervaporation (PerVap) membrane separation, 3) this is followed by additional methanol reaction to yield dimethyl carbonate (DMC) and ammonia. Ammonia is regenerated as an end product and recycled to the process for urea formation, thereby acting as a chemical path carrier. Application of HIRD with side reactors and PerVap membrane separation significantly enhanced the techno-economic viability of the process. The proposed HIRD process would replace phosgene based production of DMC. The numerous and expanding uses for DMC make it an ideal candidate, with favorable economics, for consumption of captured CO2.

Global consumption of polycarbonate stands at 5 million tons per year, with growth rate of about 7% globally. Lithium ion battery electrolytes use organic carbonates such as DMC and complexes of lithium ions. DMC is also being evaluated as oxygenated fuel additives. DMC is not corrosive and it will not produce environmentally damaging by-products, it also is a valuable commercial product finding utility as an organic solvent and in the production of other alkyl and aryl carbonates. If implemented worldwide, this production route has the potential to consume in excess of 5 million metric tons/yr of CO2.

This design study presents an AspenPlus process analysis of an integrated process of HIRD with side reactors and PerVap membrane separation. The analysis demonstrates a design methodology for rapidly scaling the batch kinetic data to a commercial plant. The methodology developed in this study is applicable to other chemical manufacturing processes where HIRD with side reactors can be applied.

351780 Thermally Efficient Short-Contact Time Catalytic Plate Reactor: Detailed Multi-Step Mathematical Model and Parametric Study for Methane Steam Reforming

Abstract:

Due to its high thermal efficiency and low pressure drop, short-contact time Catalytic Plate Reactors (CPRs) have received more attention in recent years for small scale stationary and on-board fuel cell power generators. This study focuses on detailed two-dimensional steady state modeling and simulation of a CPR for the endothermic catalytic Steam Methane Reforming (SMR) on one side of a metal plate and the exothermic methane combustion on the other. The reforming side of the plate was considered to be coated with a nickel-alumina (Ni-Al_2O_3) catalyst and the combustion side with a platinum-alumina (Pt-Al_2O_3). A multi-step surface microkinetic model was adopted to simulate the catalytic SMR process; whereas, a lumped surface kinetic model, which is based on a microkinetic model, was used to simulate the catalytic combustion process. To simulate the gas-phase methane combustion, a one-step homogeneous kinetic model was adopted. Two Langmuir-Hinshelwood Hougen-Watson (LHHW) type kinetic models of the SMR were also implemented on the reforming side for a comparative study. Simulation results obtained using the LHHW type kinetic models and the multi-step surface microkinetic model of methane steam reforming were compared with experimental results in terms of the methane conversion, and the carbon monoxide selectivity, S_{CO}. The results obtained using the microkinetic model, have showed excellent representation of the experimental data compared to the LHHW models. Base case study and parameter sensitivity study were performed using the microkinetic model, and important criterion for the CPR performance are identified. Flow arrangement between the two channels, reforming catalyst thickness, space velocities and metal plate thickness are found to have major

349652 Biogas as a Natural Gas Supplement: Consideration of Potential Safety Issues

Abstract:

Biogas is the common moniker for a flammable gas mixture that is formed from the anaerobic digestion of organic matter. Biogas is most often derived from waste streams in agricultural or chemical processing facilities. Biogas contains methane, but it is not natural gas. Biogas is used to supplement natural gas as a fuel gas for fired heaters and boilers. The integration of biogas into a natural gas header system is not overly complicated, but there are oftentimes overlooked safety concerns related to component maintenance and equipment performance when performing the integration. Pipeline natural gas is heavily processed before reaching the customer; in contrast, little preparation of biogas is performed before addition to the natural gas infrastructure of a facility. Problems may arise when maintenance and operation protocols are established for pure natural gas operations and the system is modified at a later date to accommodate biogas supplementation. In this paper we discuss the general methods for forming biogas, the expected compositions of biogas, the common methods for integrating biogas into a natural gas supply system, potential impacts biogas may have on typical gas-fired operations, followed by a review of several case studies where biogas supplement systems led to unanticipated issues with flammable gas consuming facilities.

Shredded Waste Downdraft Gasification*

Paper Number 351629

M. Cushman (mcushman@infoscitex.com), L. Sasa (lsasa@infoscitex.com), D. Montella (dmontella@infoscitex.com), S. Schlis (sschlis@infoscitex.com) M. Reed (mreed@infoscitex.com) and H. Gold (hgold@infoscitex.com)(Infoscitex Corporation, Waltham, MA); M. Young (myoung@mswpower.com) and S. Chase (schase@mswpower.com)(MSW Power, Acton, MA); and B. Pittenger (bhpittenger@jenike.com)(Jenike & Johanson, Tyngsboro, MA)

ABSTRACT

Mobile, small-scale gasifiers are finding applications in industries that require economical on-site generation of electricity and waste heat from solid biomass waste. Most waste-to-energy gasifier systems pelletize the solid fuel waste prior to processing in the gasifier to minimize problems associated with materials handling. Pelletizers are expensive, require pretreatment (e.g. metals separation and removal), and have a large footprint. The objective of this study was to develop a shredded waste downdraft moving bed gasifier that converts high volatility municipal solid waste (MSW) and biomass fuels to a low tar producer gas, minimizes waste pretreatment problems, and at the same time results in faster reaction kinetics and higher conversion efficiencies.

Shredded waste has a very high wall friction for typical gasifier refractory materials and geometries, resulting in non-uniform solid waste flow. The high surface area to volume ratio also results in low permeability flow, making it more difficult to inject secondary air into the gasifier to combust pyrolysis vapors for optimum conversion. A prototype downdraft packed bed gasifier was developed to process municipal solid waste and other biomass and tested in the laboratory. The walls of the gasifier were tapered with the cross-sectional area increasing from the inlet to the outlet of the reactor. A manually-operated grate, at the exit of the gasifier, was used to control the solid waste mass flow and to remove char and ash. The gasifier and downstream processing system were instrumented with thermocouples, pressure gauges and flow meters. The diverging tapered gasifier design resulted in bulk solids flow without any regions of stagnant flow due to bridging or arching.

Shredded waste has a low permeability to gas flow compared to pellets and the extent of the secondary air gas penetration into the gasifier can be limiting. Flow simulation studies were carried out to determine the penetration of the secondary air into the solid waste through nozzles placed in the walls of the gasifier. The results showed at very high flow resistance, corresponding to shredded waste, the secondary air only penetrated about 33% into the center of the gasifier.

Another objective of the study was to determine secondary air configurations that resulted in high producer gas energy. Secondary air injection ports were placed in three zones below the inlet to the gasifier and distributed around its periphery in each zone. In some experiments, a secondary air injection tube was also placed across the width of the gasifier. Gas samples were collected prior to the producer gas blower and their composition analyzed by gas chromatography. Vertical temperature profiles throughout the gasifier were obtained from the electronic data capture system at the time of gas sampling. High producer gas energies resulted from the injection of secondary air just below the pyrolysis zone in the gasifier and were a consequence of the combustion reactions generating CO_2 and H_2O, which in turn increased the production of CO and H_2 through the Boudouard, water gas and water gas shift reactions, reducing char to ash.

Further work is continuing and is focused on the development of other secondary air injection tube configurations, the use of feed stocks consisting of MSW (plastics, food, and paper/cardboard), and measuring the levels of tar and other contaminants in the producer gas.

INTRODUCTION

Mobile, small-scale waste to energy conversion (WEC) systems are currently finding niche applications in industries/applications that economically favor the on-site generation of electricity and heat from non-durable goods (paper and paperboard), food waste and other biomass. These applications

*Supported under Strategic Environmental Research and Development Program (SERDP) (WP-2235) under Contract No. W912HQ-12-C-0067.

include restaurants, universities, supermarkets, hospitals, sporting and entertainment events, disasters and military field applications. Downdraft co-current moving bed gasifiers are most suitable to convert high volatility fuels, such as municipal solid waste (MSW) and biomass, to low tar producer gas for use in generating electricity. In these gasifiers, the primary gasification air enters at the top of the gasifier and is pulled through the reactor by a vacuum pump. Secondary air is introduced at or above the oxidation zone in the gasifier. The shredded waste fuel flows in the same direction as the combusting air. The producer gas is removed at the bottom of the gasifier. The downdraft gasifier produces much less tar than any other gasifier because the volatiles are largely converted in the pyrolysis zone at the top of the reactor and then are cracked in the high temperature zone of hot ash (gasification zone) at the bottom of the reactor. Downdraft gasifiers have been limited in their ability to use unprocessed fuels, such as fluffy, low density shredded materials because of the high pressure drop across the gasifier. As a result, the solid fuel is pelletized or briquetted before use in the gasifier. Pelletizers are expensive and require pretreatment, e.g. metals separation, to prevent equipment breakdown and maintenance. In most downdraft gasfiers, it is difficult of maintain uniform high temperatures over a given cross-sectional area, limiting the gasifier to a smaller cross-sectional area and lower power production.

The use of shredded waste, instead of pellets, in a packed bed will reduce the footprint and cost of the feedstock conditioning pre-processing system by eliminating a pelletizer and non-ferrous metal separator. Shredded waste will also result in faster reaction kinetics, increasing the conversion efficiency because of the high surface area to volume ratio.

The development of a downdraft gasification system that uses shredded waste is not without its problems. Gasification of shredded waste has only been done successfully in fluidized bed/air blown reactors, requiring significant feedstock conditioning (sizing and drying); these systems are bulky and used primarily for much higher shredded waste flows and energy production. Shredded waste has very high wall friction for typical gasifier refractory materials and geometries, retarding the solids flow along the walls of the reactor, and resulting in non-uniform or funnel flow across the cross-sectional area and isolated regions of low permeability and excessively high non-uniform temperatures. Funnel flow is a flow pattern in which solids flow in a channel formed within stagnant material. In addition, because of the regions of low permeability, a high pressure drop is required to pull the air flow through the length of the reactor and uniform gas flow is difficult to achieve across a given reactor cross-section. Secondary air is also needed to control the gas temperatures in the reactor to produce optimum pyrolysis, combustion and gasification in the gasifier. Secondary air is usually injected into the reactor normal to the direction of the solids flow in a combustion zone just below the pyrolysis zone. The non-uniform solids flow and low permeability result in poor penetration of the secondary air into the interior of the reactor.

REACTOR DESIGN

In a downdraft gasifier, the solids flow through the reactor is controlled by a grate at the exit of the gasifier and not by the flow at the top of the reactor. Bulk solids mass flow is required both in the reactor and grate to achieve a uniform solids velocity and temperature over the cross-sectional area of the reactor. In mass flow, the entire bulk solids bed should be in motion when char is charged from the outlet of the reactor. Mass flow guarantees complete discharge of the content of the reactor at predictable flow rates. An additional requirement is that obstructions in the solids flow cannot develop. Flow property tests were performed flow on (a) shredded food waste (44.5% by weight, paper (42.2%) and plastics (13.3%) at different moisture contents and (b) char/ash from a downdraft gasifier. Data was obtained for both unpolished and polished refractory that will be used as the walls of the gasifier. Tests included:

- Cohesive strength and physical modeling - to determine minimum outlet size to prevent arching and to confirm that gas injection and control devices will not interfere with solids flow
- Wall friction - to determine wall materials and geometry to be used to allow solids mass flow
- Compressibility – to provide relationship between consolidated pressure and bulk density and provide residence time calculations

- Permeability – provides relationship between gas velocity and pressure drop for design of secondary air distribution system.

Figure 1 shows some typical wall friction data obtained for char/ash at 850°C on a polished refractory. This data was used to calculate the slope of the walls of the gasifier.

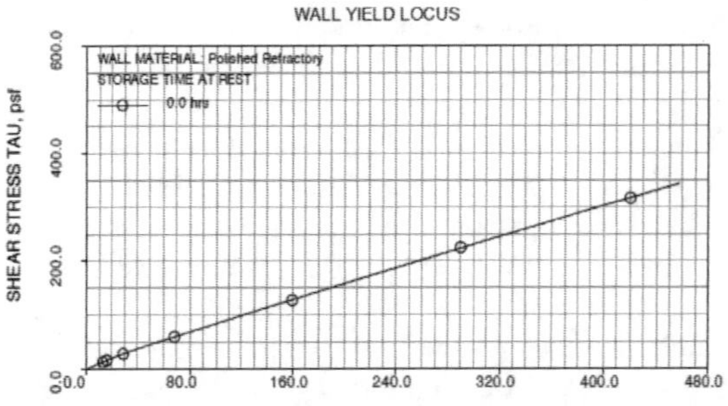

Figure 1. Wall friction data for char/ash on polished refractory

Based on the flow property test data, a prototype downdraft packed bed gasifier to process MSW and other biomass was designed, fabricated and tested in the laboratory (Figure 2). The walls of the gasifier were tapered 5 degrees with the area increasing from the inlet to the outlet of the reactor. The diverging tapered walls of the gasifier reduces the friction between the shredded waste and the reactor surfaces, reducing the potential for bulk solids flow problems, such as arching and rat-holing. The gasifier inlet had a rectangular cross sectional area of 102 cm x 15.2 cm (4" x 6") and the height of the gasifier was 61.0 cm (24"). The rectangular design reduces the distance of secondary air coverage maximizing its reach without extending nozzles into the reactor. The gasifier was designed to handle a shredded waste flow of 7.6 kg/hr (16.7 lb/hr) and a syngas flow rate of 0.61 m^3/min (20.0 ft^3/min). A solid waste feedstock consisting of paper and cardboard with a shred size varying between 15 to 25 mm (0.6 to 1") and a moisture content of 10% was used in the test program.

SECONDARY AIR SIMULATION

For high conversion efficiency, secondary air should penetrate to the center of the gasifier, reacting with as much of the waste in the pyrolysis zone as possible. Shredded waste has a low permeability (high resistance) to gas flow compared to pellets and the extent of the secondary air gas penetration into the gasifier can be limiting. SolidWorks Flow Simulation analyses [1] were carried out on the gasifier configuration shown in Figure 2 to determine the penetration distance of the secondary air into the solid waste as a function of the following operating conditions: secondary air flow as a percentage of the total syngas flow; waste permeability (resistance); nozzle diameter (flow velocity); mass flow per ring; and number of secondary air flow rings. This model does not take into account mixing between the solid waste and the secondary flows. The penetration distance was determined by the distance of the limiting streamline from the wall of the gasifier.

The solid waste was considered to be porous, with the permeability determined by the pressure drop along the length of the reactor. The gas flow in the reactor is defined by Darcy's Law [2]:

$$m = -A*\Delta p/(k*L) \qquad (1)$$

where

- m =gas mass flow, kg/sec
- A = cross-sectional area in direction of flow, m^2
- Δp = pressure difference in direction of flow, kg/m^2
- L = length in direction of flow, m
- k = porous medium flow resistance, sec/m

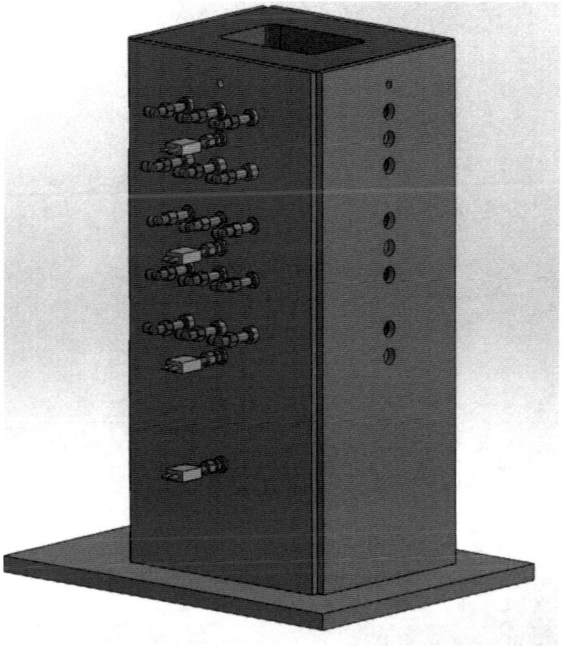

Figure 2. Downdraft packed bed gasifier

For high resistance or low permeability and for a fixed pressure gradient, the mass flow through the porous medium is very low. Conversely, for a fixed mass flow rate and high resistance, the pressure gradient is very large. The resistance coefficient was calculated as 336 sec/m for a full scale WEC unit using pelletized waste. Based on pressure drop measurements taken during testing of the shredded waste gasifier, the resistance was calculated to be 2200 sec/m, about 6.5 times that of the flow in the pelletized gasifier.

Figure 3 shows the secondary air streamlines through the center planes of the gasifier for a given flow resistance. The secondary air is coming in through the first ring of injection ports, 6.4 cm (2.50") from the top of the inlet to the reactor. There are eight 3.2 mm (0.125") injection ports in the first ring with three ports on each on the long faces of the gasifier and one port on each of the short sides. Figure 4 shows the secondary air streamlines through planes parallel to the top face of the gasifier. In Figure 4, the limiting streamline from the injection port is initially parallel to the plane of the injection ring and then bends towards the axis of the gasifier, reaching a maximum distance from the injection port slightly below the plane of the injection ring.

Figure 5 is a summary of the flow simulation results. In Figure 5a, the penetration distance is plotted as a function of secondary air (as a percent of the producer gas flow rate) for a relatively high flow resistance of $3.36 * 10^2$ sec/m. The secondary air is injected from the first ring of injection ports (2.5" from the inlet). The distance from the injection port on the long side of the gasifier to the center of the gasifier is 2.2 in. At this high resistance, there is very little penetration of the secondary air into the center of the gasifier, regardless of the secondary air injection velocity. In Figure 5b, the penetration distance is shown as a function of the resistance for a constant secondary air flow rate. At relatively high flow resistances, the penetration distance is relatively constant (~1.4 inches), about 33% of the distance to the center of the gasifier. As the flow resistance is reduced, the penetration distance increases until it reaches the center of the gasifier. At a lower flow resistance and for a constant injection flow rate, the penetration distance increases with decreasing nozzle diameter or increasing injection velocity.

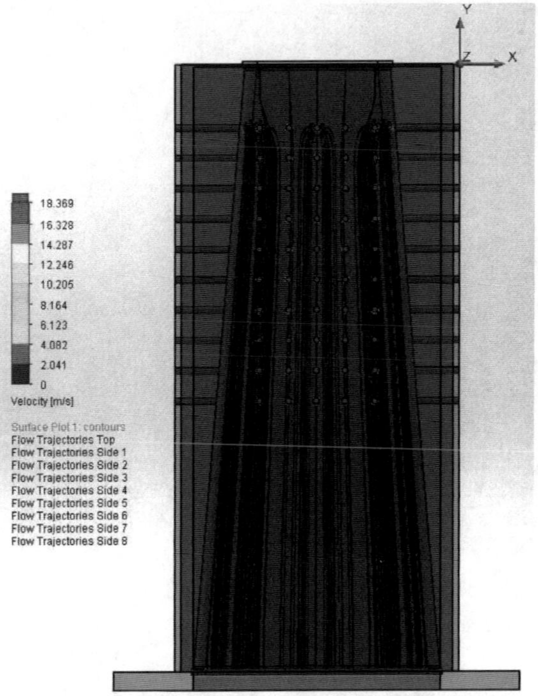

Figure 3. Secondary air streamlines along the length of the gasifier using SolidWorks Flow Simulation model. Streamlines through center plane of gasifier facing long side of reactor

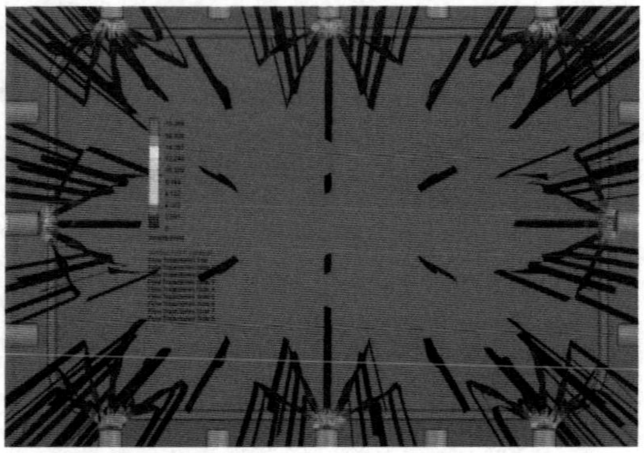

Figure 4. Secondary air streamlines through the top face of the gasifier

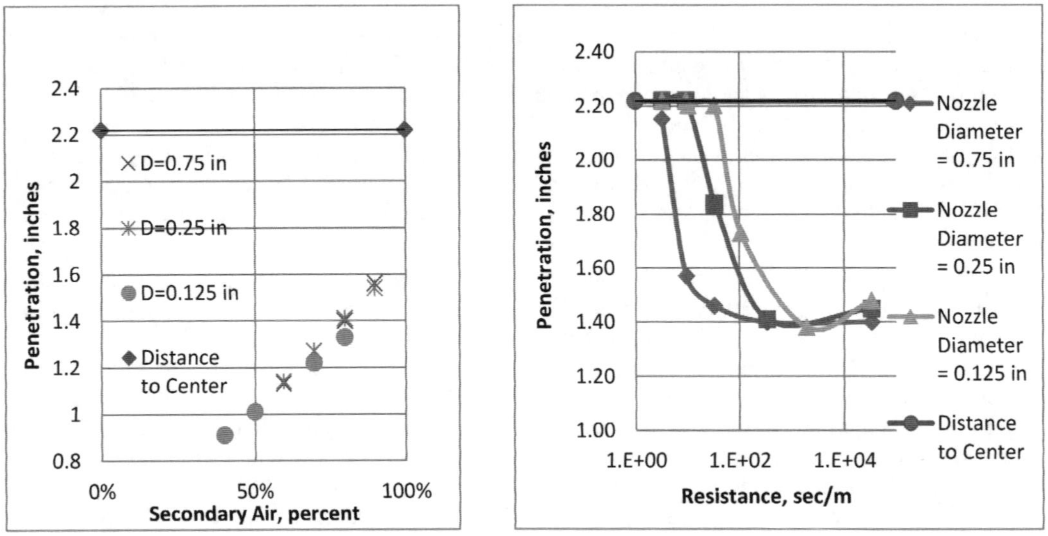

a. Resistance = 336 sec/m *b. Secondary air flow = 80% (8.2 cfm)*

Figure 5. Penetration distance as a function of secondary airflow rate and flow resistance. Producer gas flow rate = 10.3 ft^3/min

GASIFIER OPERATION

Figure 6 shows the gasifier system and downstream producer gas conditioning system. The gasifier was instrumented with thermocouples, pressure gauges and flow meters. A manual operating grate was installed at the exit of the gasifier to control the waste flow through the gasifier and to remove char and ash. A heat exchanger to cool the syngas and a filter system to remove tars and other particulates were placed downstream of the grate (Figure 6). The syngas was sampled downstream of the filters in syringes and its composition determined by gas chromatography.

A vacuum pump (blower) located downstream of the filters is the primary mover for the system. Five rings of secondary air injection ports were installed along the length of the gasifier, with the top three rings placed in the pyrolysis zone. Eight secondary air injection ports were installed in each ring; three ports each in the front and back face of the gasifier and one port each on the side faces. Secondary air was injected through the ports normal to the walls of the gasifier. The air flow rate at each ring was independently controlled, as was the producer gas flow through the vacuum pump. The differential between total gas being moved by the blower from the amount of secondary air injected as well as the producer gas produced would result in the amount of primary air added to the reactor through the top opening.

Figure 7 shows the experimental facility in the laboratory. Prior to starting a test run, char was used to start up the gasifier. When the temperature in the top part of the gasifier reached 800 to 900°C, a shredded feedstock consisting of paper and cardboard with a moisture content of 10% was manually fed into the gasifier. The weight of the shred was recorded as a function of time. A manually-operated grate was used to control the waste flow through the gasifier and to remove char and ash. The producer gas flow rate was manually set, monitored intermittently, and adjusted periodically with changes in the waste feedstock flow rate. The secondary air flow rates were also manually set and recorded on a data sheet. For the test program, secondary air was injected through ports in the walls in the top three positions of the gasifier (2.5, 5.0 and 7.5 inches). The diameter of the injection ports were 0.269 inches (6.83 mm). For two tests, air was injected into a tube placed across the width of the gasifier through two holes of diameter 0.094 inches (2.38 mm) facing downward; secondary air was also injected through some of the sidewall air injection ports. The gasifier, downstream heat exchanger, filter and blower temperatures and pressures were electronically monitored and recorded at a 1 Hz rate from system start-up to shut-down. In the event of a backdraft from the gasifier, the secondary air flows were reduced until the backdraft disappeared.

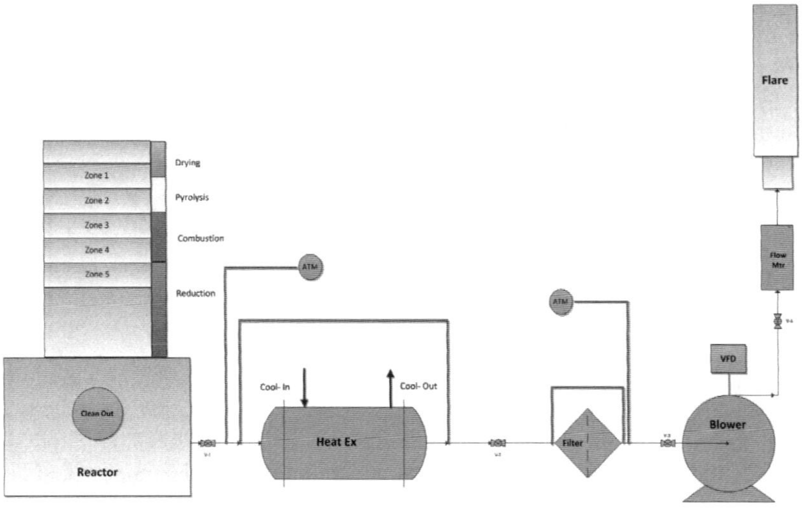

Figure 6. Gasifier and downstream producer gas conditioning system

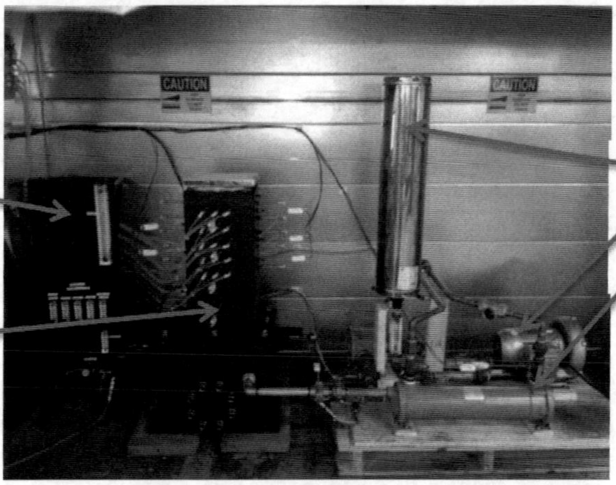

Figure 7. Photograph of downdraft gasifier facility

The producer gas was sampled via a collection port immediately upstream from a downstream flare. A syringe was connected to the port, the gas was allowed to fill the tube for approximately 20 to 30 seconds to ensure the syringe was completely filled with syngas, and then the syringe was capped. Three samples were taken from each collection port. The collected gas was injected into a SRI 8610 Gas Chromatograph (GC) to perform compositional analysis. A methanizer equipped flame ionization detector (FID) was utilized to measure the concentrations of carbon monoxide, carbon dioxide, and methane, while a thermal conductivity detector (TCD) was utilized to measure the concentrations of nitrogen, hydrogen, and oxygen. The average of the samples was used to determine the producer gas energy.

The operation of the downdraft gasifier was guided by an equilibrium analysis of the gasification process. In the gasification zone, three reactions dominate [3, 4, 5]:

Boudouard reaction: Char + $CO_2 \rightarrow 2CO$, $\Delta H = +40{,}778$ kcal/mol

Water Gas Reaction: Char + $H_2O \rightarrow CO + H_2$, $\Delta H= +32{,}472$ kcal/mol

Water Gas Shift Reaction: $H_2O + CO \rightarrow H_2 + CO_2$, $\Delta H= -8{,}306$ kcal/mol

The char gasification zone is assumed to be adiabatic and the water gas shift reaction is assumed to be at equilibrium [3, 4]. Figure 8 are the equilibrium results for a feed waste consisting of paper and cardboard. The ultimate analysis of this waste was given in [4]. The adiabatic gas temperature is 695°C with producer gas energy of 167 BTU/ ft^3 ($6.22*10^3$ kJ/m^3). This model was used to determine how closely the test results approached equilibrium.

The objectives of the test program were to determine (a) if the gasifier design resulted in bulk solids mass flow and (b) the secondary air configurations that resulted in high producer gas energy. Test data obtained to date show that the diverging tapered gasifier configuration produced bulk solids mass flow without any regions of stagnant flow in the gasifier. Steady temperatures and pressures indicated there was no arching of the waste flow in the gasifier (see Figure 9).

Table 1 shows the operating conditions and test results for a series of six tests. The producer gas energy was determined from the GC analysis for hydrogen, carbon monoxide and methane. Each time a gas sample was withdrawn from the gasifier for GC analysis, temperature profiles throughout the gasifier were obtained from the electronic data capture system (see, for example, Figures 10 and 11). The maximum temperature was determined from the data set and was taken to be the char combustion temperature just below the flaming pyrolysis zone in the gasifier [4, 5], where the solid waste and the pyrolysis vapor is combusted by oxygen, producing CO_2 and water. The adiabatic char gasification temperature was taken near the bottom of the gasifier [4, 5], where char is converted to ash. The secondary air configuration is shown in the last column of Table 1. The first number is the quantity of secondary air injected through the series of injection ports nearest the gasifier inlet (2.5 inches); the second number corresponds to the next (second) series of injection ports (5.0 inches); and the last figure corresponds to the third series (7.5 inches). High producer gas energies resulted from the injection of secondary air just below the pyrolysis zone in the gasifier and were a consequence of the combustion reactions generating CO_2 and H_2O, which in turn increased the production of CO and H_2 through the Boudouard, water gas and water gas shift reactions. The measured maximum producer gas energies are about 15% lower than the calculated equilibrium energy (167 BTU/ft^3). Table 2 lists the volume concentrations of the producer gas for each test condition shown in Table 1. The concentrations were normalized to a total of 100%. The producer gas energy was calculated from the hydrogen, carbon monoxide and methane concentrations.

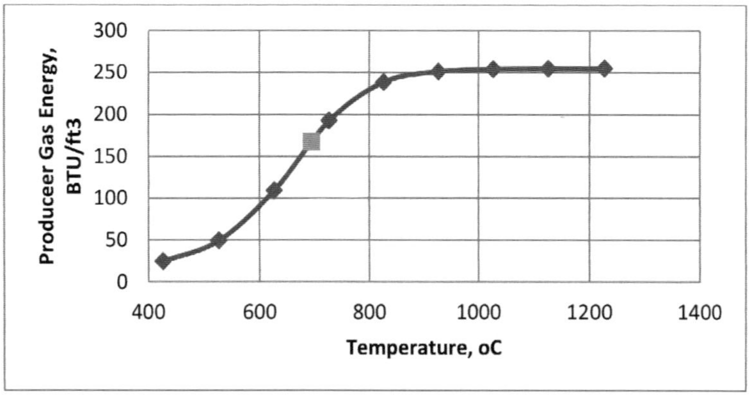

Figure 8. Equilibrium conditions for paper and cardboard waste. Square indicates adiabatic conditions

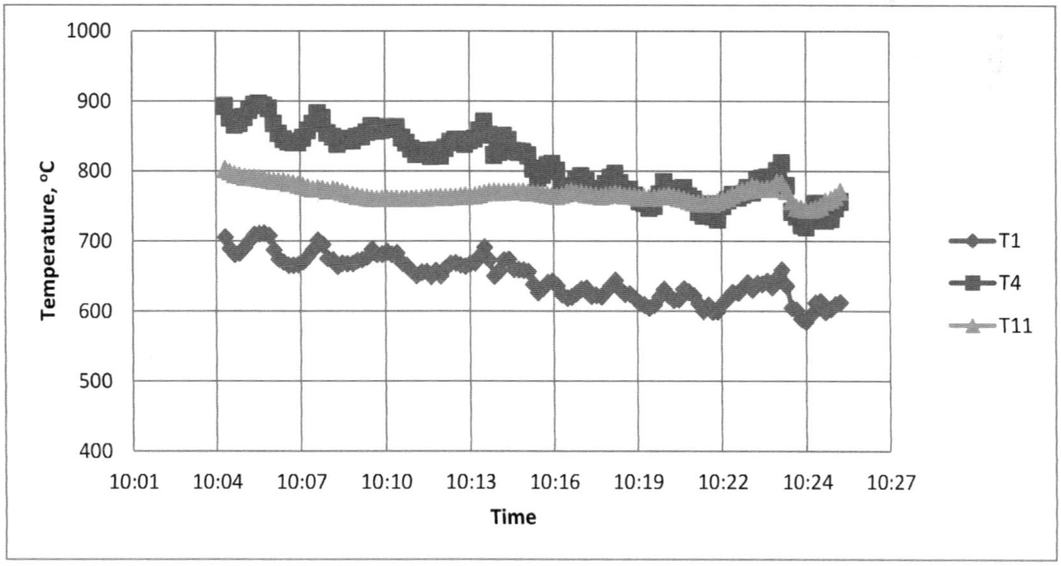

Figure 9. Typical reactor gas temperature profiles

Table 1. Test conditions and tests results for downdraft gasifier with shredded waste feedstock

Test No.	Waste Flow Rate, lb/hr (kg/hr)	Producer Gas Flow, ft^3/min (m^3/hr)	Producer Gas Energy, BTU/ft^3 (1000 J/m^3)	Maximum Temperature, °C	Char Gasification Temperature, °C	Secondary Air Flow Rate, ft^3/min
1	17 (7.7)	22 (37)	148 (5.5)	890	740	1, 4, 4
2	8.5 (3.9)	12 (20)	96 (3.6)	790	780	0, 2, 2
3	8.5 (3.9)	12 (20)	144 (5.4)	690	570	4, 0, 0
			114 (4.3)	740	620	0, 4, 0
			91 (3.4)	800	780	0, 0, 4
			127 (4.7)	740	700	0, 3, 1
			101 (3.8)	800	790	0, 1, 3
			94 (3.5)	800	780	0, 2, 2
4	14.6 (6.6)	18 (31)	143 (5.3)	860	700	3, 3, 0
			107 (4.0)	830	740	0, 3, 3
5*	16 (7.3)	20 (34)	141 (5.3)	790	770	1, 2, 5
6**	20 (9.1)	20 (34)	133 (5.0)	820	740	0, 5, 3
			113 (4.2)	750	730	0, 3, 2

* Secondary air injector tube at 3rd ring (7.5 inches from gasifier inlet)

** Secondary air injector tube at 2nd ring (5.0 inches from gasifier inlet)

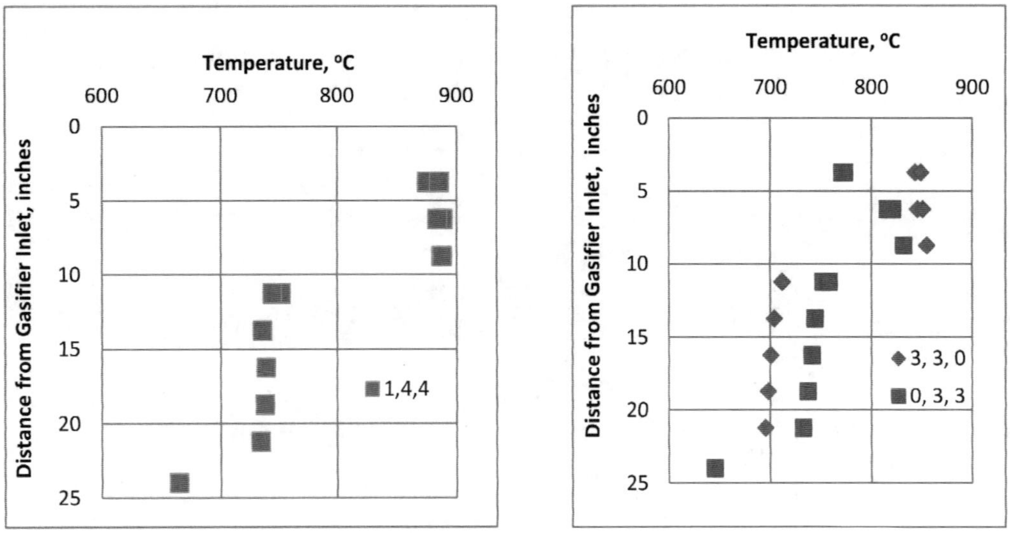

Test 1 *Test 4*

Figure 10. Gasifier temperature profiles with secondary air injection through side wall injectors

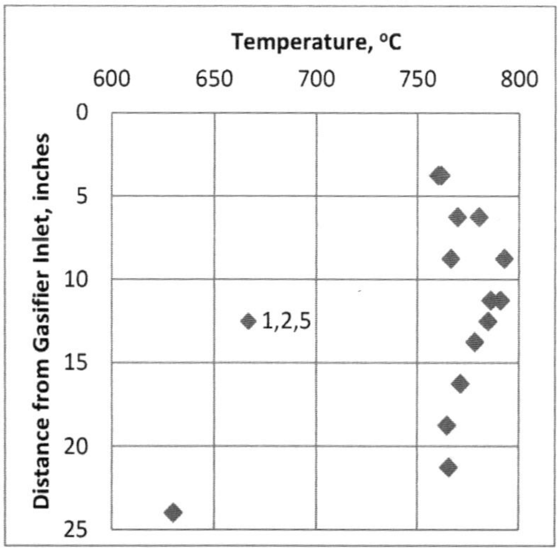

Test 5. Injector pipe at third zone (7.5 inches)

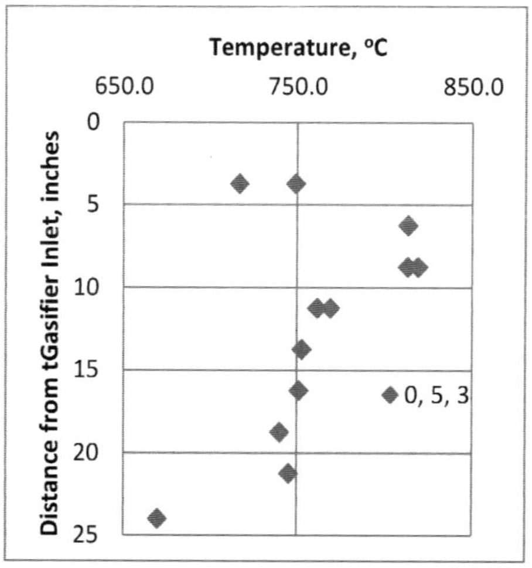

Test 6. Injector pipe at second zone (5 inches)

Figure 11. Gasifier temperature profiles with secondary air injection through pipe

Pressure drop measurements were also taken for each of the tests and showed pressure drops ranging from 2.5 to 4 inches of water. The resistivity of the secondary air in the shredded waste gasifier flow was calculated to be about 2,200 sec/meter, or about 7 times the resistivity of the full scale GEM unit that used MSW pellets. Although the low permeability of the shredded waste flow resulted in higher pressure drops, it did not inhibit the mixing of the secondary air with the solid waste, producing relatively high producer gas energies.

Table 2. Volume concentrations of producer gas

Test No.	Producer Gas Energy, BTU/ft^3	Producer Gas Concentration, vol %				
		Hydrogen	Nitrogen	Carbon Monoxide	Methane	Carbon Dioxide
1	148	9.33	47.41	29.75	2.95	10.56
2	96	8.56	57.80	18.63	1.35	13.66
3	144	11.53	50.46	27.59	2.53	7.88
	114	8.37	53.39	22.03	2.21	13.79
	91	4.79	57.49	18.21	2.12	17.39
	127	19.75	47.80	16.89	2.05	13.51
	101	11.23	54.04	16.04	2.05	16.65
	94	11.33	56.31	15.10	1.58	15.69
4	143	15.39	48.85	23.44	2.80	9.51
	107	2.98	58.81	24.07	2.35	11.79
5	141	6.71	50.76	27.83	3.57	11.13
6	133	10.16	52.09	25.04	2.71	10.00
	113	6.21	56.48	22.28	2.66	12.37

CONCLUSIONS

The diverging tapered gasifier design resulted in bulk solids flow without any regions of stagnant flow in the gasifier due to bridging or arching. Secondary air injection just below the pyrolysis zone, either through the gasifier or through an air injection pipe placed across the width of the gasifier, produced the largest producer gas energy. This is the result of more complete combustion of the pyrolysis vapors by oxygen, generating carbon dioxide and water, which in turn produce carbon monoxide and hydrogen through the Boudouard, water gas and water gas shift reactions. Although the flow simulation studies indicated very little penetration of the secondary air into the solid waste flow, mixing of the primary and secondary air with the solid waste occurred and resulted in the relatively high producer gas energy.

Further work is continuing and is focused on the development of other injection tube configurations, the use of feed stocks consisting of MSW (plastics, food and paper/cardboard), and measuring the levels of tars and other contaminants in the producer gas.

REFERENCES

1. https://www.solidworks.com/sw/products/simulation/flow-simulation.htm; http://www.solidworks.com/sw/support/808_ENU_**HTML**.htm
2. Probstein, R.F., Physicochemical Hydrodynamics, Butterworths, Stoneham, MA, 1989.
3. Probstein, R.F. and Hicks, R.E., *Synthetic Fuels*, McGraw-Hill Book Company, New York, NY, 1982.
4. Basu, P., *Biomass Gasification and Pyrolysis – Practical Design and Theory*, Academic Press, Elsevier, 2010
5. Reed, T.B., Grabowski, M.S. and Levie, B., Fundamentals, Development and Scale-up of the Air-Oxygen Stratified Downdraft Gasifier, SERI/PR-234-2571, Solar Energy Research Institute, The Biomass Energy Foundation Press, Golden, CO, 1994.

352418 Pyrolysis of Reed Canary Grass – ANFIS Modeling

Abstract:

The depletion of fossil fuel and the environmental impact of using fossil fuel as a main energy source have been subjects of intense research and policy interest in recent years. Pyrolysis of biomass to produce bio-energy is a promising process. However, with the resulting high cost, creating a cost effective chemical plant is very important. A comprehensive process model, which can be used to predict the production from pyrolysis of biomass, is therefore necessary. However, modeling is complex and challenging because of short reaction times, temperatures as high as a thousand degrees Celsius, and biomass of varying or unknown chemical compositions. As such a deterministic model is not capable of representing the pyrolysis reaction system. We propose a new kinetic reaction model, which would account for significant uncertainty. Specifically we have employed fuzzy modeling using the adaptive neuro-fuzzy inference system (ANFIS) in order to describe the pyrolysis of biomass. In this work, we firstly use this technique with Reed Canary grass. The resulting model has a good prediction of 99.8% of the pyrolysis yield at different temperatures. With this result, the pyrolysis yield can be predicted more accurate than the traditional deterministic model.

High Octane Number Gasoline Production from Lignin

Maoqi (Mark) Feng, Ph.D.

AIChE 2014 Spring Conference
April 1, 2014

Southwest Research Institute®

Agenda

- Introduction
- Background
- Objectives
- Technical Approach
- Results
- Summary and Conclusions

Introduction

- Renewable Fuels Standard (RFS2) Requirement
  - By 2022, 21 billion gallons of non-ethanol renewable fuels mandated (36 billion gallons total)
  - Fuels from lignin comply with RFS2
  - Lignin is abundant but currently only utilized as combustion heating source in industry

- Lignin Sources
  - Any biomass (in plant cell walls)
  - Woody biomass (20-35%); switch grass
  - Kraft lignin from pulp mill plants
  - Lignin extraction from black liquor
  - Cellulosic ethanol plant – coproduction of 0.5 kg of lignin from each liter of ethanol produced

Introduction

- Lignin structure – three common monolignols based on the phenylpropane core

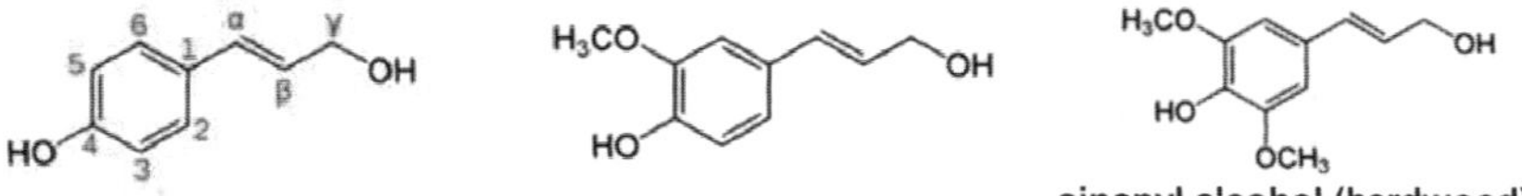

para-coumaryl alcohol (grass) | coniferyl alcohol (softwood) | sinapyl alcohol (hardwood)

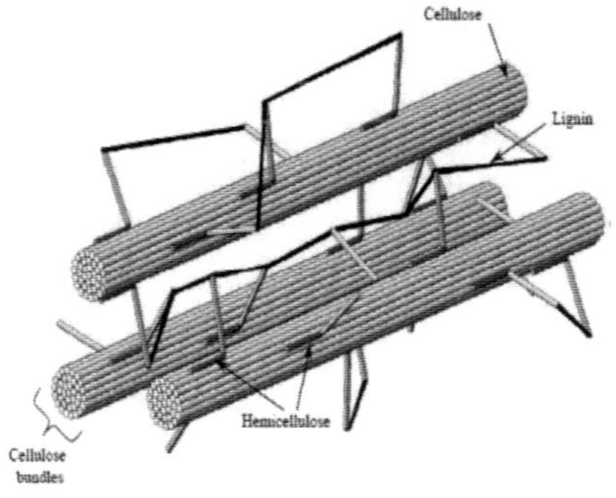

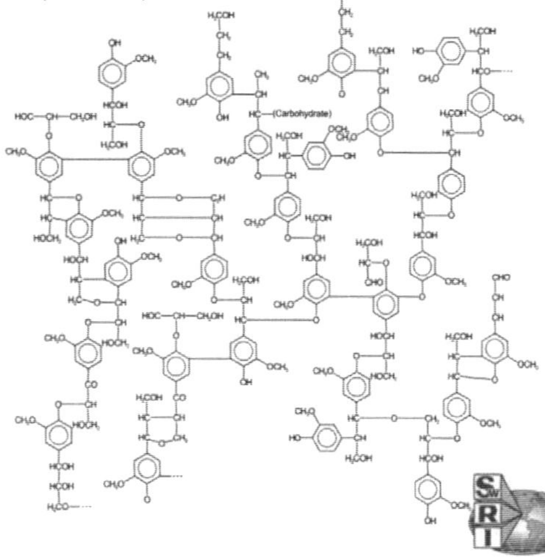

Objectives

- Develop a novel lignin conversion technology for gasoline production
  - Lignin depolymerization 80%
  - Hydrodeoxygenation catalyst development
  - Gasoline octane number $\geq$ 92

Technical Background

- Current lignin conversion processes
  - Lignin Pyrolysis
    - ✓ High temperature required (500-700°C)
    - ✓ High coking and char yield ⇒ low carbon efficiency
  - Hydrothermal conversion & Supercritical methanol conversion
    - ✓ Low solubility in water ⇒ low depolymerization conversion
    - ✓ Unstable intermediates
    - ✓ Difficult to scale up
  - Lignin gasification
    - ✓ Energy intensive for syngas cleanup, high investment cost

Technical Background

➢ Comparison of selected processes

Method	Temperature, ˚C	Intermediate Stable?	Carbon Efficiency	Relative Cost
Pyrolysis	500~700	No	40~60%	0.85
Gasification	700~950	Yes	70~80%	2.50
Hydrothermal	400~550	No	30~50%	1.75
Supercritical Methanol	350~400	No	40~60%	1.50
SwRI Method	350~400	Yes	75-90%	1.0

❖ SwRI method has higher hydrogen efficiency than other processes.

Technical Background

➢ **Key Concept** – Two-step conversion

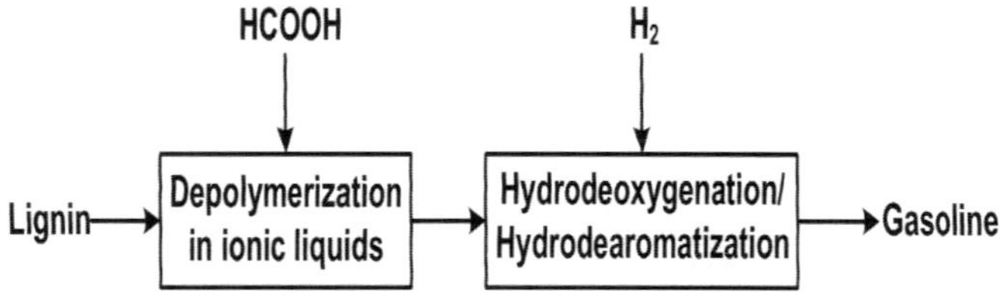

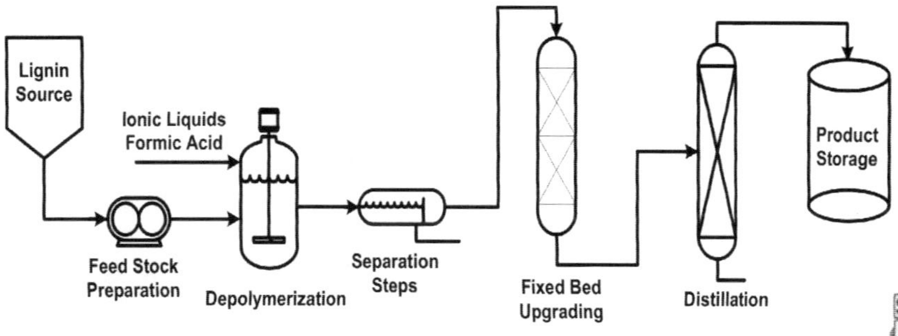

Technical Approach

- Advantages of ionic liquids + formic acid for lignin depolymerization
  - Higher solubility - Better mass transfer - Reaction efficiency
  - Intermediates stabilized by H_2
- Hydrocarbons production from hydrodeoxygenation
  - Metal carbide catalyst to replace noble metals
  - Bifunctional catalyst to maximize hydrocarbons yield

Lignin Depolymerization

- Lignin preparation – 200 mesh particles
- Ionic liquids preparation and lignin solubility measurement
- Lignin depolymerization in batch processing

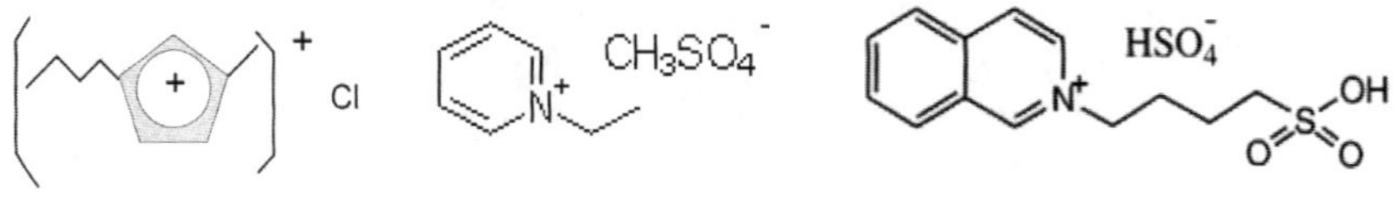

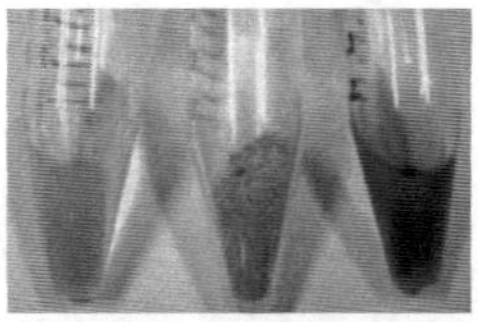

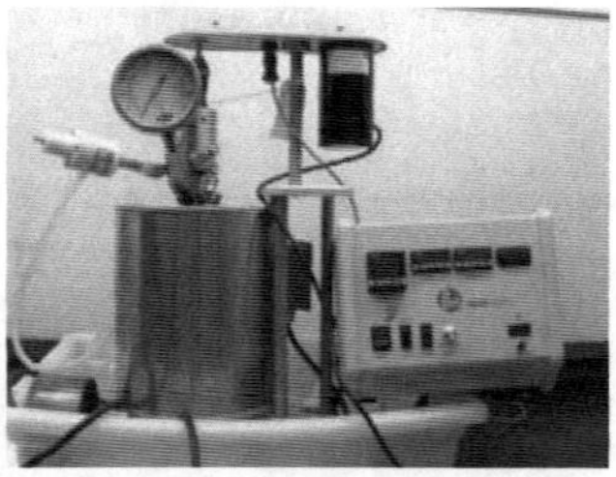

Solvent Polarity

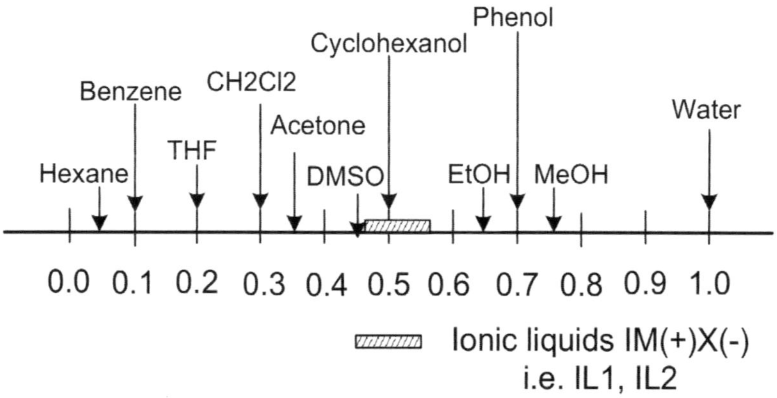

IL1: 1-n-butyl-3-methylimidazolium chloride
IL2: 1-ethyl-3-methylimidazolium chloride

Lignin Depolymerization

- Lignin depolymerization conditions

Parameter	Concentration
Ionic liquids	$[C_4MIM]Cl$ (IL1), $[EtPy]MeSO_4$(IL2), IQBS (IL3)
Temperature	120, 200, 250, 300°C
Lignin to IL ratio	1:5, 1:8
Solid catalyst	None, NiO, Co_2O_3, WC
Residence time	1 hour, 2 hours

Lignin Depolymerization

➢ Lignin depolymerization conditions

	IL1	IL2	IL3
Lignin/IL Ratio	1:5	1:5	1:5, 1:8
Temperature	120,200°C	200,250°C	250, 300°C
Catalyst	NiO	None, NiO, Co_2O_3, WC	NiO, Co_2O_3, WC
Residence time	1 hr, 2 hr	2 hrs	2 hrs

Hydrodeoxygenation

➢ Hydrodeoxygenation of depolymerized oxygenates in batch reactor

- Catalyst development – non-precious metal catalyst
- Batch process to optimize hydrogenation conditions
- Reaction progress monitoring by thin-layer chromatography (TLC)
- Product analysis by GC-MS

➢ Continuous process to provide parameters for future large scale operation

✓Continuous fixed bed reactor

✓ Guard bed to protect the catalyst bed

Reactor

- Hydrodeoxygenation in batch and continuous processing

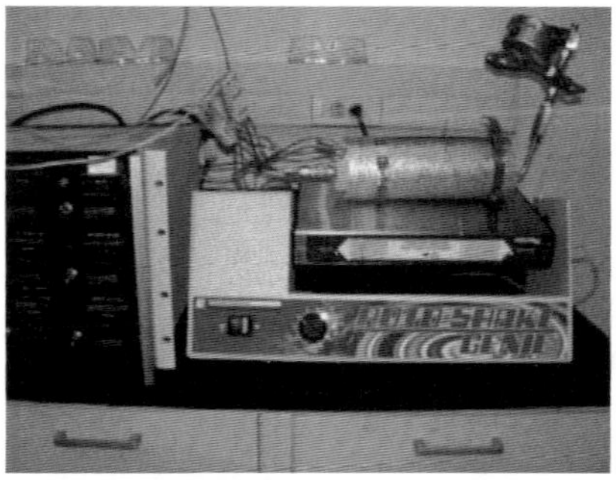

Batch reactor

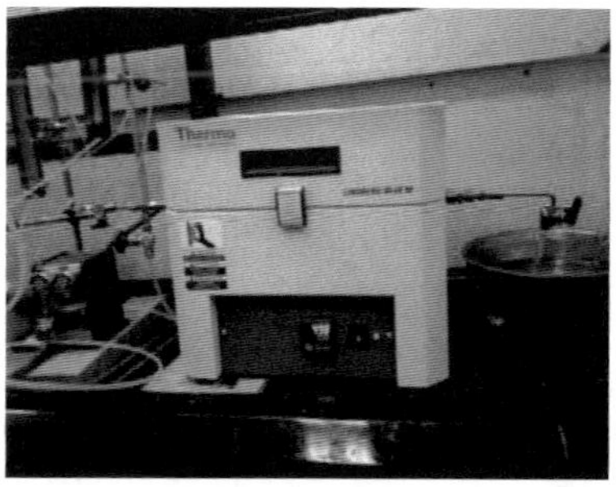

Continuous reactor

Results & Discussions

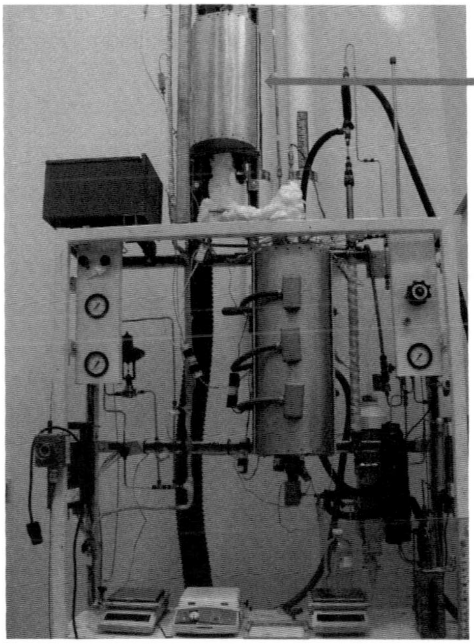

Two-Step Process

- **Key Terminologies**
  - Lignin depolymerization in ionic liquids with formic acid – facilitating depolymerization and hydrogen transfer
  - Hydrodeoxygenation with bifunctional non-precious metal catalysts

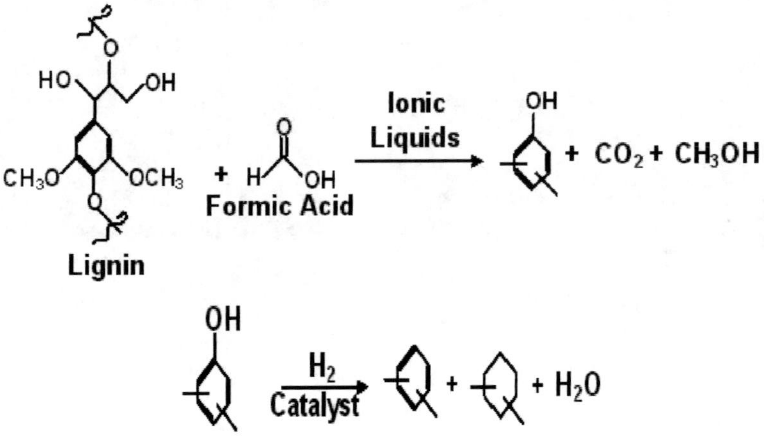

Products from Lignin Depolymerization

- Lignin depolymerization products

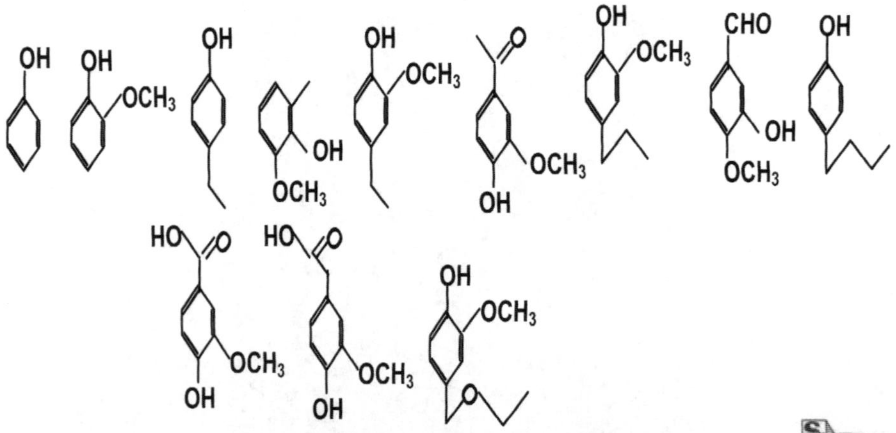

Hydrodeoxygenation

➢ Hydrodeoxygenation of model compounds

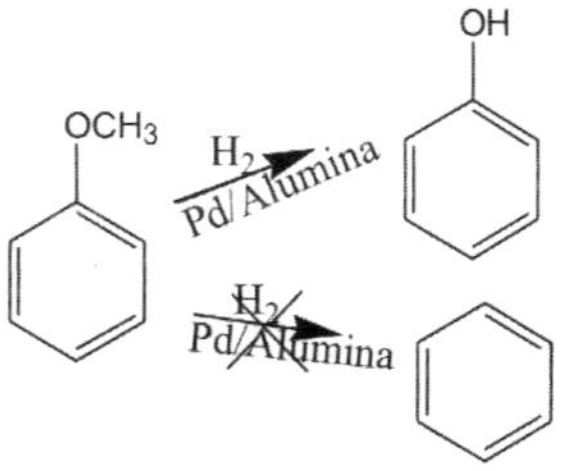

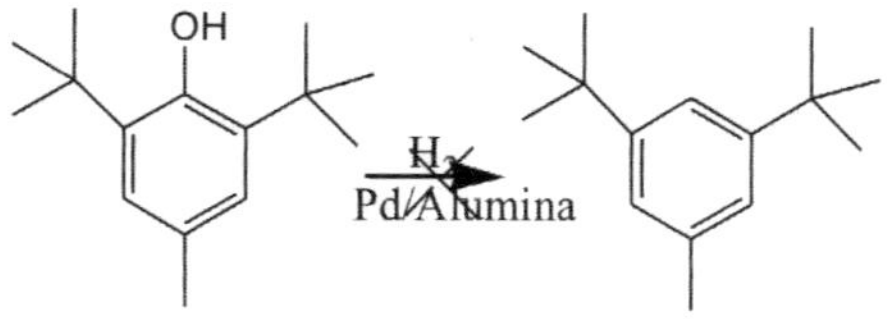

Hydrodeoxygenation

➢ Hydrodeoxygenation for hydrocarbons production

	Cat1	Cat2	Cat3	Cat4
Pressure, psi	2000	1500, 2000	2000, 3000	2000
Temp., °C	300, 350	250, 300	300, 350	300
LHSV, hr^{-1}	3, 6	2	3, 6, 9	3

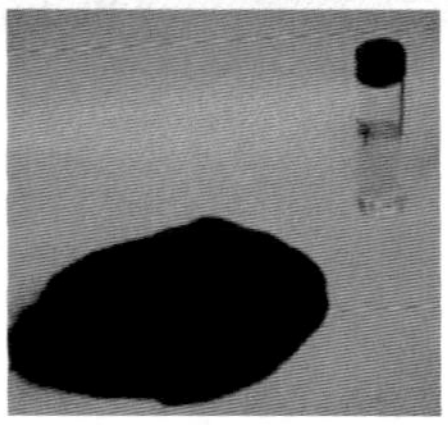

Hydrocarbons Production

- Possible reaction pathways

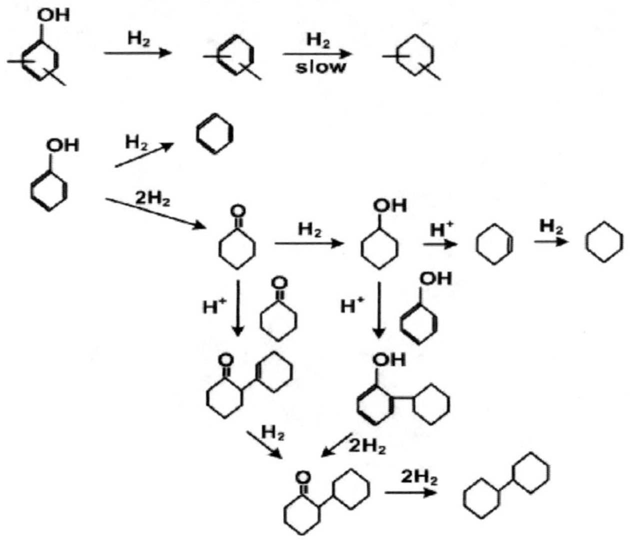

V.N. Bui, et al, Cat. Today 2009, 143, 172.

Distillation and Blending

- Distillation of hydrogenated products
  - Simulated distillation analysis (ASTM D2887)
  - Total oxygenates analysis (elemental analysis)

- Product blending study
  - Simulated distillation analysis (ASTM D2887)
  - Octane number determination

Octane Numbers

➢ Octane number of gasoline and alkylated cyclohexane

Mixture / Compound	Octane Number, RON
Regular grade Gasoline	89
Benzene	98
Cyclohexane	110
Methylcyclohexane	104
1,2-Dimethylcyclohexane	104
1,4-Dimethylcyclohexane	66
Toluene	124
Xylene(s)	162 (m), 155 (p), 126 (o)

Summary and Conclusions

➢ Lignin depolymerization in ionic liquids with formic acid for higher lignin conversion and more stable oxygenates production – formic acid facilitating depolymerization and hydrogen transfer

➢ Bifunctional, non-precious metal catalysts for hydrodeoxygenation of oxygenates to hydrocarbons

Questions?

Thank you!

Questions?

351071 Primary Benchmark, a Collaborative Tool for Water Utility Energy Planning

Abstract:

Primary Benchmark is a new collaborative energy planning tool for water and wastewater utilities. Water utilities are working hard to reduce energy consumption, increase water conservation, manage the water-energy nexus, and build sustainable infrastructure. Tracking energy consumption and understanding energy trends are two of the many activities needed for operating an energy efficient water utility. Primary Benchmark was developed to reduce the labor required for energy management, and help water utilities manage their energy use.

This presentation demonstrates Primary Benchmark, a free web-based tool that allows water utility operators, engineers, and managers to track energy benchmarks, create energy teams, manage energy goals, and compare their facilities. The presentation will use real data to show how Primary Benchmark can be implemented by water or wastewater utilities.

Primary Benchmark keeps all relevant energy data in one place and efficiently presents data for communication with operators, engineers, and managers throughout an organization. Energy data can be manually entered or uploaded directly to Primary Benchmark. Where available, data from electrical and gas utilities can be uploaded directly from the utility company.

Data uploaded to Primary Benchmark is completely private and not shared with third parties. Water utilities have the option of participating in Benchmark Share, a program that allows water and wastewater utilities to see how their facilities compare to those of other water and wastewater utilities. This feature enables greater collaboration between utilities. Creating opportunities for learning from each other helps water utilities share ideas and become more efficient.

In addition to improved communication and collaboration, Primary Benchmark improves energy management by sending notifications to users about their facility's energy trends, upcoming energy goals, and important water-energy information.

Primary Benchmark is freely available at www.primarybenchmark.com. Development of Primary Benchmark is an open collaboration of Erg Process Energy. To get involved or learn more, visit www.primarybenchmark.com.

348539 Heat Transfer and Pressure Drop Characteristics of a Phase Change Material Nanoemulsion

Abstract:

Abstract: Thermal energy storage (TES) systems using phase change slurries (PCS), are capable of resolving the frequent mismatch between energy supply and demand, because of their high energy densities as they use both the sensible heat and latent heat of the phase change material during the phase transition. As an application of phase change slurry, heat transfer and pressure drop characteristics of a beeswax nanoemulsion with 10.1% beeswax, 70% distilled water, and 19.9 % surfactants, have been investigated experimentally. This paper describes the preparation of the beeswax nanoemulsion and the experimental data on convection heat transfer and friction pressure drop in a heated test section. Those properties are useful for understanding how useful this type of nanoemulsion is in practical applications such as a latent heat thermal storage system and heat transport fluid. The test section is a straight 304 stainless steel pipe with an inner diameter of Di =11.3 mm, tube wall thickness of 0.71 mm and length L=1.04 m. Five type-K thermocouples are mounted on the test section wall at equal intervals to measure the tube wall temperature. The tube is thermally insulated in order to minimize the heat loss. Two other thermocouples are inserted at the entrance and exit of the test section to measure the bulk fluid temperatures. Ohmic heating of the test section is achieved by passing direct current from a DC power supply with a maximum heat flux of 10.4 kW/m2. At room temperature, the pressure drop of nanoemulsion was found to be about four times higher than that of water under the same flow rate. However heating of the nanoemulsion to a temperature above the melting point of the beeswax (62 oC) in the test section increased the pressure drop. The maximum recorded pressure drop was 17.6 kPa at 80°C and the flow rate also dropped sharply due to the high viscosity of the nanoemulsion. Based on the pressure drop data, the nanoemulsion viscosity at 30, 60 and 70 oC were determined to be 20, 23 and 73 times greater than that of water at the same temperature. Near the melting point of the beeswax, the wall temperatures show fluctuations which are absent at below and above the melting temperature. The average convective heat transfer coefficients calculated at a heat flux of 7.15 kW/m2 and based on the recorded bulk fluid and tube wall temperatures were 226, 435 and 935 W/m2K at flow rates of 0.15, 0.45 and 1.2 GPM. They were found to be consistent with the values predicted by a Sieder-Tate correlation.

SPERA Hydrogen Process for Large Scale H_2 Storage & Transportation

Masato Shiraga, Hironori Kawai and Tomoyuki Mikuriya
Chiyoda Corporation, Yokohama, Japan

Introduction

Renewable energy is expected to be an alternative energy as post-fossil resources. Hydrogen is a clean energy carrier since it is converted to water after utilization, and hydrogen can be produced from any primary energy including renewable energy. It is considered that hydrogen energy can be converted from renewable energy, including solar energy in the future. Hydrogen produced from water by renewable energy can be a final energy solution as "solar and water energy system".

Hydrogen production from water by electrolysis is an effective method for conversion of electricity from renewable energy to hydrogen energy. It is considered that photo-catalytic water production will be commercialized in the future which is direct conversion of solar energy to hydrogen energy in addition to the direct utilization of solar energy.

While hydrogen storage and transportation technologies are commercialized by compressed hydrogen method for fuel cell vehicles (FCV) and liquefied hydrogen method for the hydrogen fuelling stations on a small-scale, the method for a large-scale storage and transportation has not been commercialized. Chiyoda Corporation has developed the organic chemical hydride (OCH) method using a toluene (C_7H_8) / methylcyclohexane (MCH: C_7H_{14}) system as the basis for a large-scale storage and transportation of hydrogen energy, and demonstration tests were successfully carried out to verify its expected performance. We call the MCH in this system "SPERA (Latin word for hope) Hydrogen", which reflects our ambition to establish a hydrogen supply chain. This paper introduces the OCH method and its development status.

1. Hydrogen Supply Chain

Fig.1 shows the hydrogen supply chain concept which has been proposed since 2004 in Japan, by using hydrogen storage and transportation system based on OCH method[1-2]. Japan faces issues of CO_2 emission and future fossil fuel depletion. In addition, importance of renewable energy utilization has been recognized after Fukushima nuclear power plant accident due to the tsunami disastrous in 2011 in Japan. However, Japan does not have much renewable energy as same as fossil fuel energy.

The hydrogen supply chain concept will develop in two steps. In the first age, hydrogen from fossil resources, called clean hydrogen is utilized accompanied by carbon dioxide capture and storage (CCS), enhanced oil recovery (EOR) or enhanced coal bed methane (ECBM).

In the second age, hydrogen from renewable energy will be utilized. For the hydrogen supply chain, existing infrastructure used in the first age can be utilized, even when the extent of renewable energy utilization expands.

2. Organic Chemical Hydride Method

Organic Chemical Hydride method is a way to store and transport hydrogen under ambient temperature and pressure by using Liquid Organic Hydrogen Carrier. Hydrogen is fixed to aromatic compounds such as toluene by hydrogenation reaction, whereby toluene is converted to MCH. Both toluene and MCH are gasoline components and are in the liquid state under ambient conditions.

Hydrogen is generated by dehydrogenation reaction from MCH at the hydrogen user site. Toluene is recovered and recycled back to hydrogenation site and reused as hydrogen carrier.

Table 1 shows the comparison of physical properties of typical organic hydride systems. Chiyoda Corporation has employed toluene and MCH system, because these components can be kept in the liquid state under the temperature range of -95 to 101deg.C. A Solvent is not required for any conditions across the globe, while other systems need a solvent to maintain the liquid state.

Fig.2 shows the hydrogen storage density. MCH has a gravimetric hydrogen content of 6.1wt% and a volumetric content of $47kg\text{-}H_2/m^3$ theoretically. Other plots show the practical values including container in a small-scale storage for FCV. It is considered that practical value is not very different from theoretical value in the case of OCH method, since storage condition is ambient and therefore special container is not required.

In the OCH method, hydrogen gas volume is reduced to <1/500 of that under ambient condition. Natural gas volume is reduced to <1/600 by liquefaction at a temperature of -163deg.C as LNG. Although in the liquefied hydrogen (LH_2) method hydrogen gas volume is reduced to <1/800, it needs an operation temperature of -253deg.C and boil-off gas treatment. OCH method is suitable for large-scale storage and transportation, since the method has relatively high storage density in spite of ambient storage condition. Moreover, there is no loss of hydrogen during long term massive storage.

In Japan, OCH and LH_2 methods are under development as a large-scale hydrogen storage and transportation technology. It is announced that practical application of LH_2 method by large LH_2 tanker is targeted by year 2025[3)].

Both toluene and MCH have been transported commercially by middle range class chemical tanker, in excess of 50,000 tons by one ship. The amount of hydrogen transported by above chemical tanker is more than 3,000 tons which is equivalent to fuel for 600,000 FCVs with 5kg each.

OCH method was investigated in the Euro-Quebec project in 1980's using MCH to transport hydrogen from a hydraulic power plant in Canada to Europe. However, since a feasible dehydrogenation process did not exist in those days, it was not established to date[4-5)].

Fig.3 shows the scheme of OCH method. The hydrogenation process from toluene to MCH has been well established from 1970's. Since the yield of MCH is more than 99%, the hydrogenation process is well commercialized.

Conventional chemical storage tanks can be utilized for toluene and MCH storage and marine transportation is well commercialized as mentioned above. The surface transportation is also well commercialized by chemical trucks on the road and the freighter on railways. Therefore OCH method can be established by the development of a novel dehydrogenation process, and improved by the modification of the hydrogenation process based on commercialized technology.

3. Dehydrogenation Catalyst

Chiyoda Corporation has developed a commercial dehydrogenation catalyst with high performance and long life. Fig.4 shows the estimated catalyst surface model. Platinum cluster (<1nm) is impregnated to alumina carrier of which pore size is well controlled uniformly to increase the catalytic activity and the platinum cluster is partially doped with sulfur atoms to prevent the coking. Although usual platinum-impregnated alumina carrier is egg shell type in which the platinum cluster is dispersed only on the rim of the catalyst pellet, Chiyoda's catalyst is of the uniform type in which the platinum cluster is well dispersed uniformly around the inside of catalyst pellet.

Fig.5 shows the result of the catalyst performance test. The developed dehydrogenation catalyst has an excellent performance; MCH conversion >95%, toluene selectivity >99.9% and hydrogen yield more than 95% under the conditions of around 350degC, 0.3MPa and LHSV = 2.0h^{-1}. Hydrogen generation rate is more than 1,000Nm^3-H_2/h/m^3-cat. Catalytic performance is stable well over a period of 1 year continuous operation. The catalyst will be replaced after deactivation, and platinum is recovered from the spent catalyst. Since the recovered platinum will be used in catalyst production for the next charge, catalyst cost is reasonable, in line with conventional petrochemical catalysts.

4. SPERA Hydrogen Process

Chiyoda Corporation has developed the new dehydrogenation process using this developed catalyst, and also improved the hydrogenation process based on commercialized technology. Fig.6 and 7 shows simplified process flows of the hydrogenation and dehydrogenation process respectively. The remarkable feature of SPERA Hydrogen process is its simple process configuration. Each process feedstock reacts in the fixed bed tubular reactors in vapor phase, and high product yield is achieved with simple vapor and liquid separation drums.

Hydrogenation process:

The toluene feed is vaporized (Eq.1) and mixed with hydrogen including recycle gas. The feed mixture is superheated to reaction temperature (Eq.2) and enters the top of the hydrogenation reactor (Eq.3), which is a fixed-bed tubular type reactor charged with hydrogenation catalyst. In the tubes, toluene reacts with hydrogen to produce MCH. Hydrogenation is an exothermic reaction, so the generated heat is removed by cooling medium, such as steam generation system, to control the reaction temperature.

The effluent gas from the reactor is cooled and the condensed MCH is separated (Eq.6), with the hydrogen gas being recycled to the reactor. The liquid MCH product is sent to storage tank.

Dehydrogenation process:

The MCH feed is vaporized (Eq.11), and superheated in charge heater (Eq.12) before entering the dehydrogenation reactor (Eq.13). The reactor is of a fixed bed tubular type charged with dehydrogenation catalyst in the tubes. Dehydrogenation is an endothermic reaction, so a heating source is required. The hot oil system is typically applied (Eq.14) to supply this reaction heat.

The effluent gas from the reactor is cooled to separate hydrogen and toluene (Eq.16). The separated hydrogen is compressed and treated to meet product hydrogen specifications. (Eq.17, 18)

5. Demonstration Test

Since April 2013, demonstration tests are being conducted in a 50Nm^3-H_2/h capacity pilot plant located in Yokohama, Japan. The plant includes hydrogenation and dehydrogenation process units, with toluene recycle system. Fig.8 shows this demonstration plant. Both the hydrogenation and dehydrogenation reactors are installed with the actual catalytic tube to be used in the commercial plant.

Fig.9 shows the performance of hydrogenation and dehydrogenation process. As of November, we have achieved over 4,400 hours of successful and trouble-free operation with excellent performance. Through extensive and continuous data acquisition, the process modeling and simulation tools for commercial plant design are being fine-tuned.

6. Conclusion

The OCH method is suitable for a large-scale storage and transportation of hydrogen. Chiyoda Corporation has developed a novel catalyst for the dehydrogenation reaction and dehydrogenation process, and commenced the demonstration of this method at the pilot plant with a capacity of $50Nm^3/h\text{-}H_2$ in Yokohama, Japan since April 2013. As of November, we have achieved over 4,400 hours of successful operation. Thus, OCH method will proceed to the commercial stage, by which SPERA Hydrogen process opens the door to a low-carbon society.

References

1) Okada Y. et al. (2004) 15th World Hydrogen Energy Conference, 30B-5
2) Okada Y. et al. (2006) Int. J. Hydrogen Energy, 31, 1348
3) http://www.khi.co.jp/english/ir/pdf/130425_4.pdf
4) Gretz J. et al (1990) Int. J. Hydrogen Energy, 15(6), 419
5) Gretz J. et al (1994) Int. J. Hydrogen Energy, 19(2), 16

Fig.1: Hydrogen supply chain concept

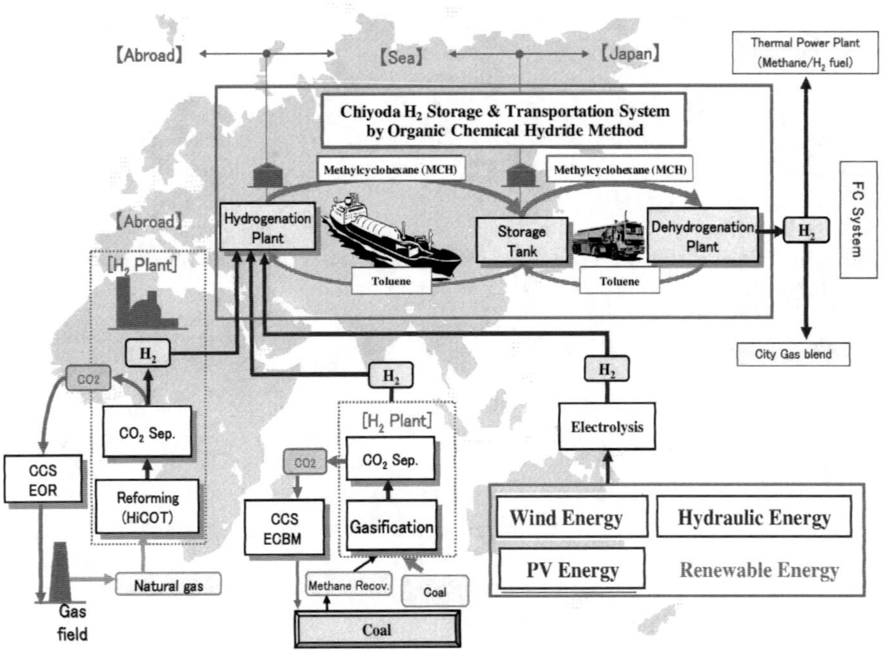

Table 1: Physical properties of typical organic hydride system

		MCH - Toluene		Cyclohexane - Benzene		Decalin - Naphthalene	
		Methylcyclohexane	Toluene	Cyclohexane	Benzene	Decalin	Naphthalene
Fomura		C_7H_{14}	C_7H_8	C_6H_{12}	C_6H_6	$C_{10}H_{18}$	$C_{10}H_8$
M.W.		98.19	92.14	84.16	78.11	138.25	128.17
State at room temp.		Liquid	Liquid	Liquid	Liquid	Liquid	Solid
Density (g/cm^3)		0.7694	0.8669	0.7791	0.8737	0.8963	0.9752
mp. (℃)		-126.6	-95.0	6.5	5.5	cis : -43.0 trans : -30.4	80.3
bp. (℃)		100.9	110.6	81.0	80.1	cis : 194.6 trans : 185.5	218.0
Hydrogen Storage Density	(wt%)	6.2		7.2		7.3	
	($kg-H_2/m^3$)	47.4		56.0		65.4	

Fig.2: Gravimetric and volumetric content of hydrogen

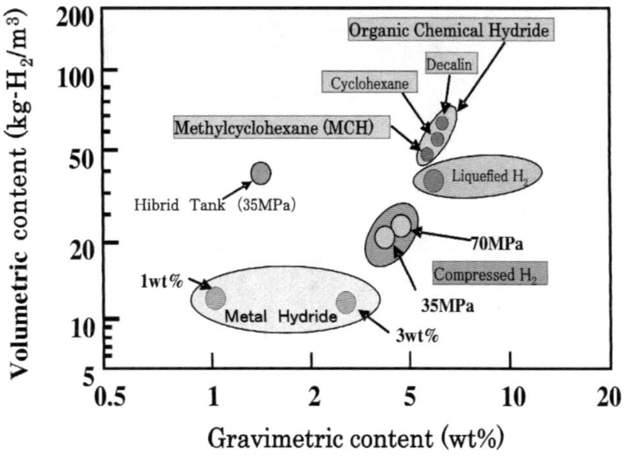

Fig.3: Scheme of OCH method

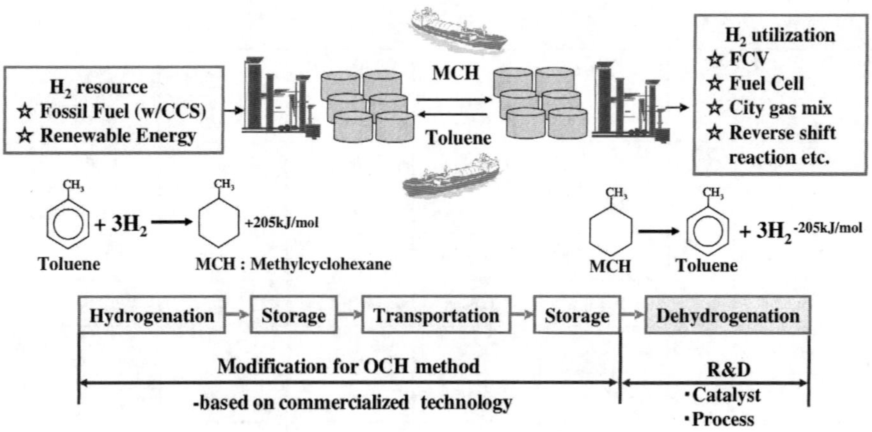

Fig.4: Estimated surface model of developed dehydrogenation catalyst

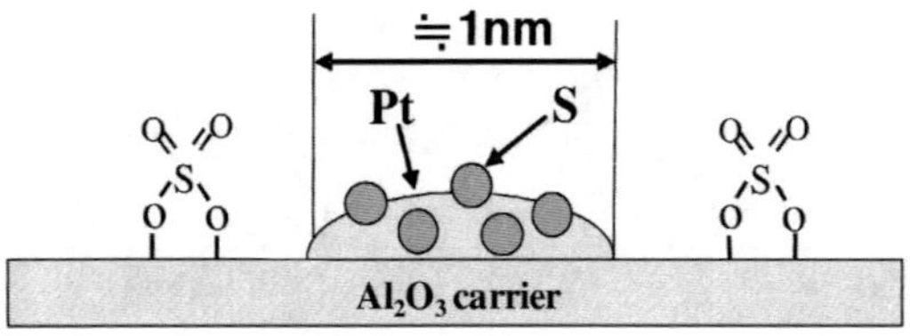

Fig.5: Result of dehydrogenation catalyst performance test

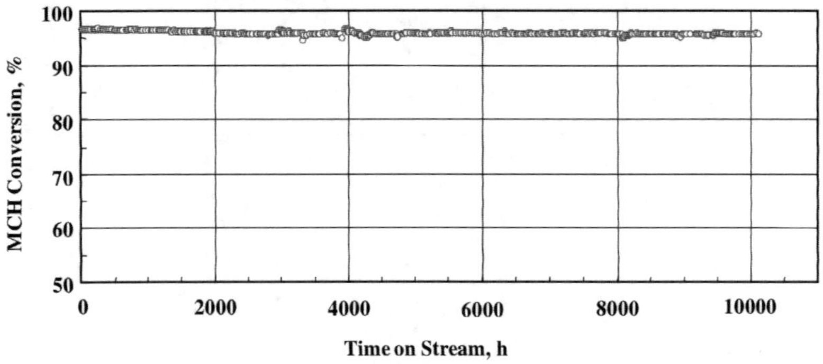

Fig.6: Hydrogenation process

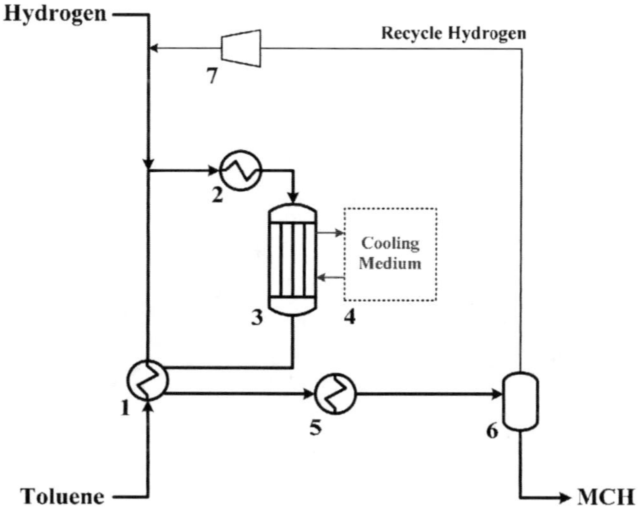

Fig.7: Dehydrogenation process

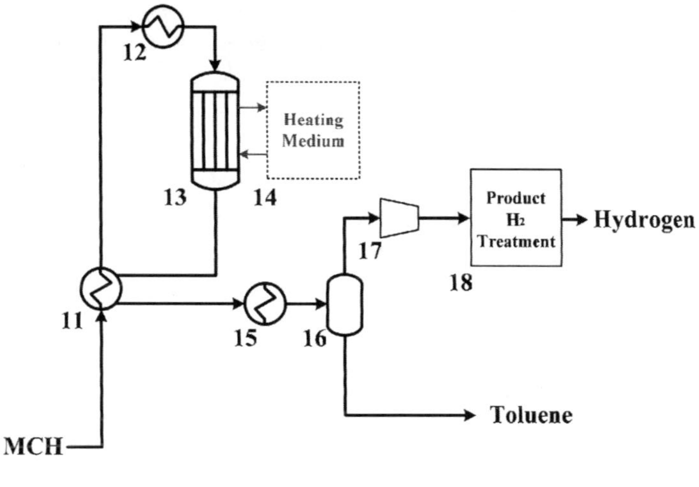

Fig.8: Demonstration plant

Fig.9: Performance of hydrogenation & dehydrogenation process

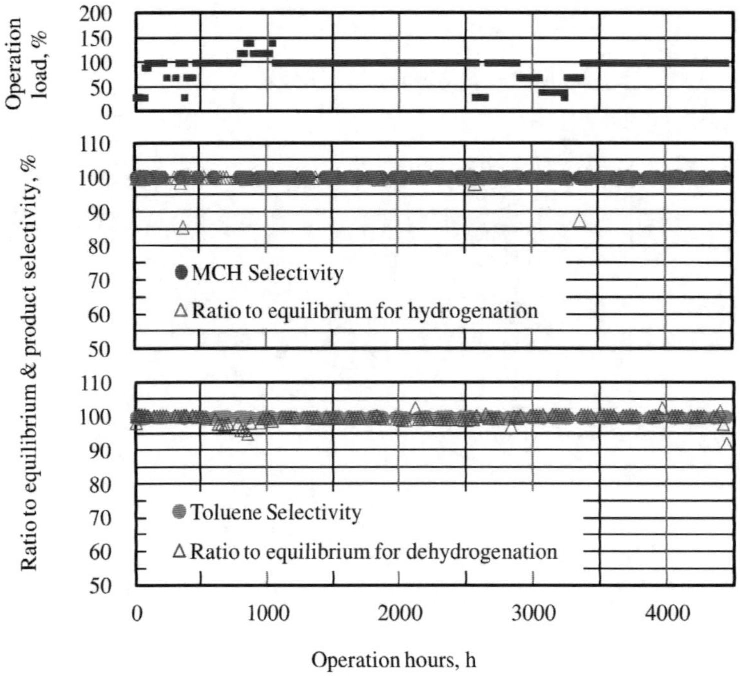

348541 Thermal Energy Storage System for Cold Engine Start up in Automobiles

Abstract:

Driving with a cold engine increases fuel consumption and greenhouse gas emissions in automobiles. Thermal Energy Storage (TES) devices could be utilized to provide heat during engine warm-up. A TES device could be charged from an engine's waste heat during normal operation. In this work, a latent heat TES device has been designed, constructed and tested to investigate waste heat recovery and reduction of engine warm-up time by using phase change materials (PCMs) as a thermal energy storage (TES) medium. Two types of paraffin waxes with melting points of 70 and 83 oC, and a 50/50 mixture have been examined to characterize their behaviors under repetitive heating/freezing cycles. The paraffin was placed in a rectangular container and hot or cold water was circulated through thin cooling plates, which were placed in the container at regular spacing. Thin aluminum fins were also placed between the cooling plates to increase heat transfer inside the paraffin. In the experiments, temperature variations in the paraffin and at the water inlet and outlet were recorded for a period of five minutes. From the experimental results, the fins enhanced heat transfer for both narrow and wide plate spacing, but the enhancement was more significant for the wider plate spacing. The efficiency of heat recovery was also positively correlated to the high circulating flow rate of the cold water and the use of a 50/50 mixture of the two paraffin waxes. An overall heat transfer coefficient was evaluated from the heat transfer rate and the temperature difference between the average water and paraffin temperatures. The product of the overall heat transfer coefficient and plate surface area was found to nearly double by inserting the thin aluminum fins. In five minutes, the total heat recovered from 3.6 kg of paraffin ranged from 600 to 1,000 kJ, depending on the water flow rate and plate spacing. Finally, experiments were conducted to heat up an aluminum block weighing 30 kg which simulated a common engine block. The aluminum block was heated up from 0 oC by water circulating through multiple TES containers. In five minutes, the aluminum block's temperature could be increased by 30, 34 and 42 oC, using 2, 3 and 4 TES containers, respectively. The results of these engine block heat-up simulation experiments demonstrated the potential of the latent heat TES in automobile applications.

350674 Optimum Operation of Complex Combined Heat and Power Systems of Two Parallel Gas Facilities

Abstract:

In the oil and gas industry, the energy demands of heat, steam, power and fuel is highly dependent to the oil or gas processed by the facility. The utilities' system design for those facilities normally built to account for the maximum operating rates. However, the plants sometimes operate at partial loading and because of the reliability constraints, boilers operate at minimum loading resulting to having some wasted steam passing over finfan condensers. Current models for optimizing the operation of utilities' systems (Combined heat and Power Models) of a facility were being developed to look at the available operational modification changes in a facility yields to better performance, but in the partial loading the operational modifications are limited and sometimes not available. Partial loading leads to high energy consumption and wasting of valuable energy even with using optimization tools sometimes. In this paper, a new optimization technique is being introduced so that the overall energy consumption for the whole system of parallel facilities would be minimized. The idea here is basically extending the boundary of the optimization problem from a facility level a lone to optimizing a system of similar facilities. In house models were being built to come up with our findings. The complete set of models developed composed of: (2) Standalone models for utilities optimization (CHP optimization models) for each facility interacting with a planning model for allocating oil and gas production to each facility to meet certain oil and gas production needed. Energy correlations to relate energy requirements of a facility with the amount of gas or crude processed. These correlations were built using historical data of actual operation of those facilities. The analysis resulted in around 3% to 5% more energy savings compared to individual plants optimization.

351307 The Shale Gale is Blowing: Plotting a Course that Avoids the Shoals and Rocks

Abstract:

Shale gas has revitalized the Chemical Industry in the U.S. The economic benefits have been widely described, but there is little discussion if the impacts of the great increase in ethane cracking. The shifting feedstock slate creates both challenges and opportunities for new technologies. The shoals and rocks caused by the shale gale will be detailed and a course described that can provide an even brighter future for the industry will be presented.

352096 Development of Shale Oil & Gas Programs - Critical Design Considerations for Economic Viability and Sustainability

Abstract:

The quantity and quality of crude oil & natural gas (from producing fields) change with time. These changes are large enough to require design and operational flexibility of the processing plants. Establishing trends for these changes are also very challenging. In addition, dynamics of the oil, gas, and gas liquids pricing bring additional demands in operational flexibility requirements. The paper will present these issues and discuss approaches to address the challenges.

352681 Future Energy and the Shale Gas Revolution: Implications and Opportunities

Abstract:

“Future Energy” will entail an unprecedented growth in demand, together with increasing stakeholder concerns for the impact of energy use on climate and the environment, and ensuring security of supply. These factors will drive diversification of energy sources as well as delivery systems and technology, to address both global and regional needs. Recent developments in shale gas and oil production via hydraulic fracturing together with increased use of bio-based and other renewable energy sources, will drive new opportunities for chemicals in delivering energy system infrastructure, improving energy efficiency, and re-optimizing process routes to minimize cost, as well as carbon and environmental footprints. In particular, low cost of gas relative to oil on an energy content basis will drive the development of new process technologies for commodity chemicals derived from C1 chemistry. Low energy costs will drive a new era of “smart manufacturing”.

This seminar will examine trends and opportunities in responsible energy and chemicals resulting from the shale gas "revolution", and end with an open discussion of what emerging chemical sciences and engineering innovations are most likely to be impactful in addressing future energy and sustainability needs.

2014 AICHE Spring Meeting

March 30 to Apr 03, 2014

Burns & McDonnell
1700 West Loop South, Suite 300
Houston, TX 77027

Authors/Presenters:
Mark W. Lockhart, P.E., Process Technology Manager, Chemicals and Petrochemicals
Burns & McDonnell, Houston, TX
832-214-1975

Elliott Robertson, P.E., Senior Process Engineer, Chemicals and Petrochemicals
Burns & McDonnell, Houston, TX
832-214-2916

Expanded Abstract:

"Shale Gas Boom – A New Era of Technology for Ethylene and Ethylene Derivatives"

There is a new era in the U.S. with the rise of the Shale Gas Boom that has emerged in this country. Large production increases are occurring from shale gas plays ranging from C1's – C4's, C5's/condensate and "tight oil". The plentiful supply of C1's – C4's versus current demand is driving prices for these molecules down. The recent result is directionally lower prices for methane, ethane, propane and butane as well as heavier products.

The plentiful supply and low prices for these building block molecules is having a marked effect on downstream industries that use the molecules either as fuel for energy or for raw materials as feeds. Steam crackers are moving to maximally change feedstock from naphtha's and gas oils to ethane to make ethylene and recently experiencing very attractive margins as a result.

Ethylene yield is very high when using ethane as feed which results in very favorable economics given the low price of ethane that has been prevalent over recent years. Ethylene producers have already made moves to feed as much ethane as possible within the limits of their existing production facilities. Ethylene producers are also moving to debottleneck existing plants including various debottlenecks that have already been announced. Additionally, as many as seven (7) new grass-roots facilities have been announced or are under consideration primarily along the gulf coast. Although, at many as two new ethylene plant projects are under review to be located in the North East to enable taking advantage of ethane production from the Marcellus Shale Play located in the states of Ohio, Pennsylvania and West Virginia.

Ethylene capacity totaling over 7 mm MTons/yr have been announced or are under study between debottlenecking of existing plants and new grass-roots ethane crackers. The debottlenecking projects can be accomplished in a shorter timeframe with the capacity increases continuing through 2013 to 2015. The new grass-roots capacities are slated to come on-line through 2016 and 2017. Several of the new ethane crackers will also install additional polyethylene capacity as part of the new facilities.

Historically, naphtha and gas oils have been used as feed to steam crackers. These feeds typically come from processing crude oils in refineries and, as such, the pricing for naphtha and consequently ethylene is directly related to the price of crude. Ethane production from shale gas plays provides crackers a new source of feed that is not related to the refining of crude oil or world-wide oil markets. This represents a significant paradigm shift in the traditional economics for ethylene production.

Ethylene is a major building block for numerous chemicals and petrochemicals. Consequently, the shale gas renaissance is affecting a broad spectrum of the chemical industry.

Shifting ethylene production from naphtha and gas oil to ethane as feed has a dramatic effect on the production of other byproducts that traditionally come from steam crackers. The yields of propylene, C4's, C5's and heavier products that are produced when feeding ethane are very low.

The reduction in propylene production is affecting the economics of producing propylene through the "on purpose" route. Numerous propane dehydrogenation projects are already in design with others being announced. This is just one example of the changing markets and the necessary transition that is taking place in the U.S. as the markets are in transition.

The total spending for the chemical/petrochemical industry for the period 2013 to 2030 is now estimated at approximately $120 billion. There is an anticipated sharp increase in spending starting in 2014 to peak in 2016/2017 at over $13 billion. The total spending for the period 2013 to 2017 totals nearly $48 billion.

With the advent of the announced ethylene plants slated to come on line by 2017, ethylene will be in greater supply. The economics of supply and demand suggest an oversupply of ethylene produced in the U.S. which will result in lower ethylene prices. A lower ethylene price has the effect of swinging economics in favor of producing more derivatives made from ethylene which will have to be exported from the U.S. to the global market place in order to maintain a supply and demand balance.

It is estimated that along with ethylene production various ethylene derivatives including polyethylene, ethylene oxide, ethylene glycol, EDC, VCM, PVC, among others will also increase. This has the effect of increasing the capital spending through 2020 - and beyond.

Where will all this production go? It is forecasted that the U.S. will be increasing net exports of the ethylene derivatives.

Lower energy cost along with cheap ethane has caused a marked shift in where U.S. stands on the global ethylene cost of production curves. An impressive result of the Shale Gas Boom and consequent lower cost of energy and cheap ethane has been the effect on U.S. cost of production for ethylene on the global cost curve. Comparing the 2007 and 2012 cost of production curves it is clear that the shift has now brought the U.S. much closer to the lowest cost producers including Canada and the Middle East.

Ethylene Dichloride (EDC) is one of the derivatives of ethylene that is forecasted to experience capacity increases in the U.S. EDC is used to produce Vinyl Chloride Monomer (VCM) which in turn is used to produce PVC. The U.S. is likely to increase export of PVC for use in home building and plumbing products world-wide. Burns & McDonnell has chosen to highlight the EDC -> VCM value chain in this presentation.

The process to produce chlorine is very energy intensive at 20,200 Btu/lb. Reduced U.S. energy costs represent a significant advantage in lowering the cost of production for chlorine which is another advantage in the EDC, VCM and PVC value chain. There are two widely accepted methods for producing EDC: direct chlorination and oxychlorination.

In direct chlorination, chlorine and ethylene are reacted to form the EDC molecule such that two chlorine molecules are used for each molecule of ethylene. The typical reaction is between the dissolved gases in liquid over an iron chloride catalyst at 15 to 75 psig pressure and 200 to 300 deg F. The use of high purity reactants is typical to avoid the formation of a broad range of impurities which may further complicate EDC purification. An excess of chlorine is typical to ensure complete ethylene conversion.

In oxychlorination, the reaction occurs over a copper chloride impregnated alumina catalyst in the liquid phase at a temperature of approximately 300 deg F and a pressure of up to 75 psig. The typical fluid bed technique offers heat transfer, prevents the occurrence of hot spots and allows more efficient catalyst regeneration. Ethylene conversion of 93 to 97% can be achieved with selectivity in EDC of 91 to 96%.

The EDC purification section removes trace amounts of impurities in the furnace feed which can inhibit the cracking reactions and cause fouling. In addition, EDC must be dry to prevent excessive corrosion in equipment and piping downstream of the reactor. Therefore the combined EDC from three different sources must be purified to greater than 99% EDC from direct chlorination, EDC from oxychlorination and EDC recovered and recycled from the cracking step. This combined EDC is washed with water and scrubbed with caustic to remove chlorine, HCl and other water soluble impurities including chloral. The crude EDC and unconverted EDC from the VCM column are combined and dried and purified in two distillation columns. In the first "Lights Column", water and lower boiling point impurities are separated as the overhead product. The bottoms product is fed to a "Heavies Column". Pure EDC of approximately 99% purity is taken as the overhead product while the heavy impurities leave the column as the bottoms product.

The pure EDC is forwarded to a furnace for thermal cracking of EDC to vinyl chloride. A typical furnace consists of a radiant section, a convection section, and a stack. The furnace fire box operates at 950 deg F. The convection utilizes combustion gases to preheat the feed before entering the firebox.

The vinyl purification section consists of two distillation columns in series which are used to separate vinyl chloride from EDC, HCl and remaining by-products. The first column ("HCL Column") distills the HCL mixture to a pure overhead product which is recycled to the oxychlorination reactor. The HCL Column is a trayed column and operates at a pressure of 135 psig. The bottoms product of the HCL Column is fed to the second column ("VCM Column") which operates at 65 psig. A VCM product of greater than 99 wt% is produced as the overhead product of the VCM Column. The bottom of the VCM Column is recycled to the Lights Column for re-purification.

A brief overview of the economic analysis for EDC/VCM plant investment indicates that the total installed cost (TIC) for a 300,000 metric ton/yr facility is in the range of $450MM to $500MM. With current EDC pricing of $245-275/ton and assuming an ethylene and chlorine cost of $1320-1362/ton and $275-310/ton, respectively and assuming the energy, labor, overhead and tax contributions are in the range of $70-85/ton results in a payback of less than five years. These rough economics do not include cash costs for land, start-up nor the capital investment required for utilities and infrastructure.

Constructing a new EDC plant at an existing facility where existing infrastructure and utilities are present will make the investment more attractive than a pure green field site. Debottlenecking an existing EDC facility is likely yield an ROR that represents another attractive investment.

Furthermore, installing or debottlenecking existing VCM and PVC facilities make this value chain one of keen interest in the coming low cost ethylene environment that is predicted.

In conclusion:

- The Shale Gas Boom is resulting in high production for C1's – C4's, which has resulted in reduced prices for these chemical and petrochemical building blocks.
- Several new grass-roots ethane crackers and ethylene plant debottlenecks to increase ethane feed capacity have been announced.
- Increased ethylene production in the U.S. is forecasted to lower ethylene prices in the long run.
- Ethylene is a central building block in producing various chemicals and petrochemicals.
- Therefore, a large sector of the U.S. chemical industry will be affected.
- Ethylene derivatives benefit from reduced ethylene prices.
- U.S. exports of ethylene derivatives are forecasted to increase to supply the world-wide market.
- The U.S. can compete well in the world chemical and petrochemical market place with the advent of the Shale Gas Plays.
- Chlorine cost of production is and will continue to benefit from the low energy prices in the U.S.
- EDC and VCM are examples chemicals that are forecasted to benefit as more ethylene production comes on line along with consequent lower ethylene pricing.
- The economics of producing EDC and VCM in a low cost ethylene environment is compelling.
- The EDC->VCM->PVC value chain will benefit. Export of these chemical products to the world-wide market is forecasted to increase in the future.

The Shale Gas Boom has shifted the U.S. energy picture as well as the outlook for chemical and petrochemical production. The U.S. is once again a key actor and leader in the world chemicals stage.

This truly is in a new era of Technology!

347676 Opportunities in Chemical Engineering Being Created by New US Shale Gas Supplies

Abstract:

Recent technical breakthroughs in extraction techniques have allowed the combination of hydraulic fracturing with long horizontal wells to enable the release of unconventional hydrocarbon sources. The ability to tap into these previously unreachable shale gas resources has completely transformed the US natural gas and crude oil supply and demand outlook. These recent technology breakthroughs have resulted in natural gas production in the United States increasing by 25% over the last 5 years, largely due to shale gas production. New US natural gas supplies are playing a key role in "turbocharging" US manufacturing over the next 10 to 20 years, creating an industrial renaissance. This paper summarizes potential opportunities in chemical engineering that could arise as a result of new shale gas production.

Shale Gas and Tight Oil

(97c) Shale Gas Boom and Tight Oil - a New Era for the Refining Industry

Lockhart, M. W., Burns & McDonnell

Varraveto, D.,

Originally presented on: 4/1/2014 15:30:00 - 16:00:00

Production of Tight Oil from the Shale Gas Plays in the U.S. is having a profound effect on the Refining Industry. Refiners are maximizing the amount of advantaged, "tight oil", they can feed their crude units some with geographic locations that allow them to capitalize the most on the sometimes attractive spreads provided by what is mostly light and sweet crude. New topping columns, crude units and condensate splitters are being constructed in various refineries in the U.S. This paper provides an overview of some of the studies and projects that are taking place and provides an overview across the spectrum of possibilities that refiners are facing. Burns & McDonnell has much experience with actual projects that will be highlighted to provide an exciting overview of this activity.

347649 Economically Treating Sour Gas from Tight Oil Formations

Abstract:

GT-SPOC – Economically treating sour-gas in the shale plays

In the shale plays, there are pockets of sour gas, or sour associated gas. Treating this gas to pipeline specifications with amine units gets the gas to pipeline specifications. However, operators have to deal with the resulting acid gas stream that cannot generally be vented due to permit limits on sulfur emissions. If the gas is not very sour, sulfur scavengers are the obvious solution, however, as the gas contains more than 0.1 ton/day of sulfur, the treating method of choice is not as clear.

GT-SPOC offers an alternative to the conventional furnace based "modified Claus" method for sulfur recovery. The technology is very suitable in the 0.5 – 20 ton/day sulfur production range, which is typical of the sulfur rates we see in the shale plays.

The approach improves upon the Claus process which is used to produce >95% of sulfur around the world, by using a patented catalyst at the front of the plant for SO_2 production. The catalyst utilizes constituents that address typical operator issues relating to heavy hydrocarbon contamination of the gas, start-up and shut-down issues, sour-gas composition variations and turndown. The presentation will discuss the capital and operating economics, as well as specific technical aspects of the technology that allow modularization, reduced footprint, and a tremendous reduction in capital cost.

A Multi-Layered Optimization Framework for Design and Analysis of Renewable Energy Systems

Aryan Geraili, Jose A. Romagnoli

April 2014

PSE @ LSU

LSU

PROCESS SYSTEMS ENGINEERING AT LOUISIANA STATE UNIVERSITY

Presentation outline

- **Motivation and objective**
- **Alternative technology options**
- **Optimization strategy**
- **Results**
  - ✓ Optimal portfolios
  - ✓ Optimal operating conditions
- **Conclusions**

2

Motivation

➤ Over the last century, world energy consumption has increased progressively

➤ Current global state of energy supply is highly dependent on fossil fuels

Owing to :

- Finite nature of fossil fuels
- Rapid increase in their prices
- Concerns about their environmental impact

→ Extensive research required to find alternative energy sources

LSU

3

Motivation

Commercialization of renewable energy technologies has been hampered by a multitude of endogenous and exogenous factors:

- Unavailability of appropriate feedstock supply
- Lack of capital and risk of investment
- Inefficient feedstock conversion systems

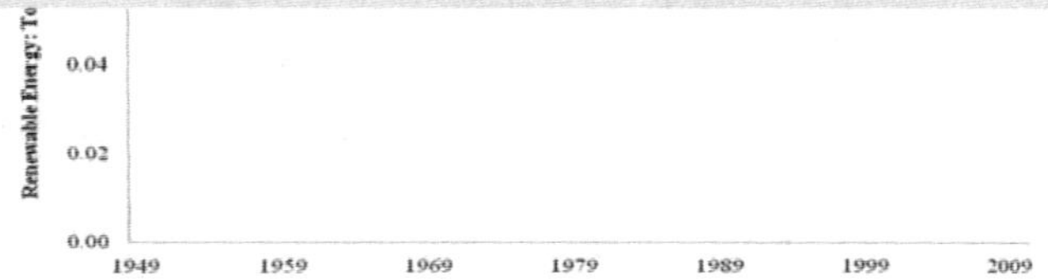

Renewable energy production as a percentage of total energy supplied in the United States; Source: EIA.gov

<u>**Garnering sufficient decision support is required to guide long and short term actions**</u>

LSU

4

Objective

➢ Overall objective:

Developing a comprehensive decision support system for sustainable design of renewable energy systems

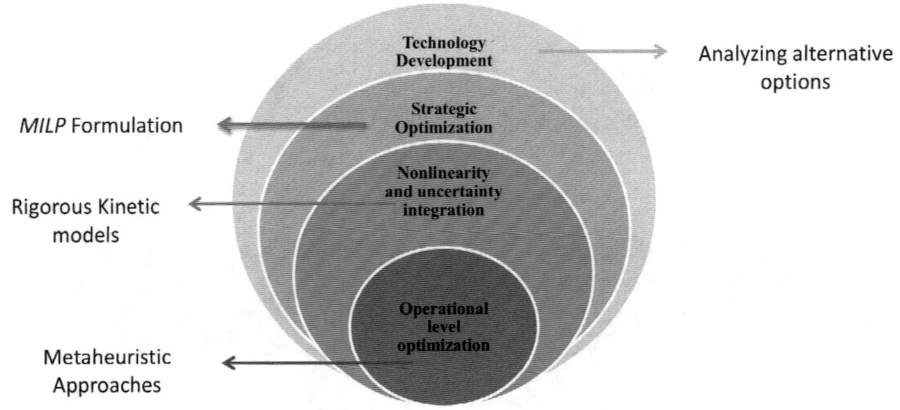

5

Biorefinery

To demonstrate the utility of our proposed framework a ***lignocellulosic biorefinery*** is considered as a case study

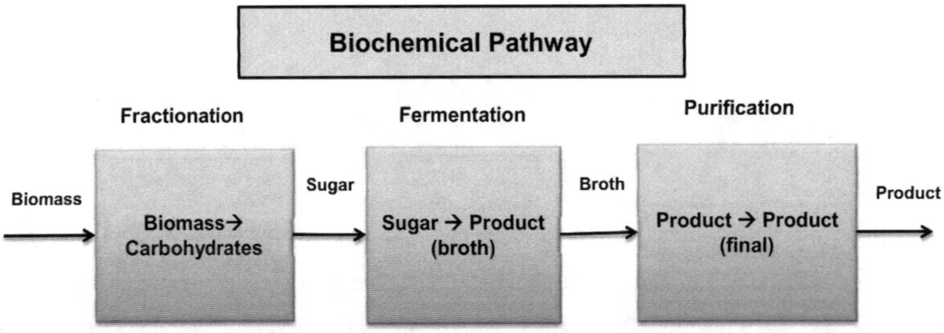

6

Dilute Acid Pretreatment

Dilute acid is selected as pretreatment technology in simulation.

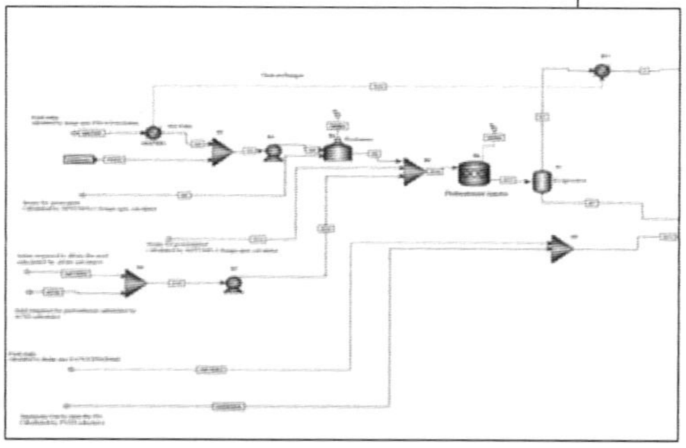

Limitations:

- ✗ Due to acid existence ⟶ This technology is very corrosive (design limitation)
- ✗ Degredation products are produced such as furfural ⟶ negative impact on downstream process sections (process limitation)

Detoxification is necessary prior to fermentation

Overliming

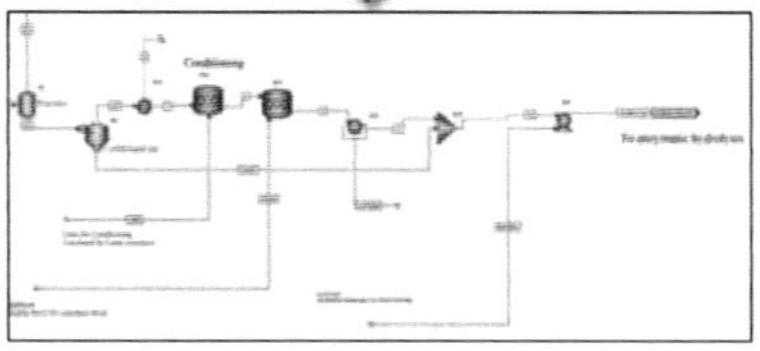

Ammonia

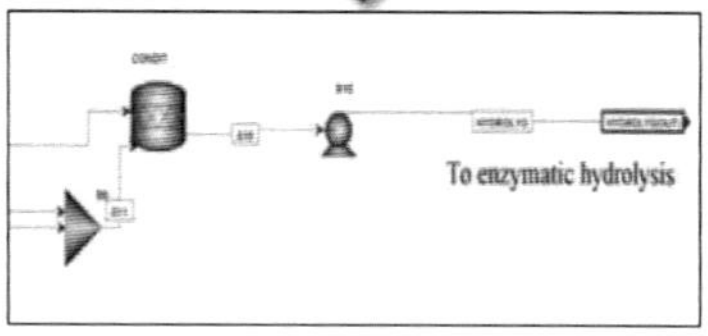

7

Hydrolysis & Co-fermentation

Kinetic model for enzymatic hydrolysis(Kadam et al., 2004)

$$R_s = \alpha \frac{C_s}{S_0}$$

$$r_1 = \frac{K_{1r}\, C_{E_{1B}}\, R_s\, C_s}{1 + \frac{C_{G2}}{K_{1IG2}} + \frac{C_G}{K_{1IG}} + \frac{C_{Xy}}{K_{1IXy}}}$$

$$r_2 = \frac{K_{2r}(C_{E_{1B}} + C_{E_{2B}})\, R_s\, C_s}{1 + \frac{C_{G2}}{K_{2IG2}} + \frac{C_G}{K_{2IG}} + \frac{C_{Xy}}{K_{2IXy}}}$$

$$r_3 = \frac{K_{3r}\, C_{E_{2F}}\, C_{G_2}}{K_{3M}\left(1 + \frac{C_G}{K_{3IG}} + \frac{C_{Xy}}{K_{3IXy}}\right) + C_{G_2}}$$

$$C_{E_{iB}} = \frac{E_{imax}\, K_{iad}\, C_{E_{iF}}\, C_S}{1 + K_{iad}\, C_{E_{iF}}}$$

$$k_{ir(T2)} = k_{ir(T1)}\, e^{\frac{-E_{ai}}{R}\left(\frac{1}{T1} - \frac{1}{T2}\right)}$$

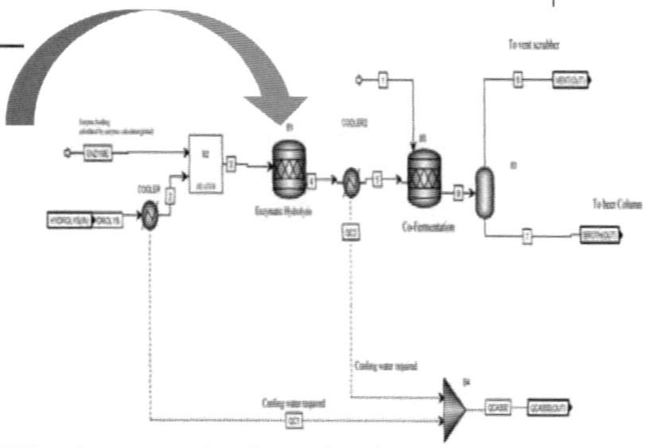

glucose and xylose simultaneously(**Co-Fermentation**)

Glucose ⟶ 2 Ethanol + 2 CO2	Conversion : 90%

To incorporate the complexity of biological reactions, experimentally-derived kinetic models are utilized to simulate the reactions in these processes.

8

Software Architecture

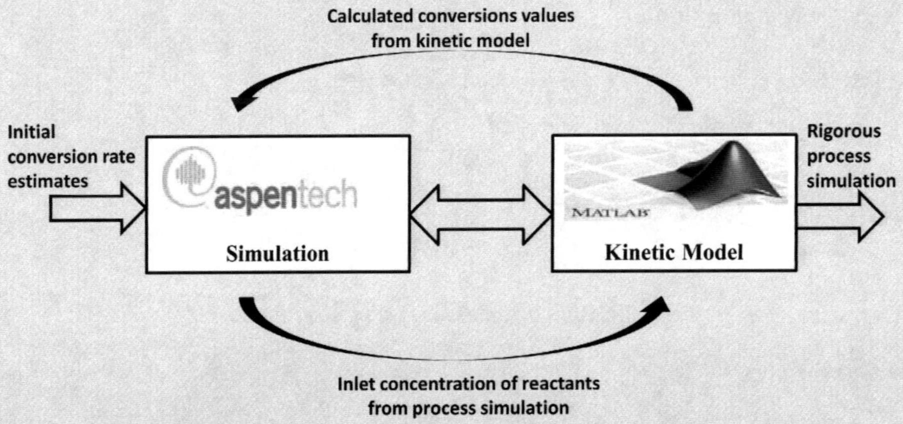

Convergence criteria:

Calculated inlet concentrations in Aspen remain the same for two consecutive iterations

9

Solid Separation Technologies

Solid Separation options

❖ **After Purification**

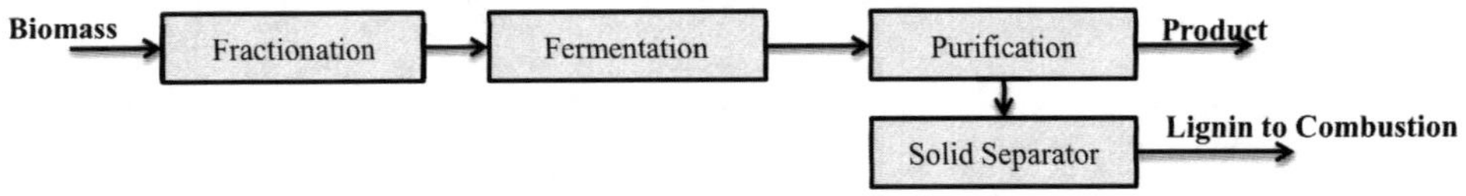

❖ **Before Fermentation**

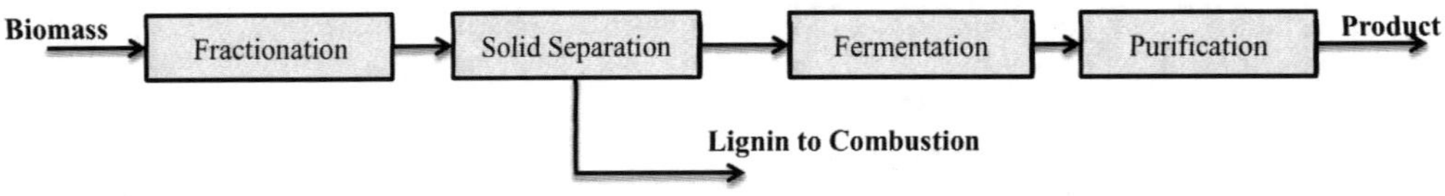

LSU

10

Alternative Process Configurations

Table 3: Alternative process configurations

Options	DA	AM	OL	SHCF	S/L sep 1	S/L sep 2	ETOH
Scenario 1	✓	✓	--	✓	--	✓	✓
Scenario 2	✓	--	✓	✓	--	✓	✓
Scenario 3	✓	✓	--	✓	✓	--	✓

DA: Dilute acid; AM: Ammonia conditioning; OL: Overliming; SHCF: Separate Hydrolysis and co-fermentation; S/L sep 1: Solid separation after hydrolysis; S/L sep 2: Solid separation after beer column; ETOH: Ethanol;

11

Optimal process configuration

Scenario 1&2 :Detoxification technologies

Annual revenue, profit and capital cost in scenario 1	
	Cost (MM$/yr)
Ethanol (1)	119.778
Succinic Acid (2)	--
Electricity to grid (3)	3.88082
Annual revenue (4) = (1) + (2) + (3)	123.659
Total production cost (5)	74.417
Capital cost (MM$)	**118.000** ⇦

Annual revenue, profit and capital cost in scenario 2	
	Cost (MM$/yr)
Ethanol (1)	119.2923
Succinic Acid (2)	--
Electricity to grid (3)	3.865069
Annual revenue (4) = (1) + (2) + (3)	123.1573
Total production cost (5)	72.278
Capital cost (MM$)	121.000

✓ **Ammonia Conditioning is selected**

12

Optimal process configuration

Sugars inhibition effect

$$r_9 = qpmax,g\frac{C_g}{K_{9g}+C_g}\left(1-\frac{C_{Et}-C_{Etip,g}}{C_{Et_{maxp,g}}-C_{Etip,g}}\right)\left(\frac{K_{9Ipg}}{K_{9Ipg}+C_g}\right)$$

$$r_{10} = qpmax,xy\frac{C_{xy}}{K_{10xy}+C_{xy}}\left(1-\frac{C_{Et}-C_{Etip,xy}}{C_{Et_{maxp,xy}}-C_{Etip,xy}}\right)\left(\frac{K_{10Ipxy}}{K_{10Ipxy}+C_{xy}}\right)$$

Sugar conversion in fermentation

$$\text{Sugar Concentration} = \frac{\text{Sugar}}{\text{Liquid} + \text{Solid}}$$

✓ **Solid separation after purification is Selected**

13

Optimization strategy

Analytical models suggested for designing decision support systems specially in the field of biorefinery:

Select best feedstock

Optimal production pathway

Strategic models

❖ Conversion mechanisms are inherently nonlinear ➡ **Nonlinearities** should be incorporated to frameworks

Operating condition optimization ➡ **Operational level optimization**

14

Multi-layered framework

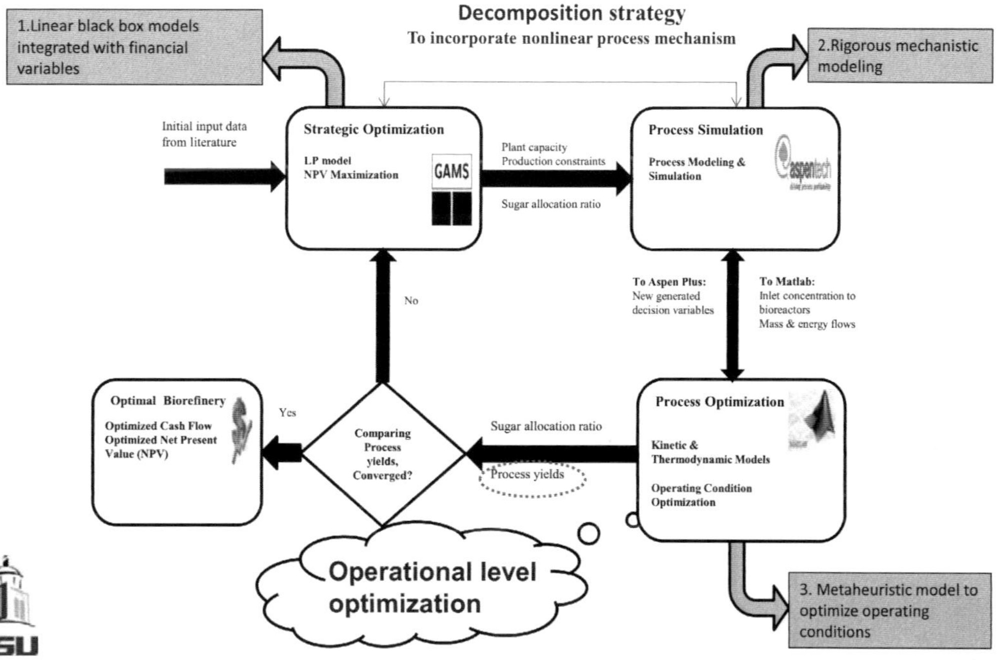

LSU

Strategic Model Description

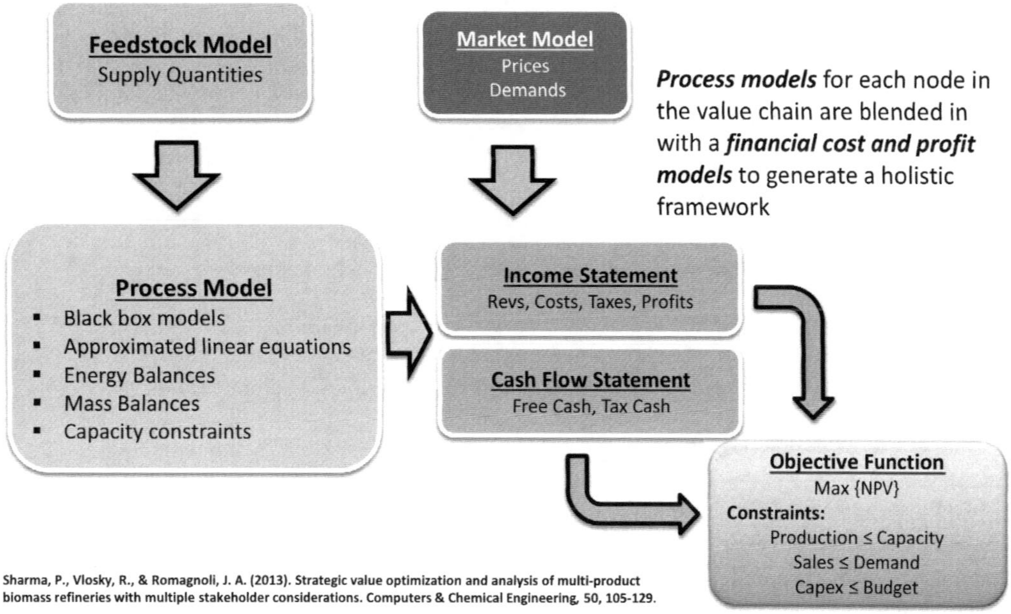

Process models for each node in the value chain are blended in with a ***financial cost and profit models*** to generate a holistic framework

Sharma, P., Vlosky, R., & Romagnoli, J. A. (2013). Strategic value optimization and analysis of multi-product biomass refineries with multiple stakeholder considerations. Computers & Chemical Engineering, 50, 105-129.

Strategic Model, portfolio Selection

- Initial Portfolio :
  - Switchgrass, Energy Sorghum, Energy Cane

- Initial Portfolio
  - **Fuels** → Ethanol, Butanol
  - **Chemicals** → Lactic & Succinic Acid

17

Strategic Model – flexible capacity design

Stepwise capacity expansion

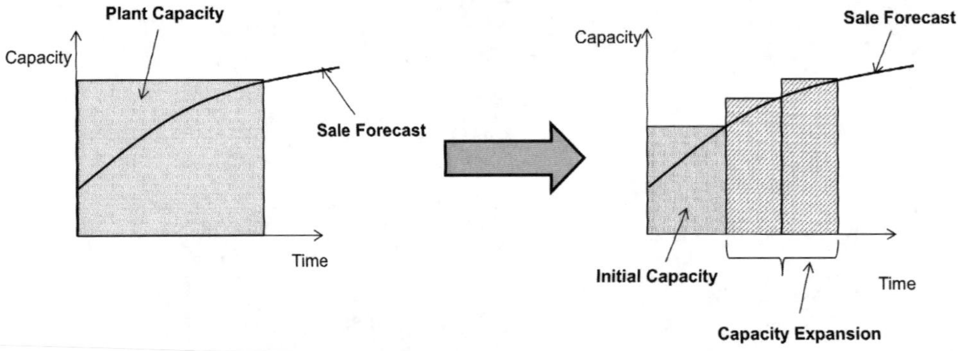

Linearized estimation of equipment cost:

Equipment cost (t) = BV(t) x Fixed cost + Expansion(t) x Variable cost

Capacity increment binary variable at each time period

18

Strategic Model – Feedstock portfolio

➢ **Switchgrass** and **energy cane** are selected as optimal feedstocks

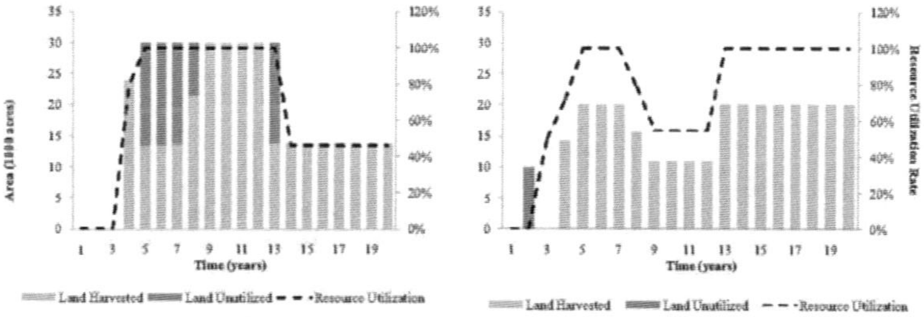

Biomass	Average Usage (tons per day)	Average Biomass Yield (tons per acre harvested)	Average Biomass Yield (tons per acre available)
Switchgrass	339	6,000	3,728
Energy Cane	559	10,680	8,195
Total	898	8,340	5,962

Fig. 3. Utilization of biomass based on total availability.

Results show that ***perennial crops with long production cycles*** are preferred over annual crops that require re-establishment every year

LSU

19

Strategic Model – Product portfolio

➢ **Ethanol** and **succinic acid** are selected as the optimal products

Butanol: low fermentation and recovery efficiencies
Lactic acid: Unitary Cost of production is higher than product price

Ethanol & succinic acid fermentations are integrated

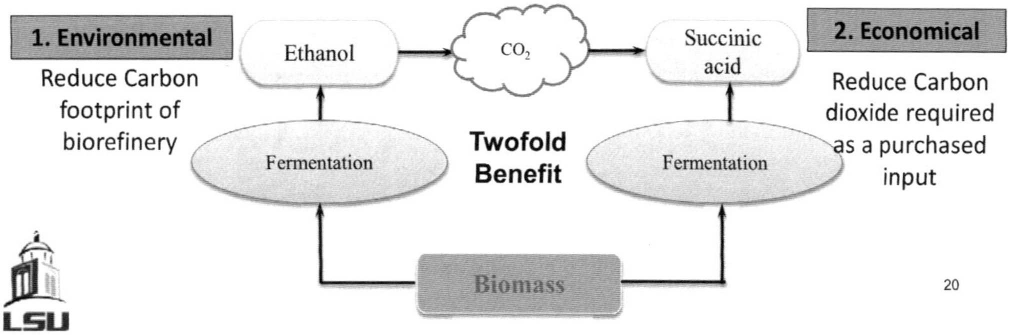

LSU

20

Framework Details: Process Optimization

Process operating condition optimization

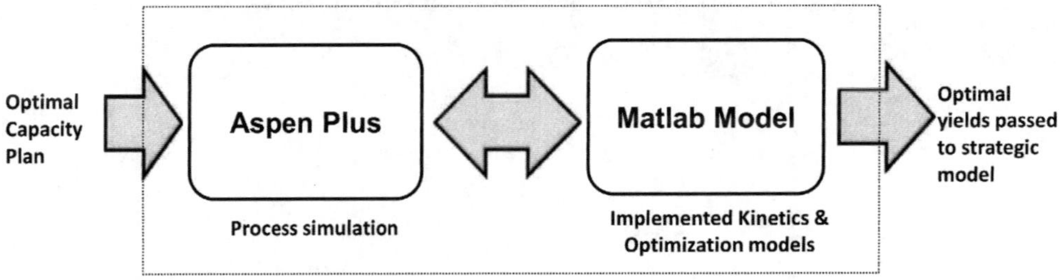

Why Metaheuristic approach:

1. To solve deterministically, constraints should be incorporated to objective function.
2. ***Metaheuristic methods*** do not require manipulation of the mathematical structure of the objective function and constraints. Hence, they can be used for any type of problems

21

Decision Variables

1. **Temperature in hydrolysis reaction (T)**

   Enzyme activity is correlated with temperature through Arrhenius model

   $$k_{ir(T2)} = k_{ir(T1)}\, e^{\frac{-E_{ai}}{R}\left(\frac{1}{T1}-\frac{1}{T2}\right)}$$

2. **Sugar allocation between ethanol & succinic acid fermentation (X)**

   Allocated sugar streams impact the fermentation reaction kinetics and consequently the product yields in the fermentation effluent

The manipulated decision variable during optimization is the fraction of total sugar that is used for ethanol fermentation (*X*)

22

Decision Variables

3. Enzyme loading during hydrolysis

Higher enzyme loading:

- ✔ Higher sugar yield in hydrolysis
- ✔ Higher ethanol yield in fermentation
- ✖ Succinic acid production decreases beyond an optimal point
- ✖ Higher direct cost of the plant

Sensitivity analysis

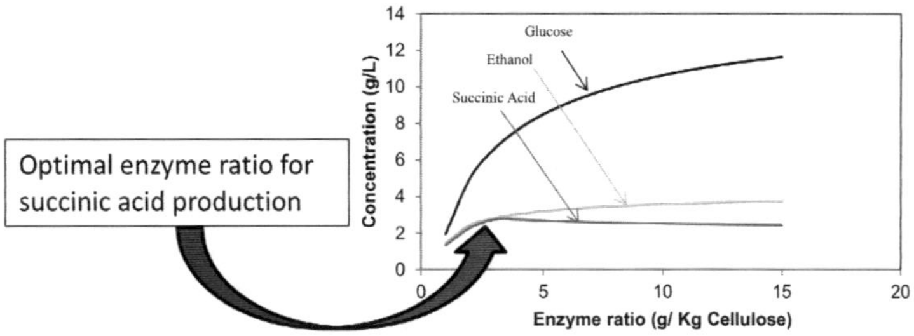

23

Results

Iteration results in optimization strategy

Parameters and Variables		Iteration1 Step 1	Iteration1 Step 2	Iteration1 Step 3	Iteration2 Step 1	Iteration2 Step 2
Capacity Constraints	Feedstock (1000 ton/yr)	--	333.3	--	333.3	--
	Ethanol (MM gal/yr)	--	11.082	--	10	--
	Succinic Acid (1000 ton/yr)	--	15.6	--	13.3	--
Process Variables	Sugar Allocation	--	0.52	--	0.40	--
Yield Parameters	Sugar (kg / kg)	0.87	--	0.81	--	0.81
	Ethanol Fermentation	0.85	--	0.71	--	0.71
	Succinic Acid Fermentation	0.25	--	0.15	--	0.15
	Ethanol Purification	0.99	--	0.98	--	0.98
	Succinic Acid Purification	0.78	--	0.78	--	0.78

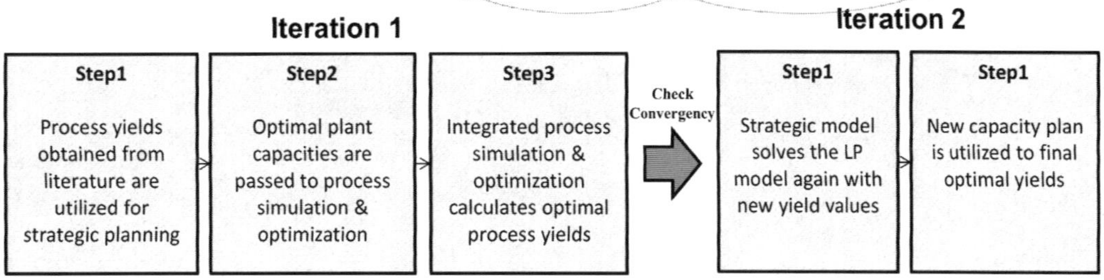

Results

Final Optimal Configuration

Optimal Framework results

	Variable	Final Value	
Optimal Capacities	Feedstock Capacity (100 ton/yr)	333.3	Long term decisions
	Ethanol production Capacity (MM gal/yr)	10.00	
	Succinic acid production Capacity (100 ton/ yr)	13.30	
	Net present value ($ MM/yr)	68.00	
Optimal Yields	Sugar yield (kg/ kg)	0.81	Nonlinear model
	Ethanol Fermentation yield (kg/ kg)	0.71	
	Succinic Acid Fermentation yield (kg/kg)	0.15	
	Ethanol Purification yield (kg/ kg)	0.98	
	Succinic Acid Purification yield (kg/kg)	0.78	
Optimal Operating Conditions	Hydrolysis Temperature (°C)	44.00	Operational level decisions
	Enzyme loading (g enzyme/ Kg Cellulose)	20.00	
	Sugar allocation ratio (sugar to ethanol)	0.40	
	Cash flow after tax ($ MM/yr)	70	

LSU

25

Conclusion

- Alternative technologies are analyzed for bioethanol production
- We have **introduced and implemented a framework to optimally design renewable energy production systems** that are governed by nonlinear process dynamics
- **Rigorous kinetic process models** were incorporated to impart greater fidelity to the simulation in Aspen Plus.
- The framework shows a deviation in process yields, and a deviation in the production capacities and operating conditions, from initial literature estimates.

LSU

26

Questions

Comments

27

Increasing the Viability of Algae-Based Bio-Fuels through Co-Product Development

William M. Chirdon, Ph.D. and Kartik Katti
wchirdon@louisiana.edu
Department of Chemical Engineering

April 1, 2014

UNIVERSITY OF LOUISIANA Lafayette

Motivation – Why Oil from Algae?

Key advantages:

- Large amounts per acre can be produced
- 10,000 gallons per acre annually or more
- 50x more productive than traditional, terrestrial crops
  - Corn: 18 gal/acre/yr
  - Soybean: 48 gal/acre/yr
  - Palm oil: 635 gal/acre/yr

Data from: Pienkos, P. "The Potential for Biofuels from Algae" Algae Biomass Summit, San Francisco, CA, Nov 2007.

Research for a reason.

UNIVERSITY OF LOUISIANA Lafayette

Motivation – Why Oil from Algae?

Key advantages:

Does not have to compete with food crops (even with land usage)

Use waste CO_2

Treat wastewater streams

Research for a reason.

UNIVERSITY OF LOUISIANA Lafayette.

Motivation –Oil from Algae

Key disadvantage:

- Economics
  - Biodiesel from LA would be approx $6/gal
  - Feedstock is 80-90% of the cost
- Value of algal cake
  - $200/ton (as fertilizer or feed)
  - $500/ton could make oil-from-algae feasible
- Plant-based glues
  - $1400/ton*
  - Formaldehyde-free, very "green" material

*Masumbu, F.F.F., Masamba, W.R.L., Fabiano, E. "Development and evaluation of maize and cassava starch-based cold-setting adhesives for case-sealing." *Malawi Journal of Agricultural Sciences* 2(1) 2003.

Research for a reason.

UNIVERSITY OF LOUISIANA Lafayette.

Motivation – Oil from Algae

What if we were to stop using petroleum?

Need replacements for a wide range of materials:

- Fuel
- Plastics
- Petrochemicals
- Asphalt

Co-products from biomass must be developed, just like they have for crude oil

Research for a reason.

UNIVERSITY of LOUISIANA Lafayette

The Challenge

Convert tons of algae cake which is "mostly protein" into value-added co-products

Chemical composition of algae cake is poorly defined

Separation/refinement may be cost-prohibitive

Solution: use very simple and unspecific chemistry to denature proteins into an adhesive state to create glues and composite materials

Research for a reason.

UNIVERSITY of LOUISIANA Lafayette

One Possible Solution

Modify the pH of the protein to denature protein

Breaks up higher level structures to convert proteins into random coil structures with exposed hydrophilic groups

Must balance pH, temperature, and time to allow for denaturation with minimal hydrolysis

Surfactants may also be used

Research for a reason.

UNIVERSITY OF LOUISIANA Lafayette.

Co-Products

Glues for paper and packaging:

Denature using mild base

Filter if clear adhesive is needed

Dilute if needed

Paper pad made by using algae proteins denatured via an alkaline solution.

Research for a reason.

UNIVERSITY OF LOUISIANA Lafayette.

Co-Products

Tough, Lightweight Composites:

Use denatured protein and

Large volumes recycled paper

Similar properties to ceiling tile or drywall

Does not require high-quality paper

Co-Products

Sand-based Composites:

Attempt to replace concrete/asphalt

Low compressive strength

Low water resistance

Experience more ductile failure

Potential Applications:

Flowable, escavatable Fill

Soil Stabilization

Temporary infrastructure

- Oilfield
- Lumber

Structures to be quickly "reclaimed" by nature

Biodegradable Mardi Gras Beads

Co-Products

Particleboard Composites:

Most promising co-product

Use algae glue with any natural filler:
particleboard filler, sawdust, rice husks, sugarcane bagasse, chicken feathers, grass clippings, oyster shell, or any fibrous agriculture byproducts

Easy to cut with a bandsaw

Formaldehyde-free

VOC free

No "chemical odor"

Potential Applications:

Interior use (similar to soy proteins)

Biodegradable wood composites

Experimental Results

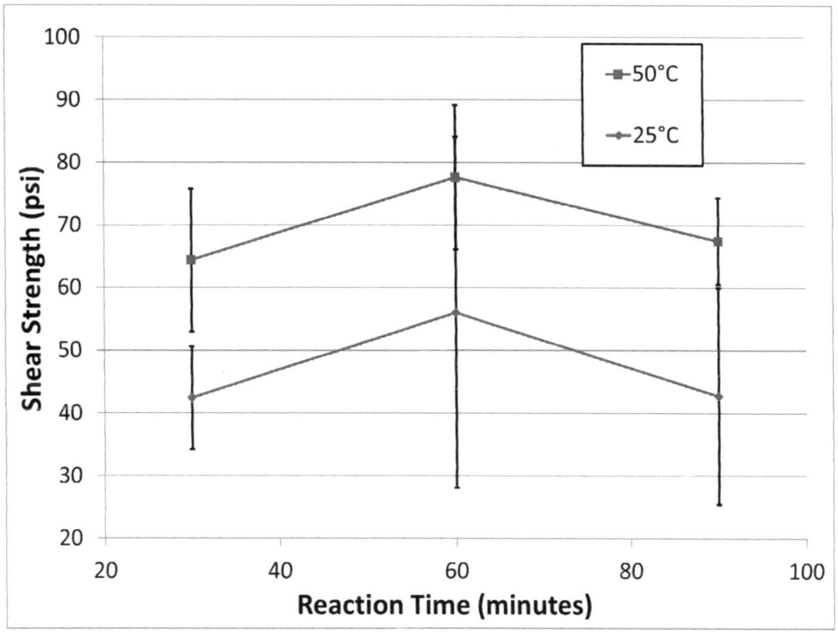

Experimental Results

Shear strength (psi) using various alkaline solutions with 90 minutes cooking time at 60 °C			
Sample	Urea	Calcium Hydroxide	Sodium Hydroxide
1	41.47	27.73	66.31
2	45.47	29.16	74.49
3	50.27	---	92.36
Average	45.73	28.44	77.72

Experimental Results

Lipid Extraction Temperature Effects on Shear Strength

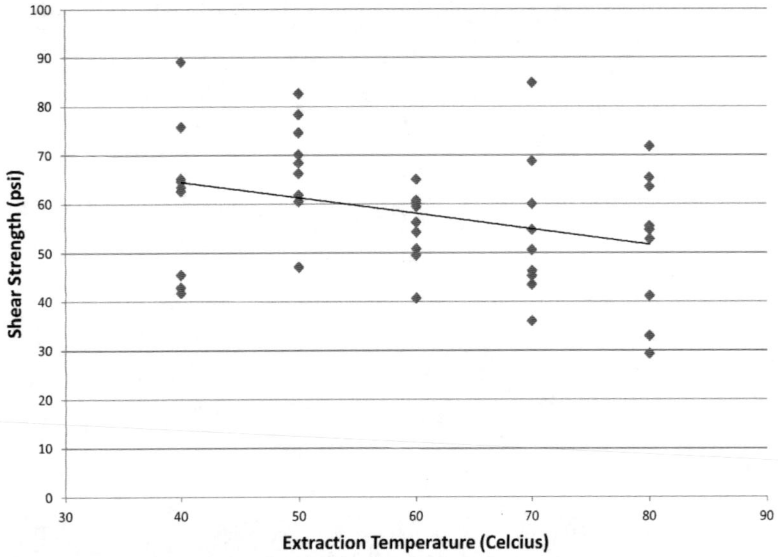

Effect of lipid extraction temperature on adhesive shear strength. p= 0.03

Experimental Results

Sand Composites

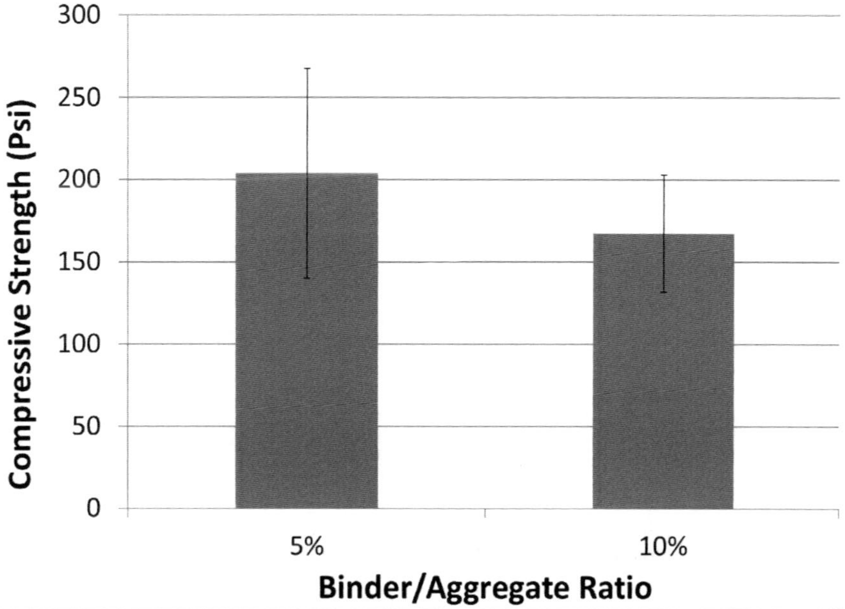

Research for a reason.

UNIVERSITY of LOUISIANA Lafayette

Experimental Results

Flexural Strength – Bagasse/Sawdust composites

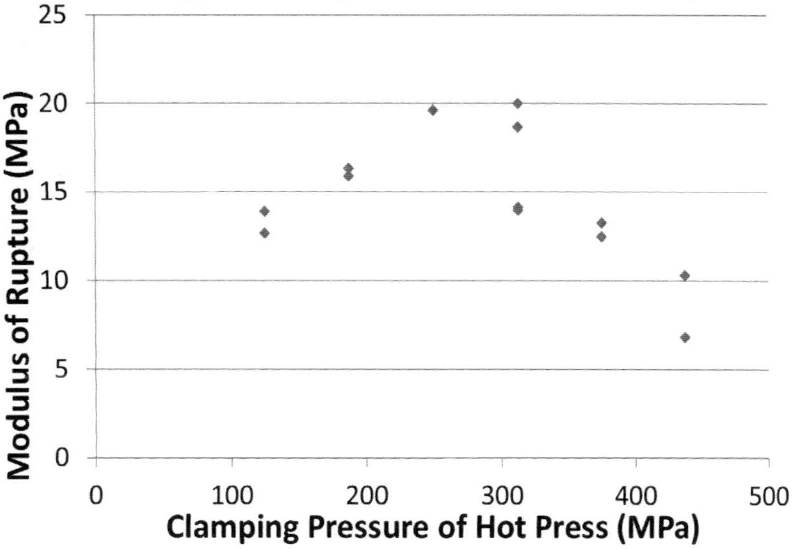

Literature values range from 2-14 MPa for UF resins with strong dependence on filler.
Rathke, et al. "Design of Particleboards," BioResources 7(3), 2970-2985.

Research for a reason.

UNIVERSITY of LOUISIANA Lafayette

Experimental Results

Flexural Modulus – Bagasse/Sawdust composites

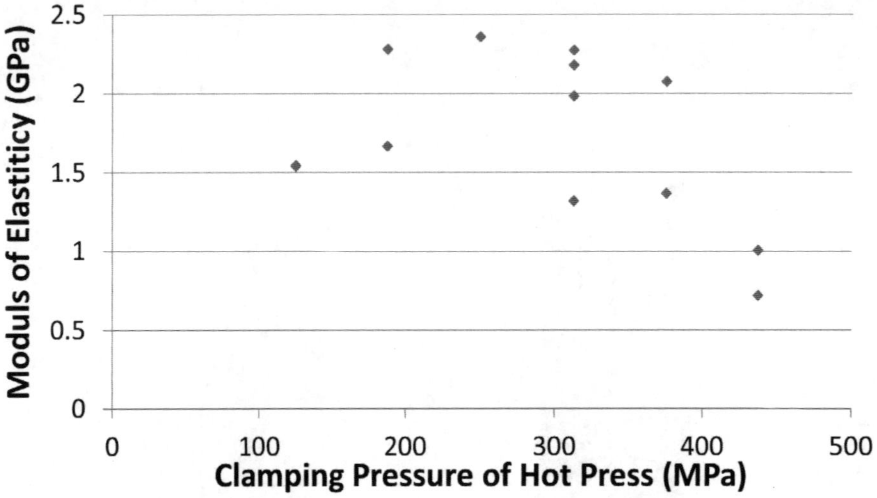

Literature values range from 0.6-2 GPa for UF resins with strong dependence on filler.
Rathke, et al. "Design of Particleboards," BioResources 7(3), 2970-2985.

Challenges:

Process:

Glue dries/cures by evaporation/boiling

Requires composite to remain porous

Limits maximum binder/filler ratio

Potential for steam explosion

Hazard can be mitigated by controlling process parameters

-Adequate hold time

-Slow decompression

-As always, use PPE including at least a face shield

-Shielding may be advisable on an industrial scale

Challenges:

Business Plan:

Need/opportunity for entrepreneurship

Difficult to startup algae culture without customers

Difficult to startup composite manufacture without supply

Requires strategic location (shipping cost for materials)

-Algae protein

-Filler source

-Processing

Conclusions:

Huge potential for algae glues and composites

Wide variety of co-products and applications

Plenty of optimization to be done

Optimization will be algae/process/product specific

Trade-off exists between lipid extraction and glue quality

Future work:

Scale-up of product size

Other microbial products (yeast, etc)

Additives

-Crosslinking agents (strength, durability)

-Anti-fungal, Anti-Microbial treatments

-Surface treatments

Fillers

Economic Analysis and Business Plan

Research for a reason.

UNIVERSITY OF LOUISIANA Lafayette

Thank you for your attention

wchirdon@louisiana.edu

UNIVERSITY OF LOUISIANA Lafayette

Effective Acidic Hydrothermal Degradation of Cellulose to Levulinic Acid through Pretreatment.

The annual energy consumption in the world is increasing each year. We mostly rely on the fossil resources for the energy product and the mass consumption of these resources, leading to the exhaustion of them, is considered as the issue. To get out of the fossil resource society, biomass would be one of the solutions. Biomass is possibly utilized as an alternative energy resource or an alternative raw material to fossil resources. However, as the source of energy, biomass has huge barriers such as the relatively low energy content, seasonality, discrete geographic availability, high moisture content or high collecting cost. As an alternative raw material, on the other hand, biomass would be beneficial resource because chemical production requires far lower volumes of biomass to satisfy demand. The US Department of Energy identified 12 chemicals which should be made from biomass. These chemicals, which could be used as building block chemicals, potentially enable biomass to be main resource of making chemicals which derive from petroleum for now. Levulinic acid is one of these 12 chemicals and is able to be obtained through the hydrolysis of some kinds of saccharides such as glucose or fructose. The use of levulinic acid is quite various, for example, it can be the materials of nylon-like polymer, artificial rubber or plastics. To obtain levulinic acid from biomass, we need the degradation of cellulose, one of the main components of biomass, via glucose. Concerning the degradation of cellulose to glucose, enzymatic saccharification is one of the well-known methods. The huge advantage of enzymatic saccharification is high selectivity. However, the bioprocess requires high cost because we need high technology of sewage treatment and complicated control of reactor. In addition, sludge produced through the process decreases the efficiency and causes long reaction time. Considering these disadvantages of biochemical technology, we need to develop thermochemical technology. Under the present circumstances, thermochemical technology has the disadvantage of low selectivity. To overcome this situation, we pretreated cellulose to convert it into desirable, degradable precursor once for the following acidic thermochemical treatment, and investigated the effect of pretreatment on the yield and selectivity of levulinic acid.

Effect of pretreatment and difference between reagents. Filter paper cellulose was pretreated in a sealed batch reactor at 50 ℃ for 24 h using H_3PO_4 or HCl as a reagent. After the pretreatment, pure water was added to the sample. As the second treatment, the reactor vessel was heated to 160-220 ℃ for 120 min. Products were analyzed using TOC analyzer and ion chromatography to identify the carbon conversion into solution and levulinic acid. To examine the effect of pretreatment, we conducted the experiment without the pretreatment for comparison. As the results, the main products through the treatment were levulinic acid, formic acid, glucose and cellobiose, and a little amount of HMF was also preduced. However, the yields of formic acid, glucose and cellobiose were very slight compare to that of levulinic acid, so that we discuss only about the yield of levulinic acid. Pretreated cellulose increased the carbon conversion into soluble component at all the hydrothermal temperature. The yield of levulinic acid was also increased through pretreatment at most by 10 C-% at 200 ℃ acidic hydrothermal degradation, reaching to 40 C-%. The selectivity also reached the highest score 55 % at 200 ℃ degradation with pretreatment, which is defined as (the yield of levulinic acid [C-%]) / (soluble fraction [C-%]). All the results indicate that the pretreatment worked effectively concerning the treatment with H_3PO_4. Then we conducted the same experiments using HCl as the reagents instead of H_3PO_4. Pretreated cellulose turned into soluble components more than untreated cellulose but the increase was quite slight compared to the case with H_3PO_4. However, the increase of the yield of levulinic acid was remarkable, especially at 200 ℃ and 210 ℃ by 17 C-% and 10 C-% respectively. The yield reached to 49.2 C-% at 210 ℃ of acidic hydrothermal degradation with the selectivity of 62.1 %. Comparing the results of H_3PO_4 and HCl, all the scores are higher when HCl was used as a reagent. As the summary of the section, the pretreatment with acid is very effective on producing levulinic acid from cellulose. As the result, two step treatment with HCl enhanced the yield of levulinic acid reaching to 49.2 C-% with the selectivity of 62.1 % at 210 ℃ acidic hydrothermal degradation.

Treatment with ionic liquid. Solubilization of cellulose with ionic liquid enables some reagent such as enzymes or acids to attack the active cites of cellulose more effectively. In addition, solubililzed cellulose is easily recovered as deposit by the addition of non-solubilizing reagent and this deposit is supposed to be changed to amorphous structure which has higher reactivity than crystalline cellulose. In this study, we chose 1-ethyl-3-methylimidazolium bromide ([EMIM]Br) and 1-ethyl-3-methylimidazolium methylphosphonate ([EMIM]P) for ionic liquid.

Treatment with [EMIM]Br. Cellulose solubilized with this ionic liquid is known to be recovered by the addition of water. We focused on this unique character which enables uncrystallization and easy separation of cellulose. As for the pretreatment, cellulose and [EMIM]Br were mixed and heated to 80 ℃ for 6 h. Immediately after heating, the products were cooled, then pure water was added for the extraction of the solute. The deposit appears as Br-cellulose hereafter. As for acidic hydrothermal treatment, Br-cellulose and 5 wt% H_2SO_4 were mixed and heated to 220 ℃ for 2 h. Products were analyzed using HPLC. As the results, after the pretreatment using [EMIM]Br at 80 ℃ for 6 h, no solid cellulose was recognized in the sample before the addition of pure water. Then, by the addition of the pure water, 99.0 wt% of cellulose was recovered as the deposit, Br-cellulose. Crystallinity of cellulose decreased by about 30 % through the pretreatment judging from the XRD curve. However, this decrease is not big considering that cellulose was completely solubilized once, so cellulose seemed to be recrystallized in the process of recovery. The elemental composition and TG curve showed little change through the pretreatment, which indicate that cellulose only changed its crystallinity through the pretreatment. Then, Br-cellulose was degraded under the acidic hydrothermal condition. For comparison, the acidic hydrothermal treatment was conducted using filter paper cellulose and amorphous cellulose as sample through the same method. More levulinic acid was produced from Br-cellulose than from filter paper cellulose. However, amorphous cellulose was turned into levulinic acid more than Br-cellulose, which indicates the crystallinity of cellulose is one of the important factors for producing levulinic acid.

Treatment with [EMIM]P. [EMIM]P has very high ability of breaking the rigid structure of cellulose, which not only solubilizes cellulose but turns it into water-soluble cellulose. As for the pretreatment, cellulose and [EMIM]P were mixed and heated to

160 ℃ for 1 h. Immediately after heating, the products were cooled, then pure water was added. This cellulose solution appears as P-cellulose hereafter. As for acidic hydrothermal treatment, P-cellulose and 5 wt% H_2SO_4 were mixed and heated to 220 ℃ for 2 h. Products were analyzed using HPLC. As the results, different from the pretreatment with [EMIM]Br, cellulose pretreated at 150 ℃ for 1 h using [EMIM]P was solubilized completely and then never recovered by the addition of pure water. To see the change in crystallinity and thermo degradability, much ethanol was added to the sample and cellulose was recovered as P-cellulose-deposit. The yield of the deposit was 130.0 wt% containing 70 % of cellulose and 60 % of ionic liquid. Cellulose was uncrystallized very well and the thermal degradability was increased by the pretreatment judging from TG analyses and XRD analyses. Then, P-cellulose-deposit was degraded under the acidic hydrothermal condition using 5 wt% H_2SO_4. Since the recovery of cellulose requires plenty amounts of ethanol, we conducted the acidic hydrothermal degradation on water-added cellulose solution, P-cellulose as well. Both pretreated cellulose remarkably increased the yield of levulinic acid, especially P-cellulose reaching to 66 C-%. The yield 66 C-% is 79.2 % of maximum yield, according to the stoichiometry.

In conclusion, the pretreatment on cellulose, which is usually hard to degrade, changed its structure and enabled it to be converted into valuable chemical, levulinic acid. Especially, through the treatment with ionic liquid [EMIM]P, levulinic acid was obtained at very high yield, 66 C-%.

347625 Preliminary Study of Waste Biomass from Palm Oil in a Bed Fluidized Gasifier

Abstract:

This research is about the design of fluidized bed reactor for the gasification process from waste materials derived of oil palm. The raw material was characterized by thermo gravimetric analysis (TGA) and total organic carbon (TOC). The calorific power was evaluated. The biomass was densified meanwhile the bed fluidized reactor were designed in a pilot plant scale.

The gases produced were determinate by gas analyzer ppppr

354013 New Learning Curves for Industrial Biotechnology: Liquid Fuels from Bioconversion of Methane and Bioelectrosynthesis

Abstract:

Biological systems are capable of performing many useful functions with applications in energy and chemical production. However, their full potential remains unrealized due to critical gaps in our ability to engineer and control them. Two programs at the Advanced Research Projects Agency – Energy (ARPA-E) address such critical gaps by engineering microbial systems for the production of liquid fuels and will be discussed in detail in my talk. The REMOTE program, whose acronym stands for Reducing Emissions using Methanotrophic Organisms for Transportation Energy, is a first-of-its-kind program that focuses on the development of advanced bioconversion technologies to produce liquid transportation fuels from methane, the main component of natural gas. The Electrofuels program, on the other hand, focuses on the development of non-photosynthetic autotrophic organisms for the conversion of CO2 and electricity to infrastructure-compatible liquid fuels.

353223 Wealth Generation through University/Industry Collaobration

Abstract:

Chemical Process Equipment Design has remained an art for a long time. With the recent advances and convergence of enabling technologies such as advanced computations and simulations, advanced measurements and advanced manufacturing the ability to explore a large design space and design towards compact, energy efficient processes are possible. Through several examples of using computational and experimental fluid dynamics, this talk will explore the fundamental paradigm shift in the design process to enable process innovation.

Multi-scale Exploration of Multiphase PHysIcs in flowS (MEMPHIS): A framework for the next-generation predictive tools for multiphase flows

P. Angeli[2], B. J. Azzopardi[3], B. Hewakandamby, G. F. Hewitt[1], C. C. Pain[1], M. J. H. Simmons[2], O. K. Matar[1*]

[1]Department of Chemical Engineering, Imperial College London, SW7 2AZ, UK
[2]School of Chemical Engineering, University of Birmingham, B15 2TT, UK
[3]Department of Chemical Engineering, University of Nottingham, NG7 2RD, UK
[4]Department of Chemical Engineering, University College London, WC1E 7JE, UK

Abstract

In this paper, we outline the framework that we are developing as part of the Multi-scale Exploration of Multiphase PHysIcs in flowS (MEMPHIS) programme to create the next generation modelling tools for complex multiphase flows. These flows are of central importance to micro-fluidics, oil-and-gas, nuclear, and biomedical applications, and every processing and manufacturing technology. This framework involves the establishment of a transparent linkage between input and prediction to allow systematic error-source identification, and, optimal, model-driven experimentation, to maximise prediction accuracy. The framework also involves massively-parallelisable numerical methods, capable of running efficiently on 10^5-10^6 core supercomputers, with optimally-adaptive, three-dimensional resolution, and sophisticated multi-scale physical models. The overall aim of this framework is to provide unprecedented resolution of multi-scale, multiphase phenomena, thereby minimising the reliance on correlations and empiricism.

1. Introduction

Systems with deformable interfaces possess extremely complex topology and encompass a very wide range of length- and time-scales. Despite this complexity, small-scale interactions within these systems are the 'building-blocks', which define the global structure and, crucially, similar phenomena appear across different industrial applications (see Fig. 1). Examples include wave formation, and the detachment of bubbles and drops from initially separate phases. The latter may occur either due to natural disturbances present in the flow[1] or via controlled forcing to deliver a precise microstructure (e.g. emulsions with controlled drop size; or break-up of compound jets to produce bespoke encapsulated structures with controlled chemistry[2], and applications to personal, healthcare and consumer goods products).

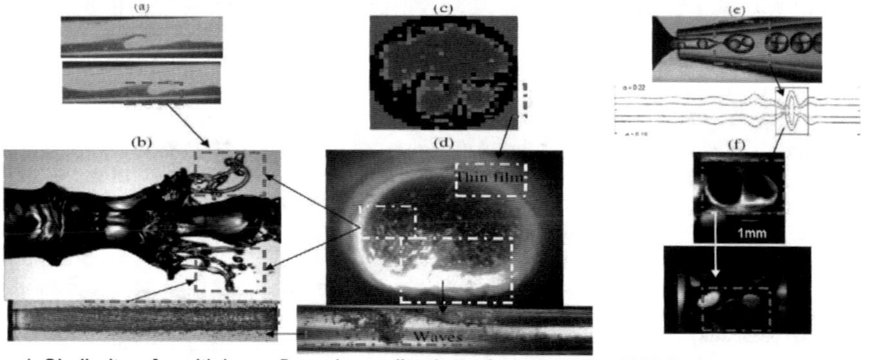

Figure 1 Similarity of multiphase flows in applications that span the FMCG and oil-and-gas sectors, amongst others, in terms of common flow features. We show the formation of waves, long filaments and droplets in wavy-stratified oil-water flows[3], (a), and gas-driven co-axial liquid jetting (b, top[4]) and turbulent water atomisation in still air (b, bottom[5]); thin film formation in gas-liquid 'churn' (c[6]) and 'annular' flows (d, top[7]), and large waves, droplet entrainment and bubble entrapment in 'slug' flows (d, bottom[8]); multiple-emulsion formation in micro-fluidics, (e, top[9] and bottom[2] show experiments and simulation, respectively), and oil-and-water flows in stirred vessels (f, top[10]) and downward pipe flows (f, bottom[11]).

* To whom correspondence should be addressed. Email: o.matar@imperial.ac.uk

The intricate dynamics of bubble and drop creation 'events' in gas-liquid and liquid-liquid systems are often controlled by micron-scale, capillary physics. These 'events' are invariably followed by entrainment and re-deposition, and the competition of these processes is influential in flow regime transitions (see Fig. 2) in oil-and-gas applications[12]. Other examples of small-scale 'events' include the coalescence of bubbles and droplets, also influential in flow regime transitions in processing, manufacturing and in oil-and-gas settings [13]. Interfacial waves, another generic feature of multiphase flows, are generated by many mechanisms which are strongly dependent on density and viscosity contrasts between different phases, and involve larger length scales: above mm, where inertial and gravitational effects become important and the flow becomes turbulent [14]. Additives (e.g. surfactants, polymers, and nanoparticles) can also have a dramatic effect on droplet generation, breakup, and coalescence due to changes in interfacial tension, and bulk and surface rheological properties; this can be manipulated to produce bespoke structured-products2[,15].

Numerous models have been proposed to simulate the complex configurations associated with turbulent multiphase flows. Direct numerical simulations (DNS) account for the presence of a large number of deformable interfaces, and are computationally expensive. Reynolds-averaged-Navier-Stokes (RANS)-based tools employ averaging of all eddies, and are generally applied to determine time-averaged quantities in steady or quasi-steady flows; they appear to work reasonably well for dispersed flows but not for separated flows, wherein the averaging procedures filter out interfacial waves, an essential 'ingredient' of these flows. These methods are also unable to predict the small-scale and rapid behaviour of 'slug' and 'churn' flows[6]. Filtering approaches, such as Large Eddy Simulation (LES)-based methods with interface-tracking have been proposed recently[16]**,** which are computationally less expensive than DNS, representing an excellent compromise, capable of simulating all the flow regimes and phenomena shown in Fig. 1.

Severe complexities also arise in laminar multiphase flows, due to conduit geometry, chemical reactions, local surface gradients due to additives and/or non-Newtonian rheology. In micro-fluidics applications, the increased surface-to-volume ratio promotes surface/interfacial and wall wettability phenomena, which significantly affect the flow patterns. The inlet configuration is also very important and has been used to form flow configurations that enhance mixing or to create multiple emulsions[9]. Simulations detailing the transition from 'dripping' to 'jetting' have only just been published[17]. For the generation of complex multiphase microstructures in the FMCG industries (see Fig. 1e), modelling of the length- and time-scales of breakage and coalescence of individual fluid filaments have only been elucidated fully for pure Newtonian fluids[18]. The added complexities of non-Newtonian rheology (e.g.[2]), surface active molecules[15], or bulk chemistry, have only been tackled to a very limited extent in idealised geometries, a major limitation for process innovation which prevents their generic use either singly or in combination in more complex flow fields.

Due to the complexities highlighted in the foregoing sections, researchers have made heavy use of empirical methods. These include 'flow regime' maps, convenient 'umbrella' descriptions and correlations of groups of behaviour with a consequent need to blend them in at regime boundaries. The erroneous predictions made via this approach can result in uncertainties in production volume, product quality and equipment failure. Prevention of film dry-out is vital to prevent boiler tube failure in the nuclear industry yet 1-D modelling approaches are pervasive and do not tackle azimuthal variations in critical parameters such as film thickness and wave shape, apart from one limited study[19]. Prediction and control of 'slug' flow is essential for oil and gas transportation and yet current models fail[20] or are unnecessarily restrictive[21] when entrainment of gas occurs in the liquid slugs. Predictions of entrainment and deposition rates of drops in two-phase flows are mainly semi-empirical and limited, particularly for liquid-liquid systems where the experimental information on wave characteristics and entrainment fractions is extremely limited[22]. Hence, it is not possible to predict reliably the pressure drop or the morphology and separation characteristics of the mixture.

2. MEMPHIS Framework

Based on the foregoing, the research challenges are:

- Accurate simulation of small-scale, laminar multiphase flows with geometrical, surface-additive, chemical reaction, and complex rheological effects;
- Elucidating turbulent interactions with interfaces; development of robust sub-grid models;
- Development of truly predictive LES-based tools (and, wherever necessary, multi-fluid models) with interface-tracking or capturing, through the use of robust sub-grid models.

The poor fundamental understanding of the physics of multiphase flows present an opportunity to bring to bear the collective expertise of a number of collaborators from Imperial College London, the University of Birmingham, the University of Nottingham, and University College London, in small- and large-scale experimentation, physical modelling, numerical methods, and computational engineering, to make a transformative impact. Our synergistic, integrated Multi-scale Exploration of Multiphase PHysIcs in flowS (MEMPHIS) Programme aims to advance the current state-of-the-art significantly by developing a unique predictive framework that offers:

- Numerical simulations that can pin-point, automatically, the largest sensitivities of the predictions to experimental data and/or equations used to model physical phenomena;
- Feedback loops from the simulations to the experiments, altering their design optimally to maximise prediction accuracy; this is absent in all previous and present simulation tools;
- Unparalleled simulation accuracy and efficiency, through parallelisation with 10^5-10^6 cores, to provide unprecedented resolution of the smallest length- and time-scales; this will minimise the dependence on empiricism and 'closures', as far as possible, and along with targeted experiments, aid in the rational development of the most reliable sub-grid models;
- Simulation of industrially-relevant flows on industrially-relevant length- and time-scales;
- Insight into the fundamentals of multiphase flows at all relevant scales leading to novel manufacturing routes for multiphase structured fluids, and equipment and process design;
- Ready-for-use predictive tools that can be assimilated easily within existing industrial codes/software, minimising the time-to-impact.

In tackling a given multiphase flow, the above framework involves the following steps:

1. Identify the small-scale features that govern the large-scale behaviour of a given flow;
2. Use a DNS-based numerical predictive tool with high accuracy, through optimal, fully-adaptive numerical resolution, to simulate the features identified in 1 with high-fidelity;
3. Develop and use of sensitivity analysis in the tool utilised in 2. to identify error sources;
4. Use the output from 3. to develop optimally-designed experiments aimed at maximising prediction accuracy;
5. Exploit, continuously, the output from 1.-4. to develop state-of-the-art sub-grid models for use with a truly predictive LES tool (and, wherever necessary, multi-fluid models with robust closures); equip these with sensitivity analysis capabilities;
6. Use the numerical tool from 5. to simulate the large-scale features of the flow in 1. and use its sensitivity analysis capabilities to direct optimal refinement of models, and small- and large-scale experiments, revisiting steps 1.-5. as often as necessary.

The ability to predict accurately, reliably and efficiently the behaviour of complex multiphase flows will have an impact on:

- Small- and large-scale manufacturing of structured fluids for fast-moving consumer-goods (FMCG) products including bespoke, personalised products, catalyst manufacture and cleaning of process plants;
- Design and operation of oil-and-gas transportation pipelines and sub-sea installations;
- Efficient downstream processing of crude-oil;
- Safe operation of nuclear reactors;
- Simulating plumes and oil slicks following a spillage and their environmental impact.

The energy efficiencies introduced as a result of a step-change improvement in the engineering science, in each of these cases, will result in an associated reduction in CO_2 emissions.

3. Sample results

Here, we describe some of the results produced as part of the MEMPHIS programme. Below, we present a discussion of our experimental results, followed by a description of our numerical results.

3.1 Experimental results

At UCL, particle image velocimetry (PIV) was used to examine the velocity field near the interface in liquid-liquid systems. An example of this is provided in Fig. 1 in which we show the flow field in an oil-water stratified flow regime in a 37 mm ID test pipe; PIV was used to study the velocity profiles in the water phase. Here, the volumetric flow rate ratio of water to oil is 2.25. A methodology was also developed to reconstruct the wavy interface. It was found that the velocity profiles are not greatly affected by wave contribution but are mainly governed by shear-generated turbulence in the lower pipe boundary. These findings will be used in order to validate and develop the MEMPHIS modelling capabilities in terms of the near-interface behaviour of turbulent fields.

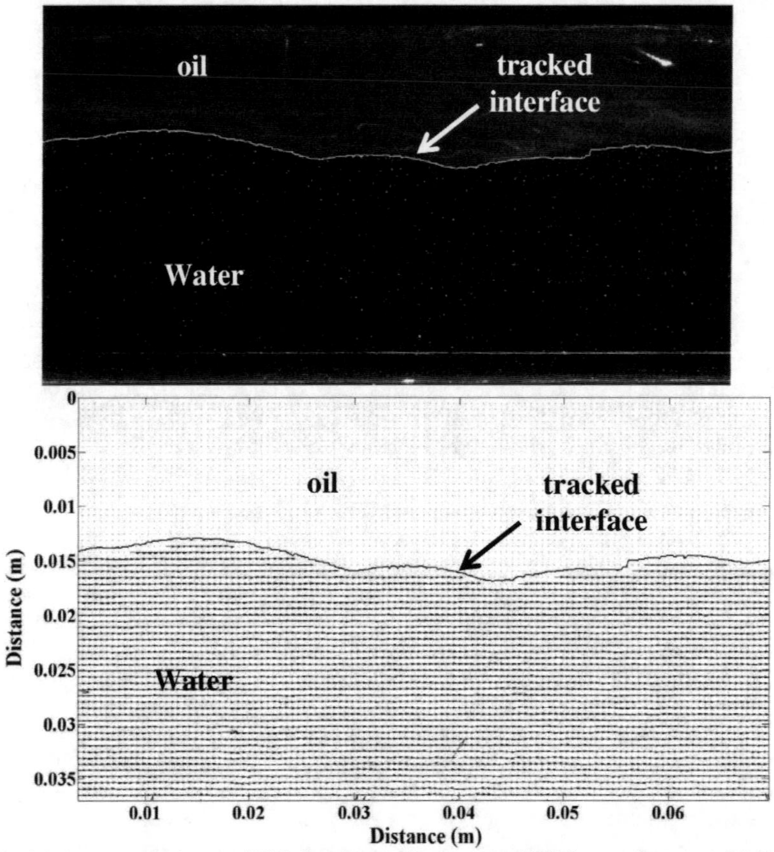

Figure 2 Particle image velocimetry (PIV) data in an oil-water stratified flow regime; top: PIV image with seeded water phase and tracked interface; bottom: instantaneous velocity vectors in the water phase.

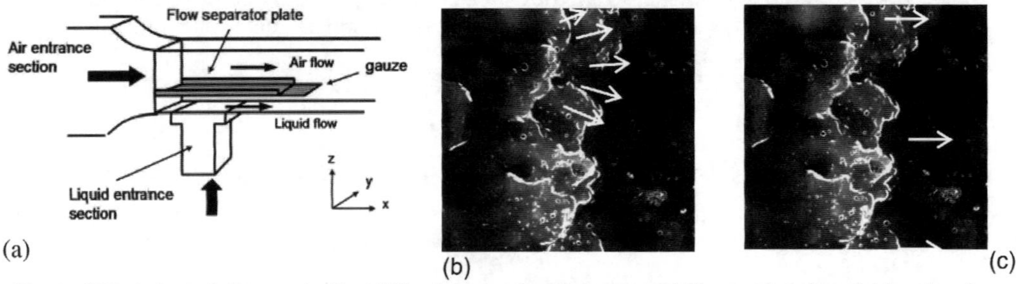

Figure 3 Turbulent air flow past a liquid film; (a) experimental setup; (b) filament breakup; (c) bag breakup.

At Nottingham, experiments were carried out involving the flow of turbulent air over a thin water film in a rectangular channel (161 x 25 mm, viz. Fig. 3a) with the following flow conditions: superficial gas velocities in the range 16-35 m/s; liquid Reynolds numbers in the range 150-520. Rhodamine-6G (15 mg/L) was injected into the water phase and laser-induced fluorescence was used to investigate the dynamics of wave formation, the transition to three-dimensional behaviour, and the onset of droplet entrainment. This work has so far shown that the two modes of droplet entrainment associated with filament breakup (see Fig. 3b) and 'bag breakup' (see Fig. 3c), which previous researchers had thought would occur under different flow conditions, actually could take place as part of the same flow. This work has also furnished a considerable amount of information regarding the genesis and evolution of disturbance waves; these results will be part of future publications, and will inform the development of the numerical models.

The Nottingham work discussed above also captured the generation of entrained droplets which then were observed to impact on the air-liquid interface giving rise to the formation of smaller droplets. This phenomenon was explored further at Imperial where the case of a droplet impacting on a falling film was studied. The results from this work, which are shown in Fig. 4, indicate that if the droplet impact speed is sufficiently large, during the impact, the droplet propels the liquid in the falling film to form a 'crown', which disintegrates into many droplets by jetting. This 'splashing' phenomenon happens because the kinetic energy of the droplet cannot be fully-absorbed by the liquid film during the impact process: the liquid in the falling film gains kinetic energy, rises around the crater, and forms a crown. Then many small droplets are formed at the rim of the crown due to instability of the crown (see, for instance, Fig. 4k).

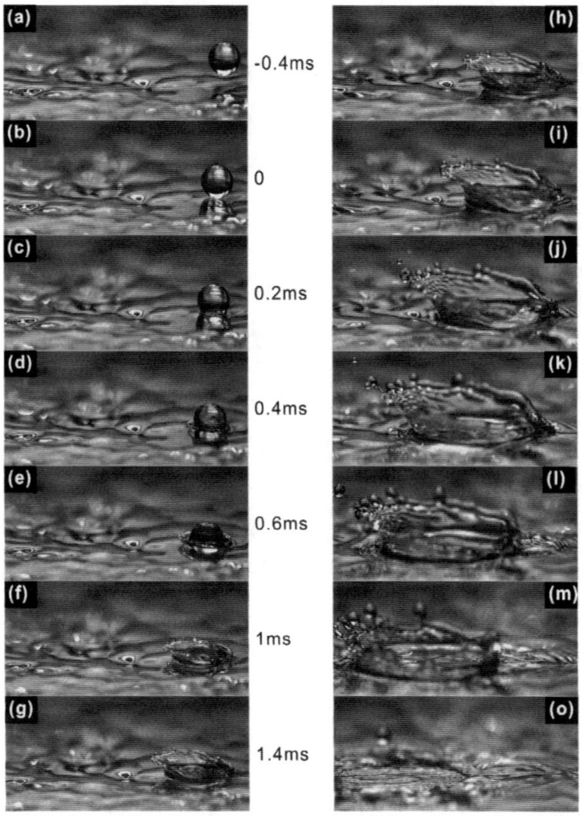

Figure 4 Splash during droplet impact on a falling film.

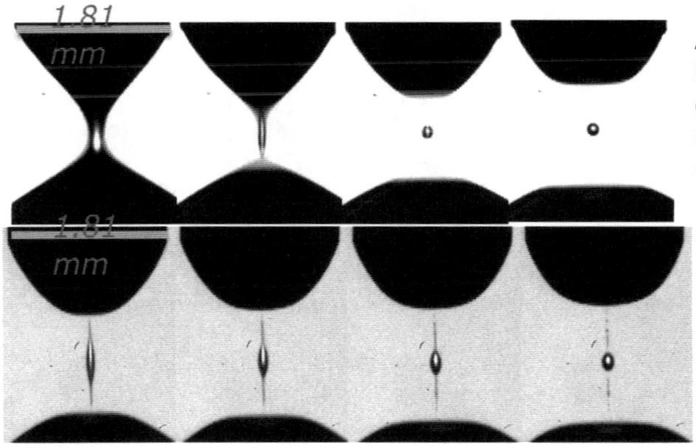

Figure 5 Breakup of a water filament in silicone oil in the absence (top) and presence (bottom) of surfactant (SLES, 0.5 CMC).

A study of the phenomena of filament breakup and droplet creation in isolation is currently being carried out at Birmingham. Fig. 5 (top) shows the production of a single droplet from a thread of water in silicone oil (η_{oil}/η_{water} =10) formed from the separation of a pendant drop from a nozzle. The addition of a surfactant, SLES (sodium laureate sulphate) at 0.5 CMC leads to a dramatically different break up behaviour, as shown in Fig. 5 (bottom), with a single drop and several smaller satellites, both above and below, identified. These simple experiments illustrate how alteration of the surface properties, in this case by introduction of surface tension gradients (Marangoni stresses) due to the distribution of surfactant on the newly formed interface, greatly influences the interfacial topology. Since control of droplet size is central to many industrial processes, models which incorporate the multi-scale physics of these problems are a critical necessity.

3.2 Numerical results

The core of the simulation tool is based on *Fluidity*, which is an open-source flow simulation tool with an adjustable, unstructured mesh, advanced numerical methods, and potential for massive parallelisation and high-performance computing. *Fluidity* has so far been used in large-scale geophysical and environmental applications e.g. ocean modelling, and flooding but the potential for its use as the base of our capabilities is enormous and game-changing. *Fluidity* has advanced anisotropic mesh adaptivity methods parallelized with hybrid OpenMP-MPI and uses our novel conservative mesh interpolation routines [23]. *Fluidity* also has tetrahedral elements capability for use with control-volume finite-element methods, which offers geometrical flexibility and mesh adaptability, making them ideal for use in simulating complex multiphase flows. *Fluidity* has a number of options for decreasing the cost of 3D pressure solvers[24], one of the main bottle-necks in computational fluid dynamics (CFD) simulations, and novel algebraic multi-grid solvers which account for the high aspect ratio of the domain and physics to accelerate convergence[25]. Our approach to course-grain parallelisation is based on domain-decomposition methods which constantly change their size and shape to balance the computational load as our meshes adapt to the flow. These methods have been shown to scale well to 8,000 HECToR cores. The combined OpenMP-MPI linear solver technology will enable the solvers to scale to million-core parallel supercomputers.

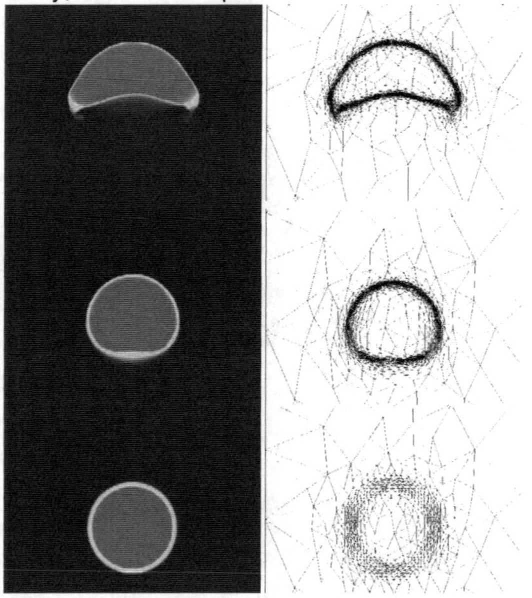

Figure 6 Numerical simulation of bubble rise in an unconfined fluid.

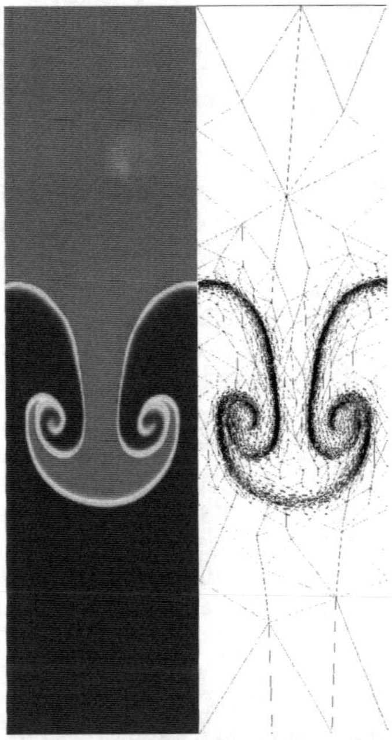

Figure 7 Numerical simulation of the Rayleigh-Taylor problem.

In order to demonstrate the capability of adaptive unstructured mesh modelling of multiphase flows, two classical examples are presented here, one is an air bubble rising in the water (Fig. 6) and the other is the Rayleigh-Taylor instability (Fig. 7). It is shown that the interface is well captured with the anisotropic mesh adaptivity, which allows the grid resolution to be concentrated in relatively important regions such as the vicinity of interfaces while lower resolution can be used in other regions. In addition, it can better represent the underlying physics of multiphase problems and reduce the computational effort without sacrificing accuracy.

A further illustration of the power of *Fluidity* is provided by Fig. 8, which depicts a snapshot in time of a numerical calculation for pipe flow through a static mixing device for a system of two miscible, shear thinning complex fluids of differing rheologies, (both satisfying a power law), modelling experiments with a polyacrylate polymer in solution at various concentrations. The uppermost image shows a cutaway through the boundary of the pipe (blue) and the six elements of the helical static mixer (grey). The centre image shows a volumetrically ray-traced plot of fraction of the more viscous of the two fluids, initially injected down the centre line of the pipe, with areas of yellow indicating high concentration and green lower. The boundary of the mixer is plotted for reference, as are streamlines (red) for a number of particles seeded at the inlet of the pipe. The lowermost image is a quiver plot with arrows indicating the local instantaneous fluid velocity, plotted at various nodal points on the unstructured computational mesh. The calculation was performed in parallel on 12 processors.

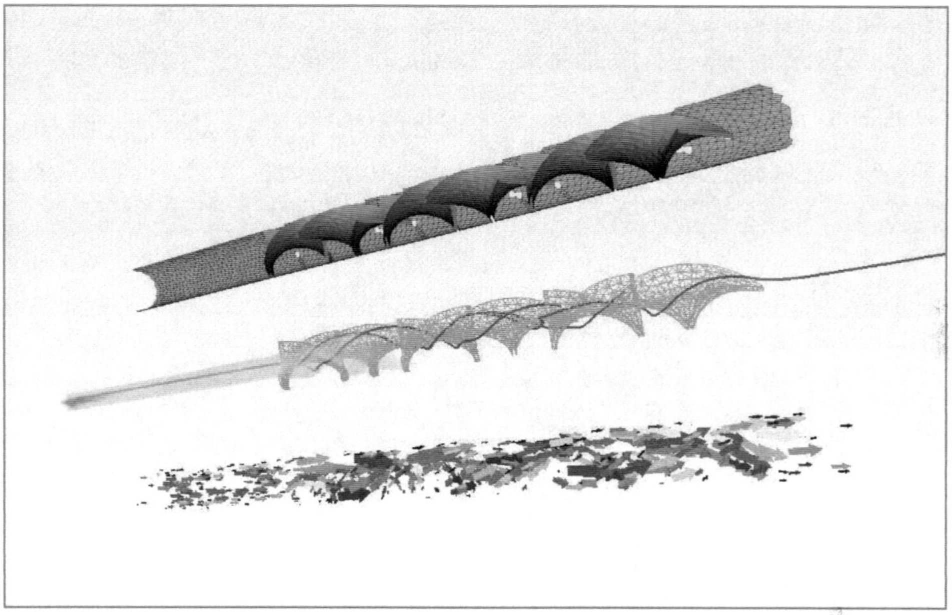

Figure 8 Numerical simulation of the mixing of two miscible, power-law fluids in a static mixer.

4. Concluding remarks

We have presented a novel framework for the development of the next-generation predictive tools for the simulation of complex multiphase flow problems. This is a joint collaboration between Imperial College London, and Birmingham, Nottingham, and UCL. The framework outlined above promises to lead to a paradigm-shift in terms of the approach adopted in the solution of multiphase flow problems, which will minimise the dependence on empiricism and the use of correlations.

Acknowledgements The authors acknowledge the support of the Engineering and Physical Sciences Research Council, UK, through the MEMPHIS Programme Grant (Grant number EP/K003976/1).

---

[1] Brauner, N., Moalem Maron, D., *Int. J. Multiphase Flow* 18 (1992), 123-140.
[2] Mohsin, M., Uddin, J., *et al., IMA J. Applied. Math.* doi:10.1093/imamat/hxr075 (2012); Gurney, C.J., Simmons, M.J.H., Hawkins, V.L., Decent, S.P., *Chem. Eng. Sci.* 65 (2010) 3474-3484; Wang, W., Xie, R., Ju,X. *et al., Lab Chip* 11 (2011), 1587-1592.
[3] Al-Wahaibi, T., Smith, M., Angeli, P., *Chem. Eng. Sci.* 62 (2007), 2929-2940.
[4] Marmottant, P., Villermaux, E., *J. Fluid Mech.* 498 (2004), 73-112.
[5] Hoyt, J.W., Taylor, J., *J. Fluid Mech.* 83 (1977), 119-127.
[6] Sharaf, S. PhD Thesis, University of Nottingham (2011).
[7] Badie, S., Lawrence, C.J., Hewitt, G.F., *Int J. Multiphase Flow* 27 (2001), 1259-1269.
[8] Davies, S.R., PhD Thesis, University of London, (1992).

[9] Shum, H. *et al., Angew. Chem. Int. Ed.* 50 (2011), 1648 –1651.
[10] Liu, L., Matar, O.K., Lawrence, C.J., Hewitt, G.F., *Chem. Eng. Sci.* 61 (2006), 4007-4021.
[11] Liu, L., Matar, O.K., Hewitt, G.F., *Chem. Eng. Sci.* 61 (2006), 4022-4026.
[12] Brill, J.P., Mukerjee, H., *Multiphase flow in wells*, SPE (1999).
[13] Kaji, R., Hills, J. H., Azzopardi, B. J., *Multiphase Sci. Tech.* 21 (2009), 1-12.
[14] Azzopardi, B.J., *Gas Liquid Flows* Begell House (2006).
[15] Uddin, J. *et al., Fluid Dyn. Res.* 40 (2008), 827-851.
[16] Liovic, P., Lakehal, D., *J. Comp. Phys.* 28 (2007), 127-144.
[17] Vu, T.V., *24th Europ. Conf. Liq. Atom. Spray Systems*, Estoril, (2009).
[18] Eggers, J., Villermaux, E., *Reports on Progress in Physics* 71 (2008), 3: 036601.
[19] Kunugi, T., Kino, C., *Computers and Structures* 83 (2005), 455-462.
[20] Pioli, L., Bonadonna, C. *et al., Geophys. Res.– Solid Earth* doi:10.1029/2011/JB008392.
[21] Fernandes, R.C., Semiat, R., Dukler, A.E., *AIChE J.* 29 (1983), 981-989.
[22] Al-Wahaibi, T., Angeli, P., *Chem. Eng. Sci.* 64 (2009), 2817-2825.
[23] Farrell, P.E. *et al. Comp. Meth. Appl. Mech. Eng.,* 198 (2009), 2632-2642.
[24] Piggott, M.D. *et al. Int. J. Num. Meth. Fluids,* 56, (2008),1003 – 1015.
[25] Kramer, S. C., Cotter C. J., Pain C.C. *Ocean Model.*, 35, (2010), 253-263.

Extended Abstract

353216 Symmetry as the Primary Driver of Process Efficiencies

Mike Kearney, Amalgamated Research LLC, Twin Falls, ID
arimk@mindspring.com

In general, the description of a process will contain factors which represent inefficiencies with respect to energy use, equipment size, waste production, modeling complications, human workload etc. Introduction of symmetries will eliminate or minimize these factors. Because symmetry considerations allow widely diverse characteristics to be addressed in this general manner, we suggest symmetry as the primary driver of process efficiencies.

As an example of this reasoning, symmetric fluid transporting fractals have been designed which perform as functional alternatives to the energy dissipative and uncontrolled characteristics of turbulence (a particularly asymmetric phenomena). The use of symmetry operations for designing this type of fluid processing equipment can be illustrated by the construction of fractal based fluid distributor/collectors, mixers and reactors wherein symmetries are engineered within each size scale and between scales of different size. The completed devices impose engineered symmetries upon the fluids being processed.

Introducing deep symmetries into the control of fluid processes can lead directly to novel equipment configurations which are quite different from those used conventionally. Resultant equipment geometries include processing vessels which approach a two dimensional plane, toroid processing columns and other types of equipment wherein fluids can be introduced, managed or collected via complex surfaces and shapes. Recently installed symmetry optimized separation processes have demonstrated a near elimination of fluid distributor and media bed pressure drop, decreased product concentration requirements and decreased turbulent energy loss.

353225 CFD Simulation and Experimental Investigation of a Fractal Fluid Distributor

Abstract:

Uniform flow distribution is critical for many unit operations in chemical engineering as it often defines hydrodynamics in process vessels. Flow maldistribution is also the major factor affecting equipment scale-up. Design of conventional fluid distributors is typically based on pressure drop with non-equivalent hydraulic pathways leading to different outlets. Fluid distributors based on fractal geometry that are based on pathway symmetry rather than pressure drop offer precise control of fluid properties. Fractal-based distributors have been used in the industry for the last 20 years, especially in the processes with granulated media. Although the flow behavior in the fractals is well understood, the interaction of specific design features and flow patterns within the resin bed needs to be additionally studied. A fractal distributor with one inlet and 256 outlets has been studied for an ion-exchange device (one distributor, resin bed, collector). Results of CFD modeling and experiments of residence time distribution will be presented. The influences of flow rate, resin depth, flow orientation, and outlet density, are investigated. CFD Simulation of this ion-exchange device has been performed with ANSYS WORKBENCH. Key evaluation parameters include COV (coefficient of variation) and pressure drop. Three different fluid distributors with different channel cross-section area have been tested and the result shows with higher flow rate, the COV increases and the middle sized distributor has the lowest COV.

353312 Effects of Stress-Dependent Permeabilities and Non-Darcy Flow Parameters on the Productivity Index of the Hydraulically Fractured Completions in Gas Reservoirs

Abstract:

Non-Darcy flows are expected to be ubiquitous in near wellbore regions, completions, and in hydraulic fractures of high productivity gas wells. Further, the prevailing dynamic effective stress in the near wellbore region is expected to be an influencing factor for the completion conductivity and non-Darcy flow behavior in it. In other words, the properties (fracture permeability and beta-factor) can vary with the time and location in the reservoir (especially in regions close to the wellbore). Using constant values based on empirical correlations for reservoirs/completions properties can lead to erroneous cumulative productivity predictions. With the recent advances in the imaging technology, it is now possible to reconstruct pore geometries of the proppant packs under different stress conditions. With further advances in powerful computing platforms, it is possible to handle large amount of computations such as lattice Boltzmann (LB) simulations faster and more efficiently. The objective of this paper is to broaden the knowledge of completion performance with the hydraulic fractures by investigating the variations of the permeabilities and beta-factor due to near wellbore stresses using the image-based simulations in proppant packs.

The two technology drivers viz. imaging of pore spaces and high performance simulations, can help in simulating fluid flows in the realistic microscopic pore spaces in order to obtain inertial flow parameters for the porous media. Further, this approach is less time-consuming, repeatable, and cheaper compared to the other alternatives. In this study, absolute permeability and Forchheimer coefficient (beta-factor) of the imaged proppant packs under varying confining stresses are calculated based on the LB simulations at varying Reynolds number.

Calculated properties show reasonable agreement with the reported values for both permeability and beta-factor. These predicted stress-dependent permeability and beta-factors corresponding to the effective stress fields around the hydraulic fractured completions is included in a 2D gas reservoir simulator to calculate the productivity index. Presented results have beneficial applications in the high production rate gas reservoirs with hydraulically fractured completions.

Hydraulic Fracturing

Water Reuse Challenges & Solutions

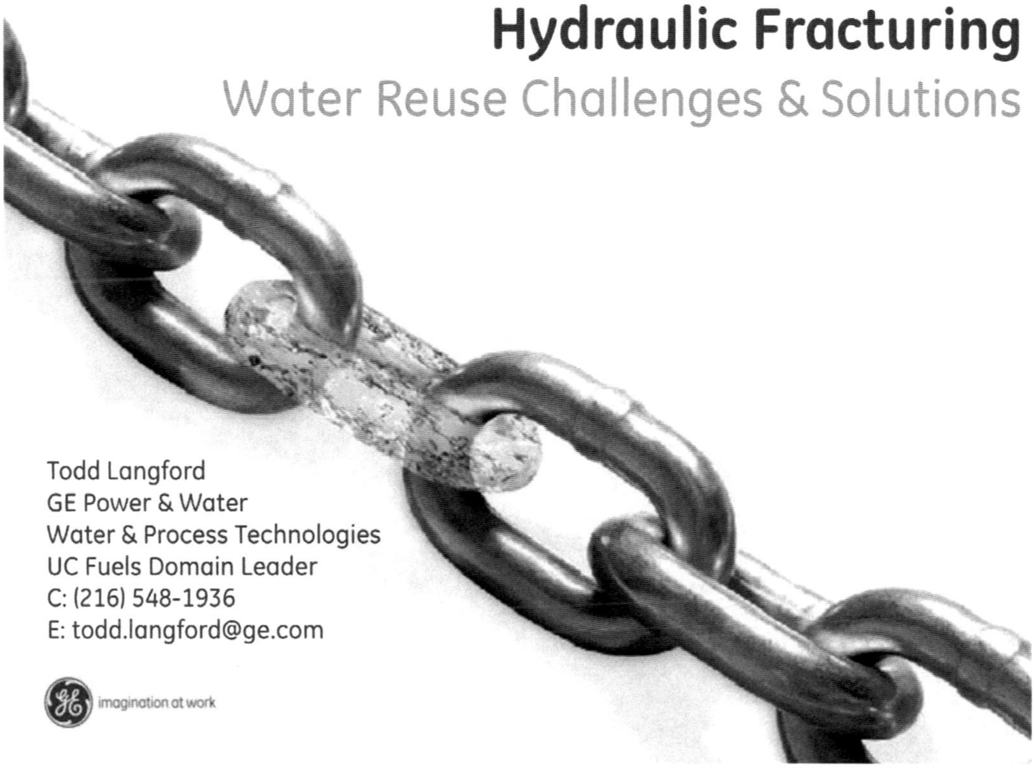

Todd Langford
GE Power & Water
Water & Process Technologies
UC Fuels Domain Leader
C: (216) 548-1936
E: todd.langford@ge.com

imagination at work

Hydraulic fracturing

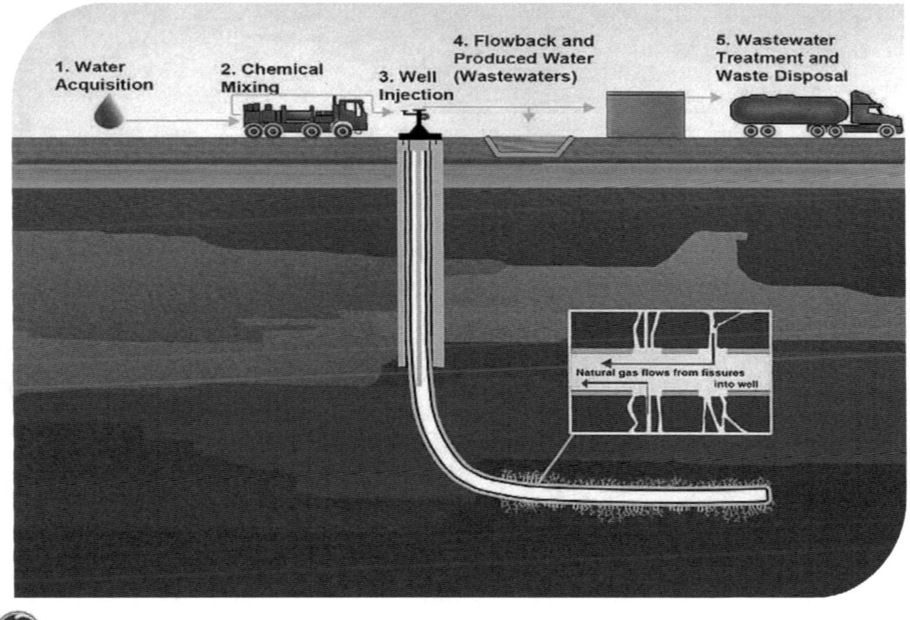

imagination at work

Source: US EPA

Shale gas *Evolution*

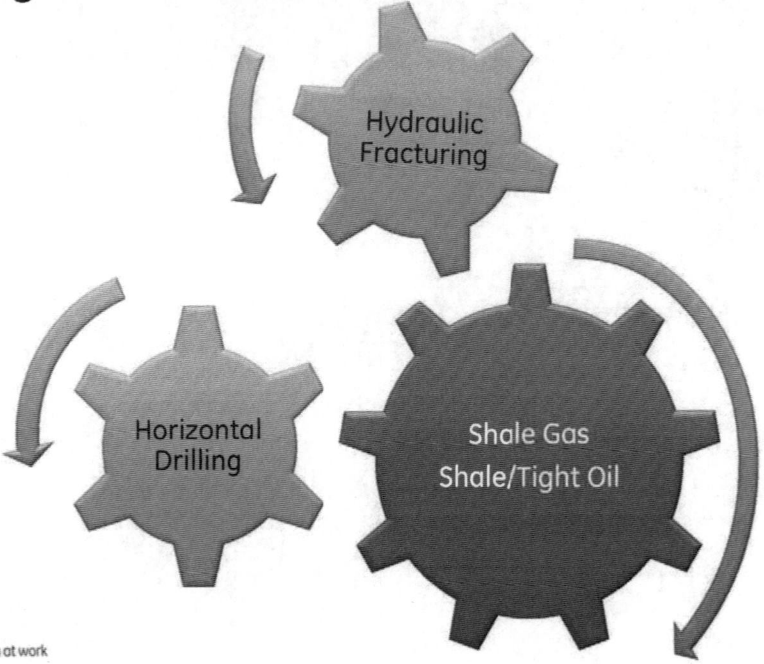

imagination at work

3
Langford
March 2014

Active shale and tight oil plays

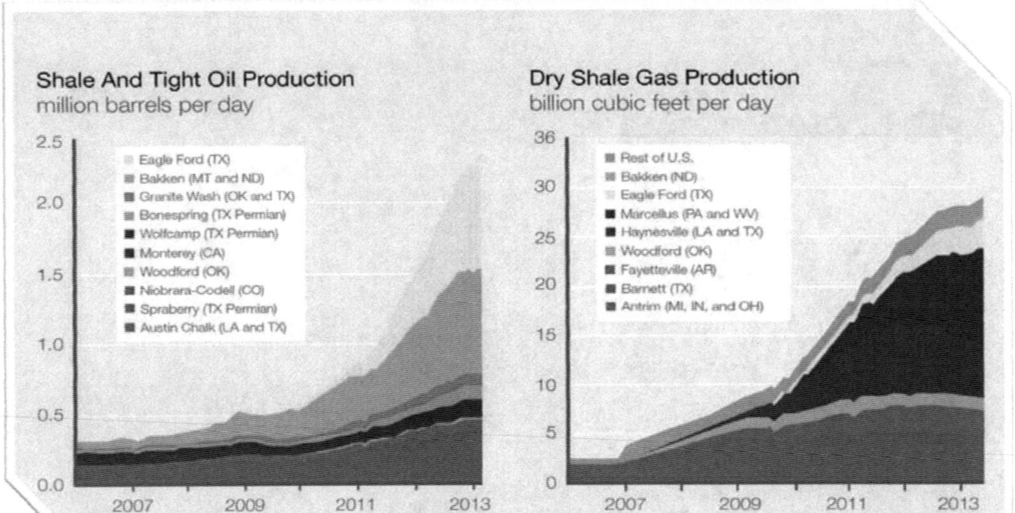

imagination at work

4
Langford
March 2014

Source: American Petroleum Institute

Energy demands on water resources

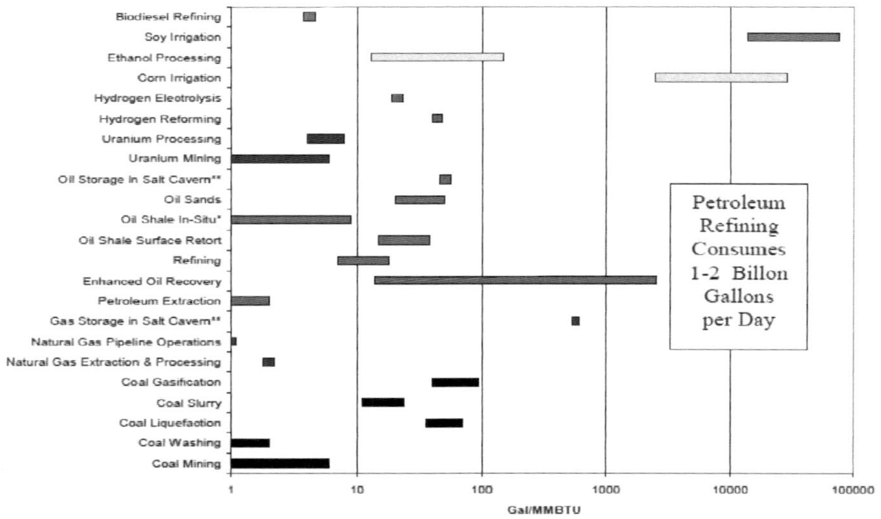

Figure V-4. Water Consumption Per-Unit-Energy and Current Water Use for Fuel Extraction and Processing
See Appendix B for Data References

imagination at work

Source: U.S. Department of Energy, 2006
Report to Congress: Interdependency of Energy and Water

5
Langford
March 2014

What's driving recycle/reuse?

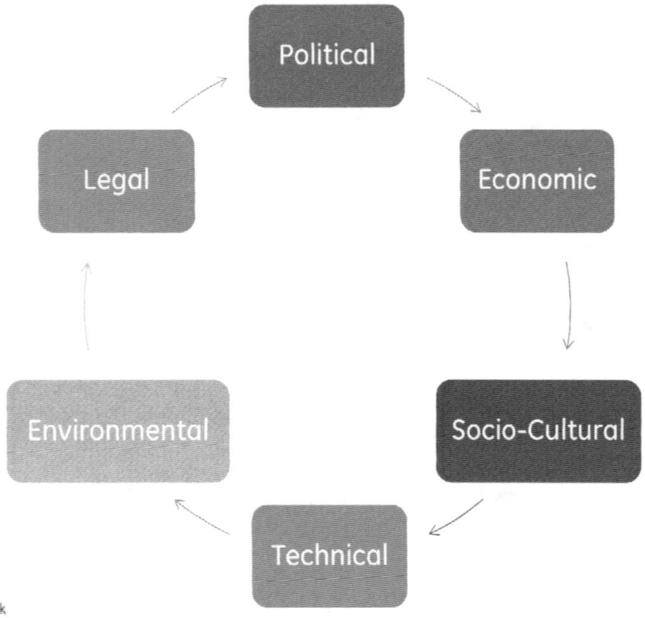

imagination at work

6
Langford
March 2014

Enviro concerns in unconventional O&G

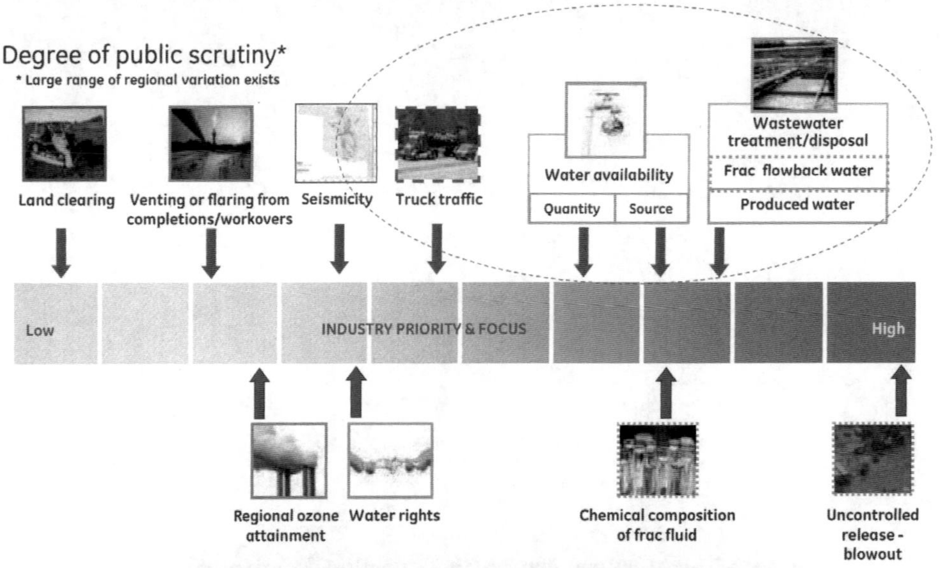

Water issues elevated in national debate
Water issues are LOCAL

imagination at work

7
Langford
March 2014

Source: GE Energy, Global Strategy & Planning; based on customer discussion not a statistical survey

Key drivers for projects to move forward

Regulatory *...limited disposal + discharge standards*

Economic *...expensive trucking to disposal*

Sustainable *...multi-generational planning*

imagination at work

8
Langford
March 2014

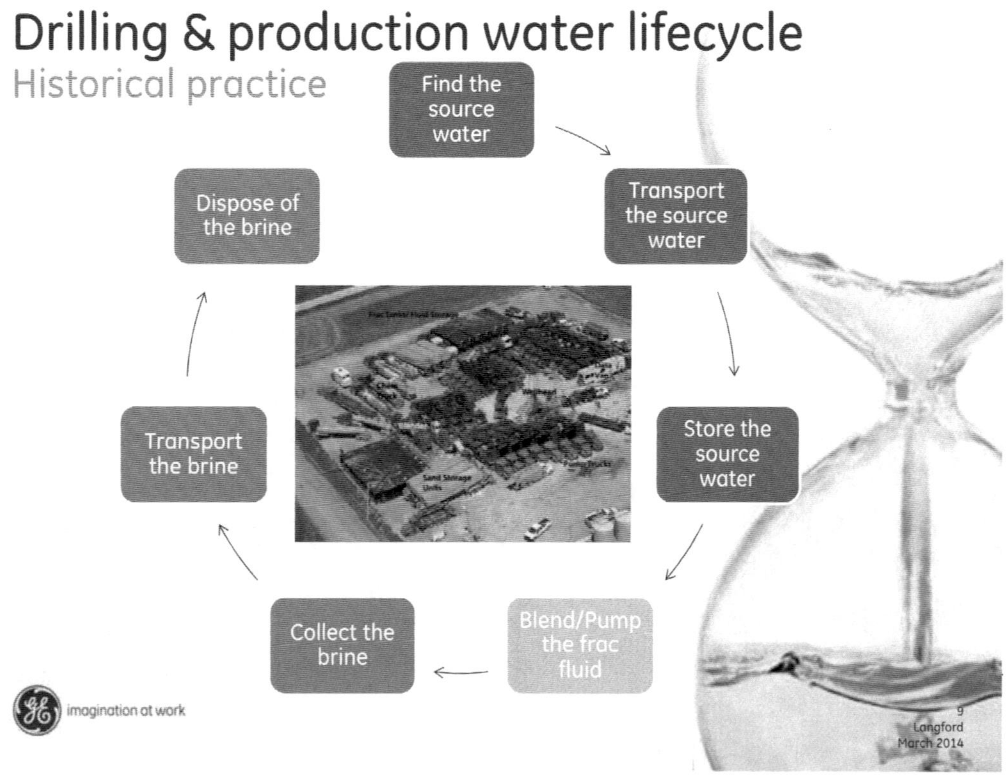
Drilling & production water lifecycle
Historical practice
Find the source water
Transport the source water
Store the source water
Blend/Pump the frac fluid
Collect the brine
Transport the brine
Dispose of the brine
imagination at work
9
Langford
March 2014

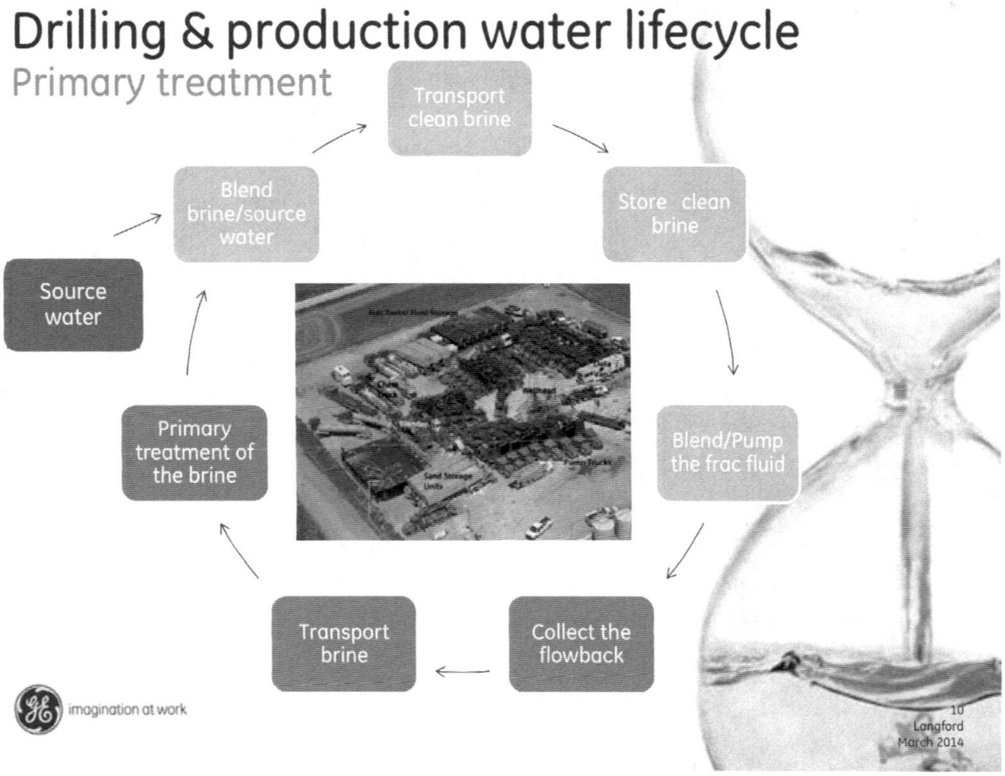
Drilling & production water lifecycle
Primary treatment
Transport clean brine
Store clean brine
Blend/Pump the frac fluid
Collect the flowback
Transport brine
Primary treatment of the brine
Source water
Blend brine/source water
imagination at work
10
Langford
March 2014

Drilling & production water lifecycle

Advanced treatment

Source water

Blend brine/source water

Transport clean brine

Store clean brine

Blend/Pump the frac fluid

Collect the flowback

Transport brine

Advanced treatment of the brine

Discharge

Beneficial Reuse

Field operations

imagination at work

11
Langford
March 2014

imagination at work

12
Langford
March 2014

Source: FracFocus.org

Flowback/produced water economics

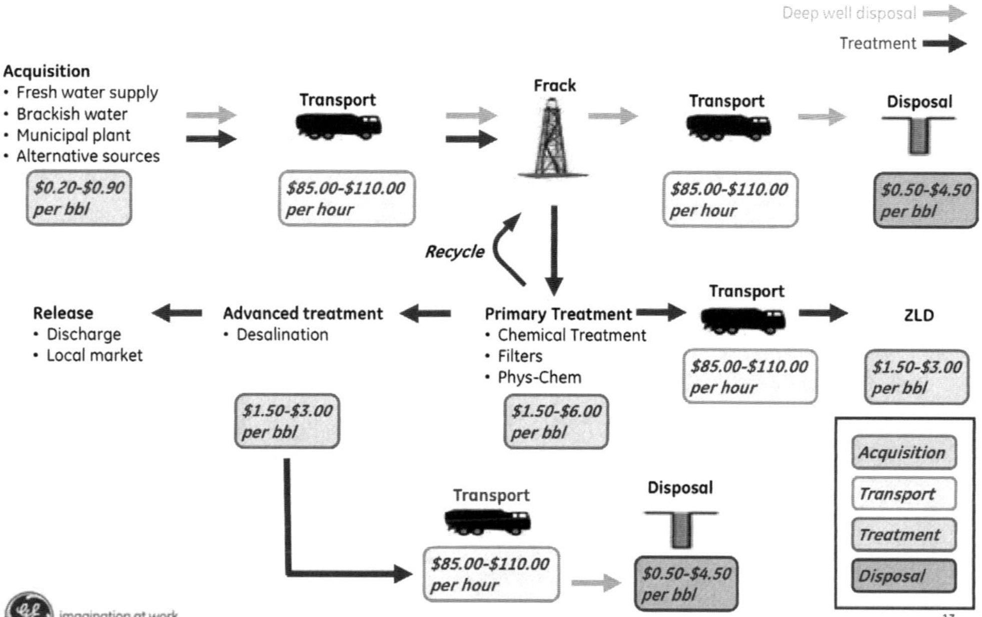

imagination at work

13
Langford
March 2014

Water spend

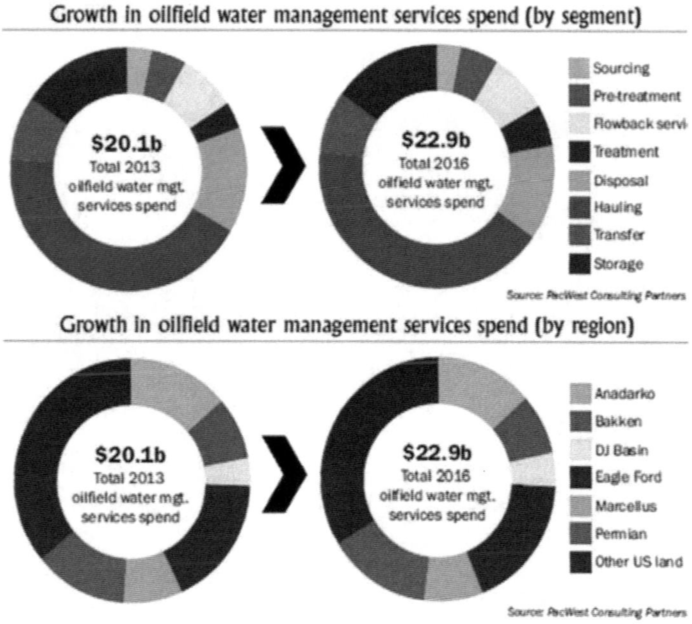

imagination at work

14
Langford
March 2014

Source: PacWest Consulting Partners/GWI March 2014

Example frac flowback & produced water

(SINGLE WELL)

Day	Percent of Fracture	Barrels
1	7.56%	6124.0
2	7.56%	6119.0
3	7.50%	6071.4
4	7.44%	6023.8
5	7.26%	5881.0
6	4.56%	3690.5
7	2.88%	2333.3
8	1.21%	976.2
9	0.88%	714.3
10	0.65%	523.8
11	0.44%	357.1
12	0.35%	285.7
13	0.26%	214.3
14	0.18%	142.9
15	0.12%	95.2
16	0.07%	57.1
17	0.04%	28.6
18	0.04%	33.3
19	0.03%	23.8
20	0.03%	26.2
21	0.04%	28.6
22	0.04%	29.8
23	0.03%	27.4
24	0.03%	27.1
25	0.04%	30.5
26	0.04%	28.6
27	0.04%	33.3
28	0.03%	23.8
29	0.08%	61.9
30	0.04%	28.6
31	0.04%	29.8

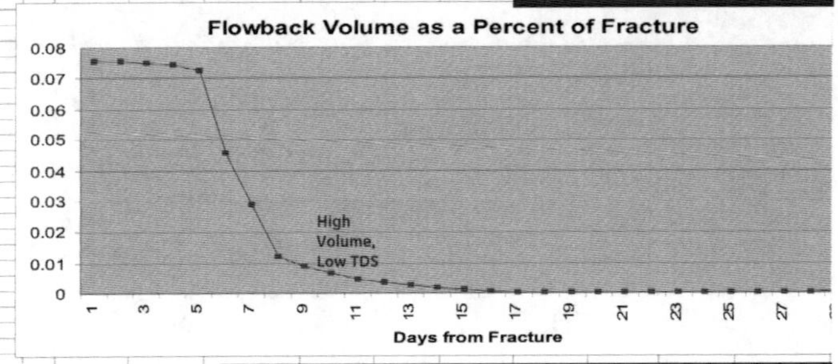

Day	Specific Gravity	Ionic Strength	TDS	pH	Ca (mg/L)	Mg (mg/L)	T
1	1.002	0.0083	559.09		40.00	3.72	
1.5	1.005	0.1025	6437.60		379.75	31.26	
2	1.009	0.2087	12773.03		935.18	77.34	
3	1.017	0.3727	23818.18		1312.51	101.69	
4	1.028	0.6434	40803.17		2447.30	190.57	
5	1.031	0.7192	45758.81		2679.66	209.47	
12	1.040	0.9371	59633.81	5.9	4395.06	381.2	1
15	1.052	1.2300	78681.52		4462.10	372.03	
21	1.102	2.4501	156570.39		10528.19	872.28	3
25	1.111	2.6452	170014.09		12084.61	1013.11	3
31	1.121	2.9331	185728.6		14575.36	1141.79	4

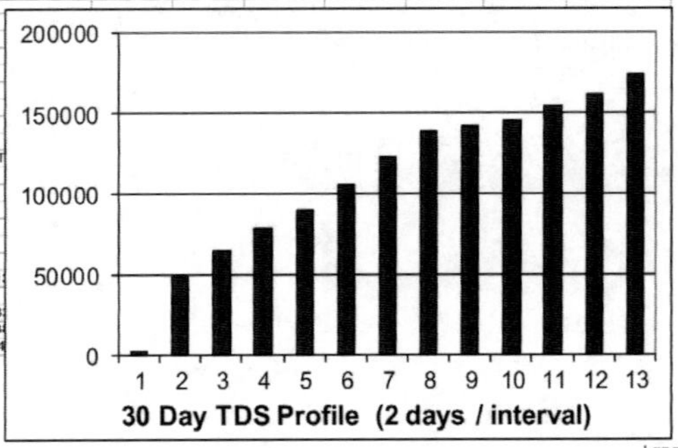

imagination at work

15
Langford
March 2014

Source: GE Water & Process Technologies

Multiple wells…

Assumptions	
# wells	250
Days drilled apart	15
Volume per frack (gal)	4,000,000
Volume per frack (bbl)	95,238
% returned on days 1-5	5.00%
% returned on day 6-7	2.50%
% returned on day 8-9	1.00%
% returned on day 10-30	0.25%
% returned on day 30-5475	0.005%
% that returns over time	64%

No reuse + fresh water

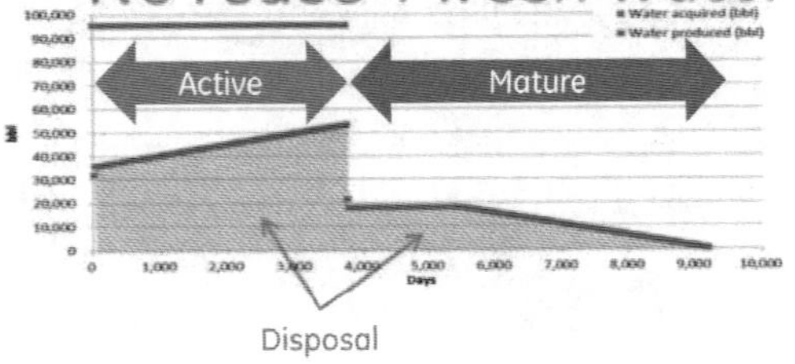

Reuse + clean brine

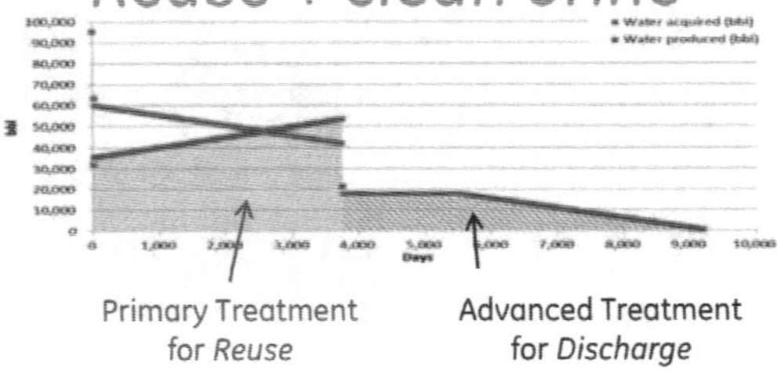

Reuse + fresh water

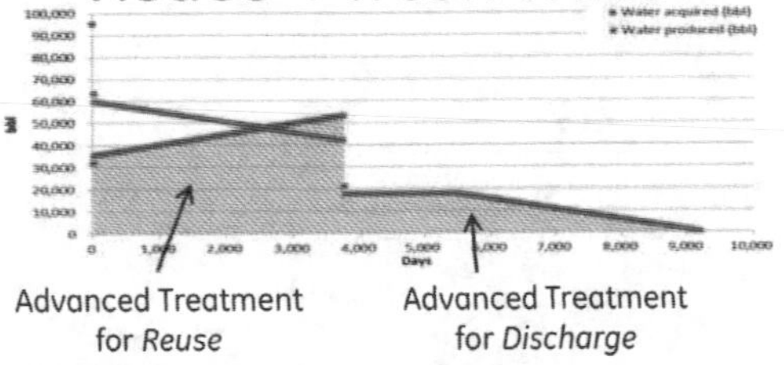

imagination at work

16
Langford
March 2014

Recycle/reuse items to consider

Target quality

- Fracture fluid
- Discharge

Source quality

- Fresh water
- Flowback water
- Produced water
- Blended quality

Process issues

- Varying feed quality
- Varying feed quantity
- System mass balance
- System mobility
- System footprint
- Power supply
- Operations
- EHS
- Permitting

Economics

- Storage
- Transport
- Processing charge
- Process waste disposal
- Beneficial reuse

imagination at work

17
Langford
March 2014

For illustration only

Customer economics to determine value

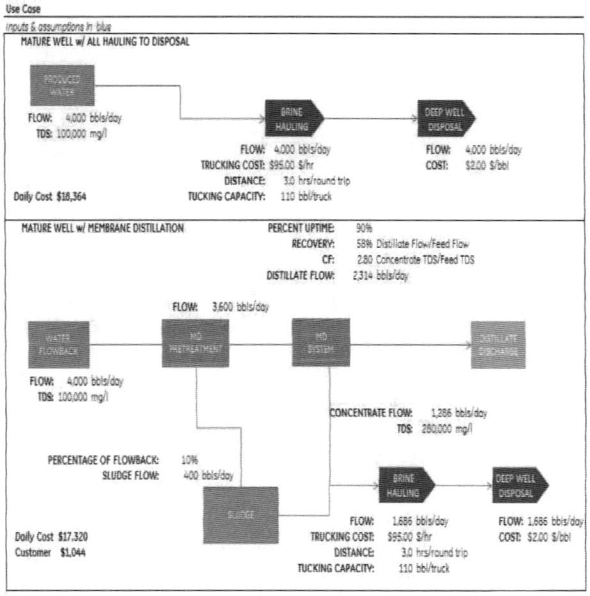

Summary

Annual customer savings			**$381,065**

Disposal Pathway

Hauling & disposal			
Brine hauling cost	$/bbl	$2.59	
+ Deep well disposal cost	$/bbl	$2.00	
= Total hauling & disposal cost	$/bbl	$4.59	
x BBL sent to disposal	bbl/day	4,000	
= Daily disposal costs			**$18,364**

VS.

Treatment Pathway

Treatment			
Volume to treat	bbl/day	4,000	
x Cost to treat	$/bbl	$2.26	
= Daily treatment cost			$9,039
Distillate (revenue)/cost			
Volume of distillate	$/bbl	2,083	
x Distillate (value)/cost	$/bbl	($0.25)	
= Total distillate (revenue)/cost	$/bbl		($520.71)
Hauling & disposal cost			
Pretreatment brine	bbl	360	
+ Concentrate flow	bbl	1,157	
+ Average daily volume caused by downtime	bbl	400	
= Average daily hauling & disposal volume	bbl	1,917	
x Total hauling & disposal cost	$/bbl	$4.59	
= Average annual hauling & disposal costs			$8,801
= Average daily cost-to-treat (w/ disposal)			**$17,320**

imagination at work

18
Langford
March 2014

Technology

	Primary Treatment			Advanced Treatment
	Preliminary	**Secondary**	**Tertiary**	**Desalination**
Application	• **TSS** • **Free/suspended oil** • **Bacteria**	• **TSS** • **Emulsified oil** • **Metals**	• **Dissolved organics**	• **Dissolved solids**
Technology	• Collection/Decant • API Separator • Hydrocyclone • Clarifier/Thickener	• Phys-Chem • MMF • Ultrafiltration • Walnut shell • Electrocoagulation • DAF, EAF • Ceramic membranes	• MBR • GAC • Optipore (DOW) • Biological Treatment • Advanced Oxidation	• EDR • BWRO • SWRO • HERO • Membrane Distillation* • Evaporation • Crystallization • Forward osmosis • Humidification-dehumidification

* Commercial scale in development

imagination at work

19
Langford
March 2014

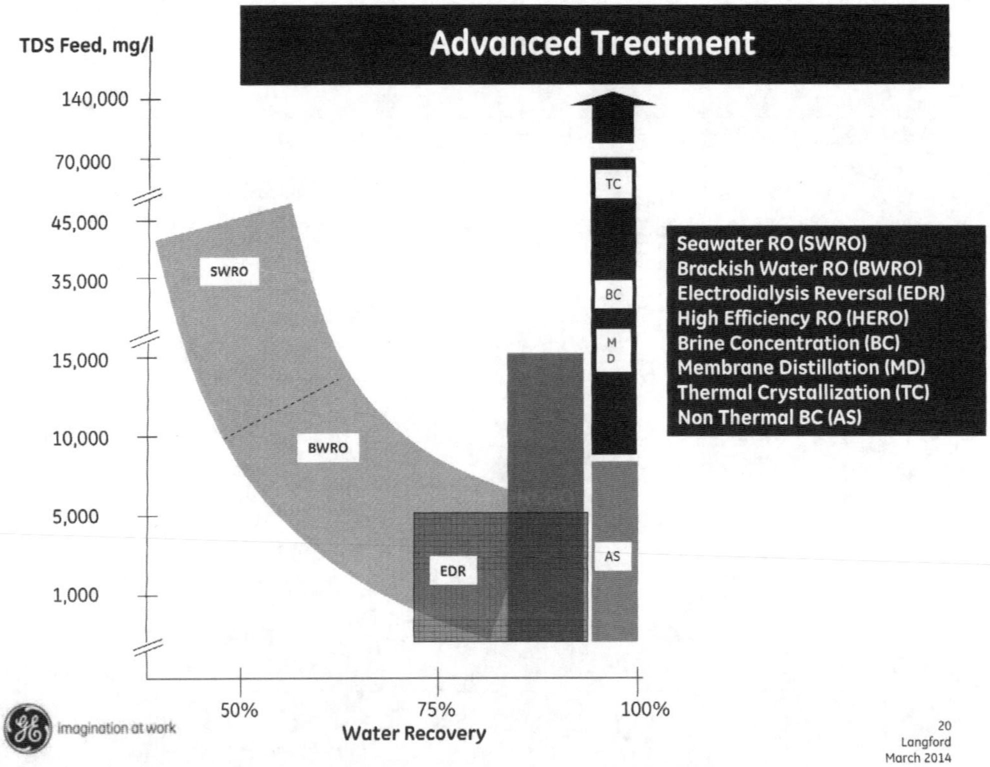

imagination at work

20
Langford
March 2014

Summary

- ➢ Evolving market
- ➢ Challenging water
- ➢ No well like its neighbor
- ➢ No silver bullet
- ➢ Evolving solutions

Need solutions tailored to today and tomorrow

imagination at work

21
Langford
February, 2014

351667 An Integrated Approach to Water Treatment and Recycle in Oil and Gas Production and Processing Via Thermal Membrane Distillation

Abstract:

Thermal membrane distillation (TMD) is an emerging technology in the field of water treatment and recycle. It is driven primarily by heat which creates a vapor-pressure difference across a porous hydrophobic membrane. It gained interest because of many advantages; it can work properly with feed temperature below boiling point and can provide high level of separation of the treated water. Because TMD relies on heating the wastewater feed stream, one of the key benefits is that it allows thermal integration with low-quality sources of heat. This thermal coupling can significantly enhance the treatment and recycle costs. The paper presents an integrated approach to the design of TMD networks. First, TMD is modeled based on key design and operating variables. Next, a framework is developed for cost-effective thermal coupling. Several applications in upstream production and downstream processing of oil and gas are presented. The results are compared with conventional separation techniques and with systems lacking heat integration.

SIMULATION AND SUSTAINABILITY ANALYSIS FOR PRODUCED WATER TREATMENT DRIVEN BY ON-SITE WASTE HEAT

Liwen Chen[1], Parag Saxena[1], Helen H. Lou[1], Lily Xu[2] and Danny Reible[3]
[1.] Dan F. Smith Department of Chemical Engineering, Lamar University
[2.] Environmental and Water Resources Engineering, University of Texas at Austin
[3.] Department of Civil & Environmental Engineering, Texas Tech University

Topics

- Produced Water Management
- On-site energy source
- Simulation of several thermal evaporation technologies (MSF, MED, MD) in Aspen Plus
- Sustainability analysis in terms of performance, economics and environmental criteria

What's Produced Water?

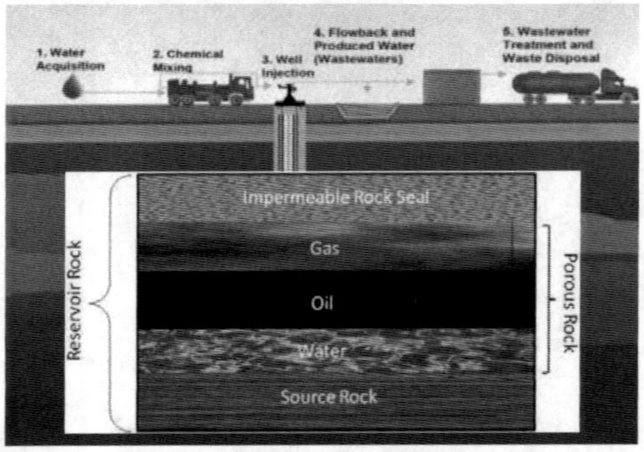

- Water that comes to the surface with oil and gas;
- Rich components (Salt, Oil and grease, Chemical additives, NORMs, HCs, etc);
- Largest volume waste stream from oil and gas production;
- Ratio of water-to-oil in the production
  - ➢World-wide estimate – 2:1 to 3:1
  - ➢U.S. estimate – 7:1
  - ➢Many older U.S. wells have ratios > 50:1!!!

source: http://www2.epa.gov/hfstudy/hydraulic-fracturing-water-cycle; E.T. Igunnu and G.Z. Chen Produced Water Treatment Technologies, International Journal of Low-Carbon Technologies 2012.

PW Composition

√ Salts, Medals,
√ Oil, Grease, HCs
√ Radioactive materials
...

Total Dissolved Solids (TDS): 8000 ~ 360,000 ppm

Parameter	Minimum value	Maximum value	
Density (kg/m^3)	1014	1140	
Conductivity (μS/cm)	4200	58 600	
Surface tension (dyn/cm)	43	78	
pH	4.3	10	
TOC (mg/l)	0	1500	
TSS (mg/l)	1.2	1000	
Total oil (IR; mg/l)	2	565	
Volatile (BTX; mg/l)	0.39	35	
Base/neutrals (mg/l)	—	<140	
Chloride (mg/l)	80	200 000	
Bicarbonate (mg/l)	77	3990	Anions
Sulphate (mg/l)	<2	1650	
Ammoniacal nitrogen (mg/l)	10	300	
Sulphite (mg/l)	—	10	
Total polar (mg/L)	9.7	600	
Higher acids (mg/l)	<1	63	
Phenol (mg/l)	0.009	23	
Volatile fatty acids (mg/l)	2	4900	

Heavy metal	Minimum value (mg/l)	Maximum value (mg/l)	
Calcium	13	25 800	
Sodium	132	97 000	Cations
Potassium	24	4300	
Magnesium	8	6000	
Iron	<0.1	100	
Aluminium	310	410	
Boron	5	95	
Barium	1.3	650	
Cadmium	<0.005	0.2	
Copper	<0.02	1.5	
Chromium	0.02	1.1	
Lithium	3	50	
Manganese	<0.004	175	
Lead	0.002	8.8	
Strontium	0.02	1000	
Titanium	<0.01	0.7	
Zinc	0.01	35	
Arsenic	<0.005	0.3	
Mercury	<0.005	0.3	
Silver	<0.001	0.15	
Beryllium	<0.001	0.004	

Rich components, Wide concentration range, High Salinity!

1% = 10,000 ppm (mg/L)

Source: Igunnu et al. 2012 Produced water treatment technologies

5

Federal & State Regulations

Federal:

- U.S. Environmental Protection Agency (EPA);
- Department of the Interior's Bureau of Land Management (BLM);
- Bureau of Ocean Energy Management, Regulation and Enforcement (BOEMRE, formerly MMS);

EPA requirement for oil and grease in PW (NPDES):

Offshore [40CFR§435.13]:

- Daily Maximum Discharge (EPA): **42** mg/L
- Monthly average limit: **29** mg/L

Onshore [40CFR§435.32]:

- Zero discharge

State:

- Regulations varied widely;

For example, Texas Commission on Environmental Quality (TCEQ)

Rule 3.8 (Water Protection);Rule 3.9 (Disposal Wells)

Require permits to discharge into surface/subsurface water

source: http://www.netl.doe.gov/research/coal/crosscutting/pwmis/fed-state-regulations

6

PW Management

Three hierarchy:

Minimization

- **Prevent PW entering well** (Mechanical blocking, Water shut-off chemicals)
- **Prevent PW reaching the surface** (Downhole separation, Sea floor separation)

Recycle/reuse

- **Injection for enhanced oil recovery** (TX: ~38540 wells (5.3 billion bbl/yr))
- **Injection for disposal** (TX: ~12000 wells (1.2 billion bbl/year), PA: 8 wells!)

Treatment

- **Dilution** (blending with freshwater) → ~$1.5-2/bbl
- **Settling** (allow enough residence time for solids to settle) → ~$2-2.5/bbl
- **Filtration** (sand filters, bag filters, disk filters) → ~$2-3/bbl
- **Chemical Precipitation** (pH adjustment, flocculants) → ~$2.5-4/bbl
- **Evaporation** (Highest effluent quality with highest energy demand. Energy cost can be mitigated by waste heat) → ~$5.5-8/bbl

Permeate Standard (WHO): TDS ≤ 500mg/L

source: Veil J.A. 2009 Produced Water Management Options –One Size Does Not Fit All

Gas Flaring

Why Flare
- Safety concerns:
  - ➢ Equipment malfunction;
  - ➢ Off-spec production;
  - ➢ Depressurizing equipment;
  - ➢ Start ups and emergency shut-downs;
- Lack of infrastructure (pipeline)

1/3 flared in Bakken Shale, $100 million/month!

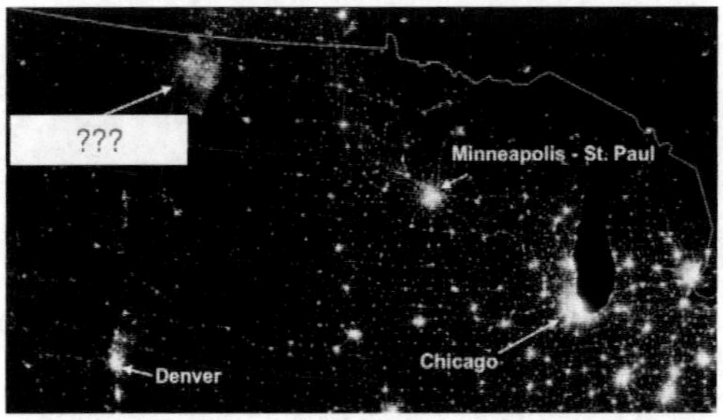

source: http://www.pennenergy.com/articles/pennenergy/2013/07/bakken-shale-natural-gas-flaring-tops-100-million-each-month.html

Energy Savings

- 5% of global gas production (150 Billion m^3/yr)
- 23% US NG use, 30% EU NG use
- 160 MMCFD NG in Bakken Shale flared out of 480 MMCFD production

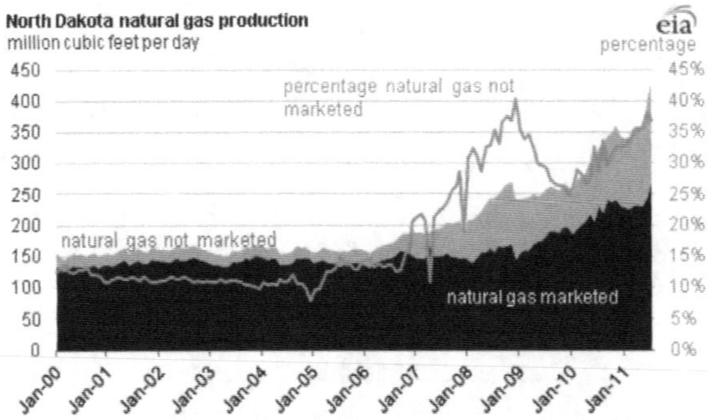

Source: U.S. Energy Information Administration, based on the North Dakota Department of Mineral Resources.

9

Flare Minimization Strategies

Legislation
- Kyoto Protocol
- Global Gas Flaring Reduction (GGFR);
- Economic Incentives

Reduction
- Reinjection EOR
- Compress to liquid;
- Dynamic Control System

Utilization
- Heat source for co-generation

10

Energy-Water Nexus Scheme

Pros:
Cost becomes profit;
Reduce environmental impacts;
On-site water treatment supply;
…

Cons:
Unsteady supply;
Safety issue;
Infrastructure;
Capital cost;
…

PW Management Desalination Technologies

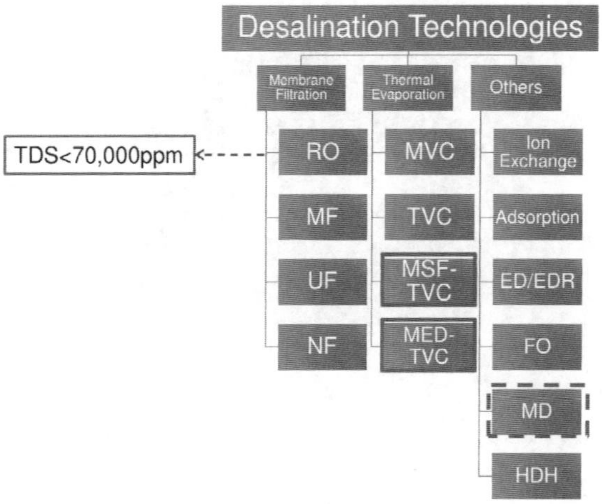

source: Shaffer et al. Environ. Sci. Technol. 47 (2013)

Sustainability Analysis

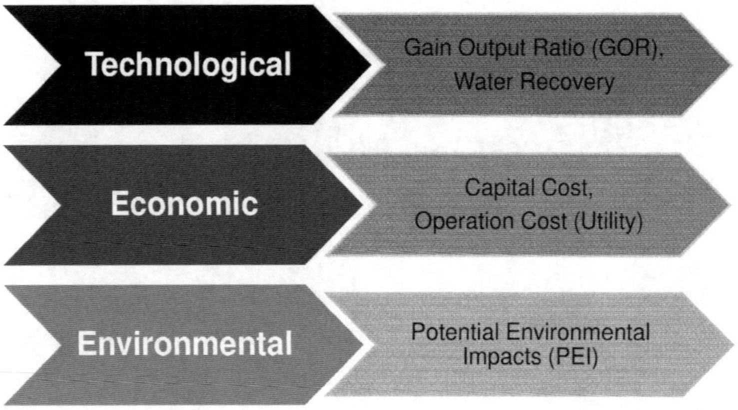

Environmental Impact Analysis

- WAR (Waste Reduction) Algorithm
  - Developed by Young & Cabezas (EPA), 1999
  - Database on environmental impacts of chemicals
  - 4 general categories, 8 sub-categories
  - http://www.epa.gov/nrmrl/std/war/sim_war.htm
- PEI (Potential Environmental Impact) index to represent environmental friendliness

  "A low PEI score indicates a lower environmental impact."
- Green house gas emission by CO_2 is the major potential environmental impact

WAR Algorithm: Impact Categories

General Impact Category	Impact Sub-Category	Measure of Impact Category
Human toxicity	Ingestion	HTPI
	Inhalation/Dermal	HTPE
Ecological toxicity	Aquatic toxicity	ATP
	Terrestrial toxicity	TTP
Global atmospheric impacts	Global warming potential	GWP
	Ozone depletion potential	ODP
Regional atmospheric impacts	Acidification potential	AP
	Photochemical oxidation potential	PCOP

Potential Environmental Impact (PEI)

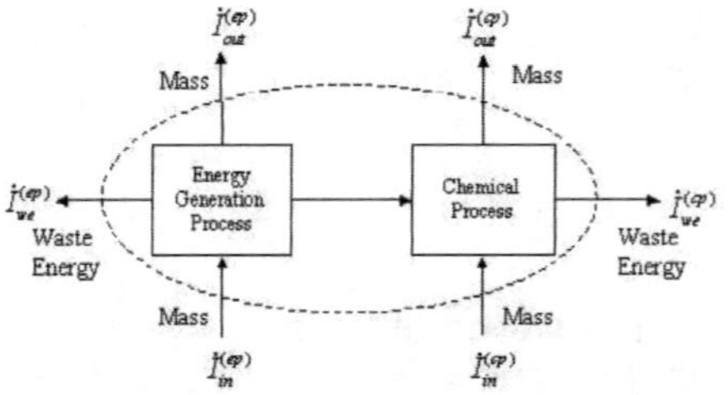

The potential environmental impact balance is as follows:

$$\cancel{\frac{\partial I_s}{\partial t}} = \dot{I}^{(cp)}_{in} + \cancel{\dot{I}^{(ep)}_{in}} - \dot{I}^{(cp)}_{out} - \dot{I}^{(ep)}_{out} - \cancel{\dot{I}^{(cp)}_{we}} - \cancel{\dot{I}^{(ep)}_{we}} + \dot{I}^{(t)}_{gen}$$

$$\boxed{0 = \dot{I}^{(t)}_{in} - \dot{I}^{(t)}_{out} + \dot{I}^{(t)}_{gen}}$$

The total input rate of PEI:

$$\dot{I}^{(t)}_{in} = \sum_{i}^{EnvCat} \alpha_i \dot{I}^{(t)}_{i,in} = \sum_{i}^{EnvCat} \alpha_i \sum_{j}^{Streams} \dot{M}_{j,in} \sum_{k}^{Comps} x_{kj} \psi^{s}_{ki}$$

The total output rate of PEI:

$$\dot{I}^{(t)}_{out} = \sum_{i}^{EnvCat} \alpha_i \dot{I}^{(t)}_{i,out} = \sum_{i}^{EnvCat} \alpha_i \sum_{j}^{Streams} \dot{M}_{j,out} \sum_{k}^{Comps} x_{kj} \psi^{s}_{ki}$$

To evaluate the PEI output index on a production basis,

$$\boxed{\hat{I}^{(t)}_{out} = \frac{\dot{I}^{(t)}_{out}}{\sum_{p}^{ProdStreams} \dot{P}_p}}$$

17

Multi-Stage Flash (MSF) Process

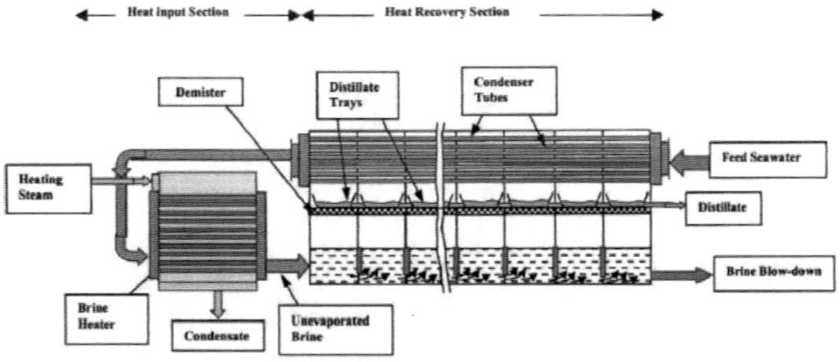

MSF-OT (once-through w/o brine recirculation)

Source: El-Dessouky et al./ Chemical Engineering Journal 1999

18

Specification

Feed Components: Na^{+}, K^{+}, Ca^{2+}, Mg^{2+}, Cl^{-}, HCO_3^{-}, SO_4^{2-} ($CaSO_4$ ↓)
Feed Flowrate: 10,000 ton/day

Three Cases: (Low TDS, Medium TDS, and High TDS)

Location	Powder River (MT)	Barnett Basin (TX)	Bakken Basin (ND)
Compnents (mg/L)	Low	Medium	High
Na+	6022	20935	71466
K+	75	61	512
Ca++	327	5441	8993
Mg++	93	1734	1467
Cl-	9074	45989	129461
SO4--	1529	1394	1885
HCO3-	1329	408	322
TDS	18447	75962	214106

Low TDS (Powder River) (<35,000ppm)

Medium TDS (Barnett) (35,000~200,000ppm)

High TDS (Bakken) (200,000~400,000ppm)

Source: US Geological Survey Produced Water Database

Multi-Stage Flash w/ Thermal Vapor Compression (MSF-TVC) Simulation in Aspen Plus

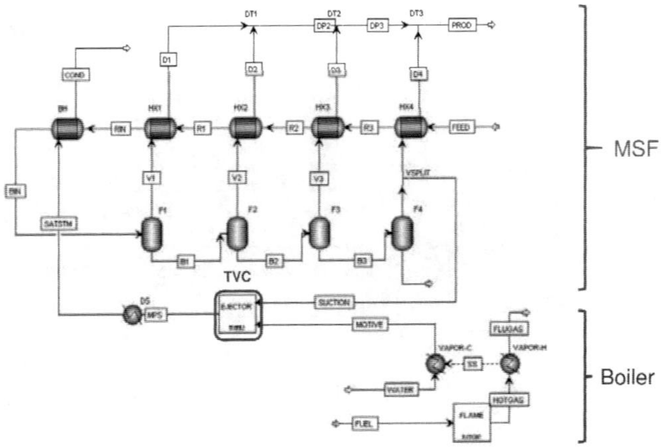

Simulation Results of MSF-TVC

MSF-TVC stages	F1	F2	F3	F4
Temperature (°C)	90	88	86	84
Pressure (bar)	0.671	0.618	0.568	0.521
Vapor (ton/day)	970	873	786	707
Prod (ton/day)	2840			
Steam Supply (°C)	133.6			
Steam Supply (bar)	3			
Steam Supply (ton/day)	440			
GOR (Prod/Steam)	6.5			
Product Recovery (Prod/Feed)	28.4%			
Energy Consumption (MMBTU/day)	1099			
Utility Cost ($/day)	4130			
CO_2 Emission (ton/day)	62			

5-year (2009-2013) average Henry Hub NG Price = $3.76/MMBTU (http://www.eia.gov)

21

Case Results

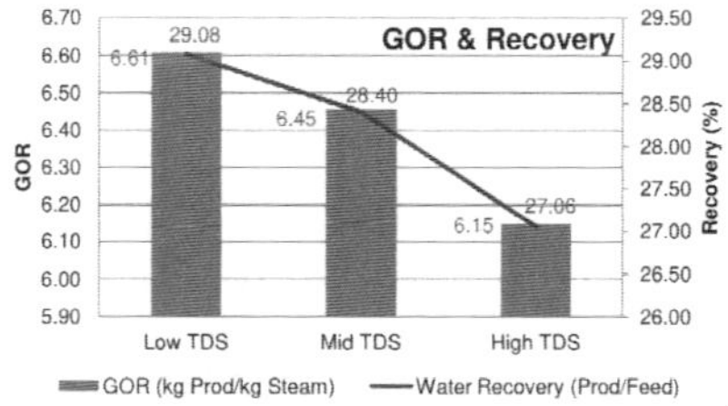

	Utility Cost
$/ton	1.42-1.53
cents/gal	0.54-0.58
cents/bbl	16.94-18.21

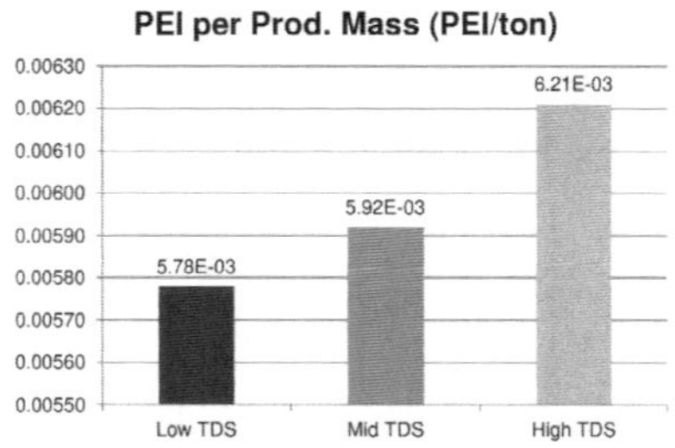

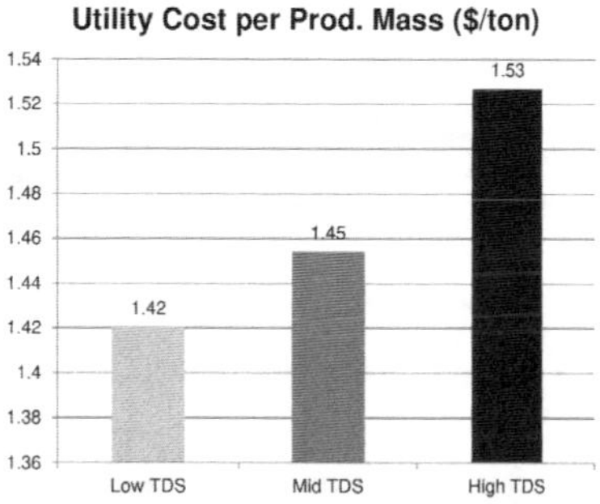

22

PEI in each categories (MSF-TVC)

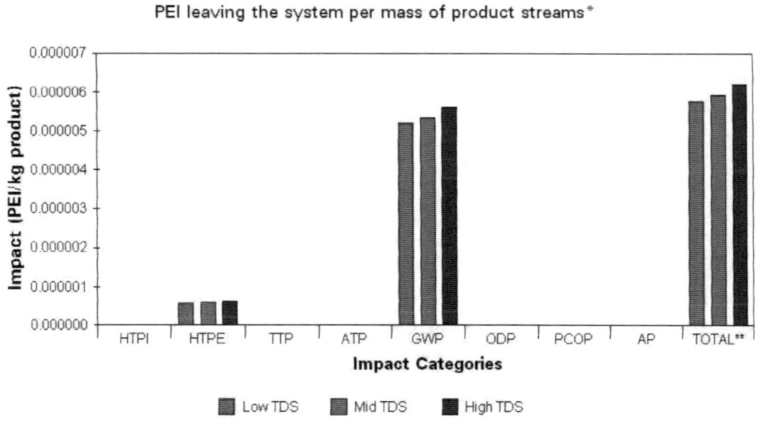

HPTE: Human Toxicity Potential by Inhalation or Dermal Exposure
GWP: Global Warming Potential

23

Multi-Effect Distillation (MED)

Features:

- Higher heat transfer coefficient;
- Max Brine Temperature 70°C (avoid scalants);
- Economic and durable material (aluminum alloy vs. copper);
- Possibility of using low grade heat;

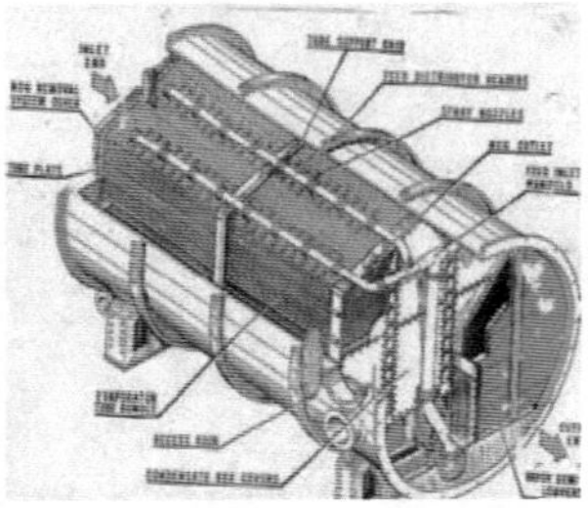

Horizontal tube assembly
A. Ophir, F. Lokiec et.al. Desalination 182 (2005)

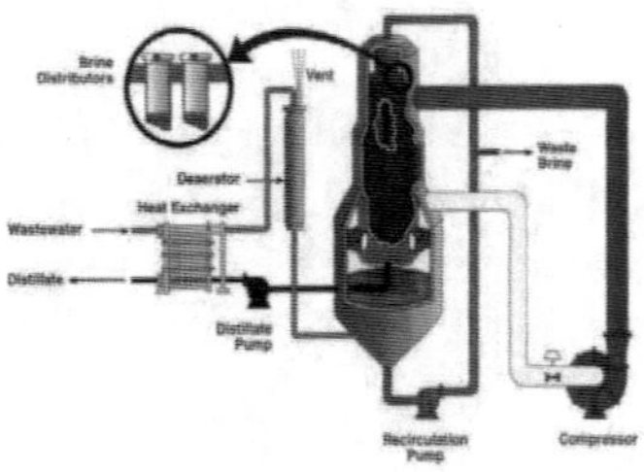

Falling Film Evaporator
(GE Power & Water 2012)

24

MED-TVC Schematic Diagram

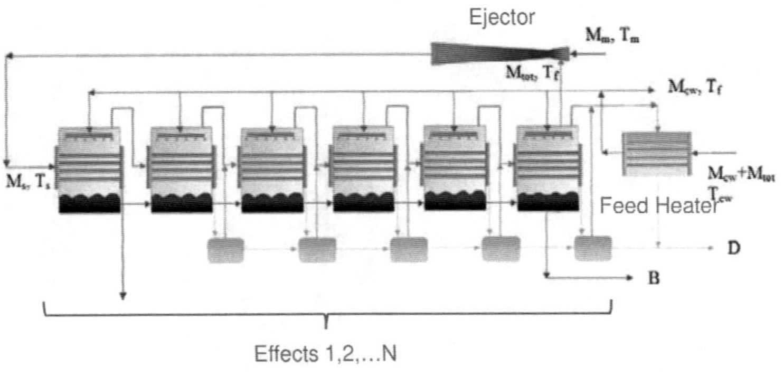

source: Manesh et al. Desalination 316 (2013)

25

MED-TVC Simulation in Aspen Plus

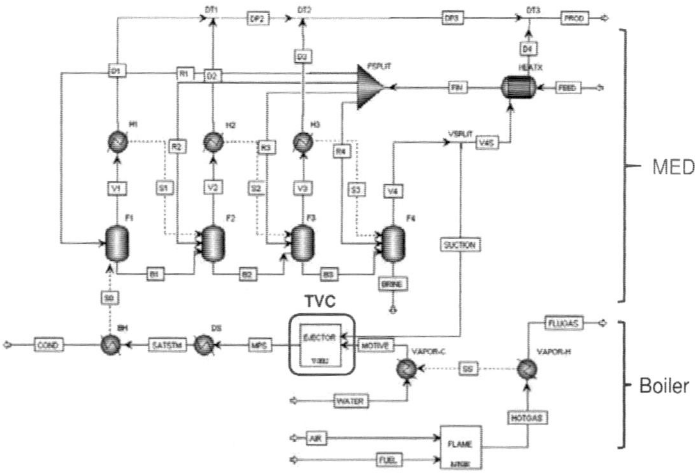

26

Results for MED-TVC

MED-TVC stages		F1	F2	F3	F4
Temperature (°C)		60	56	52	48
Pressure (bar)		0.182	0.152	0.125	0.103
Vapor (ton/day)		1156	1099	1067	1059
Prod (ton/day)		3640			
Steam Supply	(°C)	133.6			
	(bar)	3			
	(ton/day)	500			
GOR (Prod/Steam)		7.3			
Recovery (Prod/Feed)		36.4%			
Energy Consumption (MMBTU/day)		1249			
Utility Cost ($/day)		4693			
CO_2 Emission (ton/day)		71			

5-year (2009-2013) average Henry Hub NG Price = $3.76/MMBTU (http://www.eia.gov)

MED-TVC Results Summary

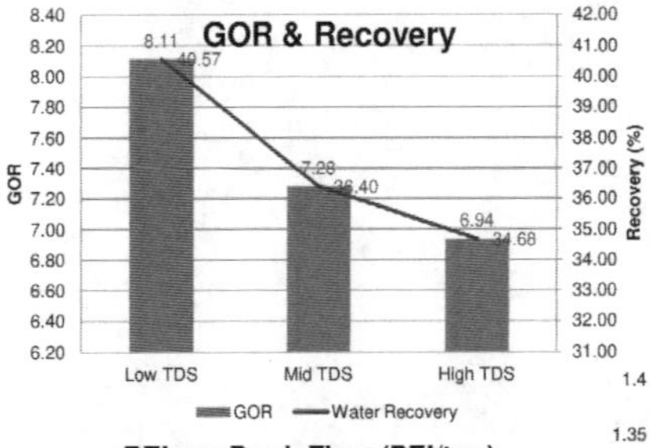

	Utility Cost
$/ton	1.16-1.35
cents/gal	0.44-0.51
cents/bbl	13.80-16.14

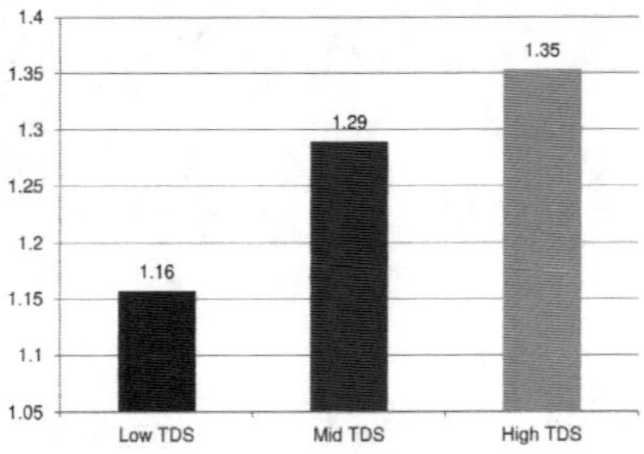

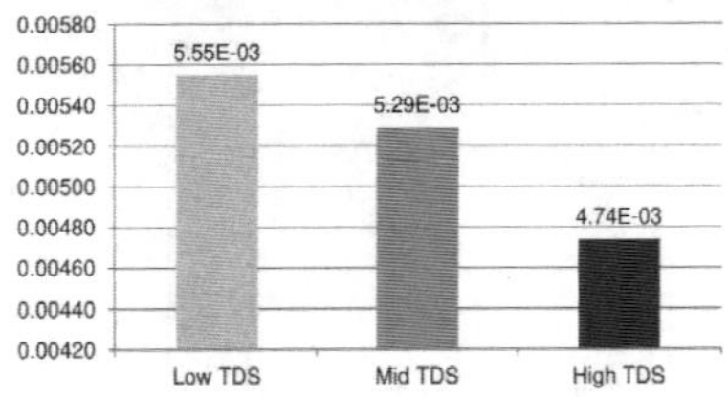

Membrane Distillation (MD)

Advantages:

- Lower operating temperatures and vapor space than MSF & MED;
- Lower operating pressure than RO;
- 100% rejection of non-volatile solute;
- Performance not limited by high osmotic pressure or concentration polarization;

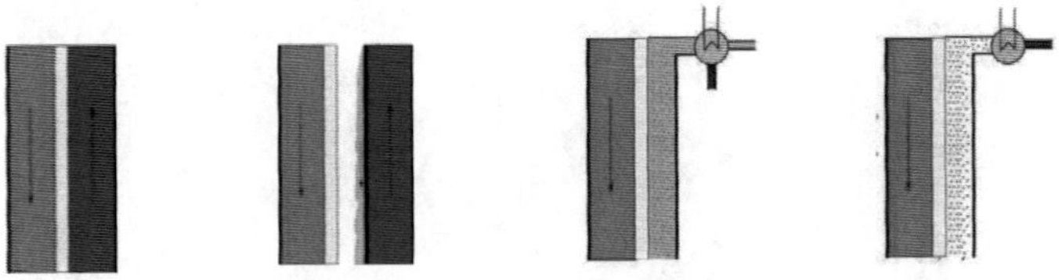

Direct Contact | Air-gap | Sweeping Gas | Vacuum

Source: Wikipedia

29

DCMD Concept

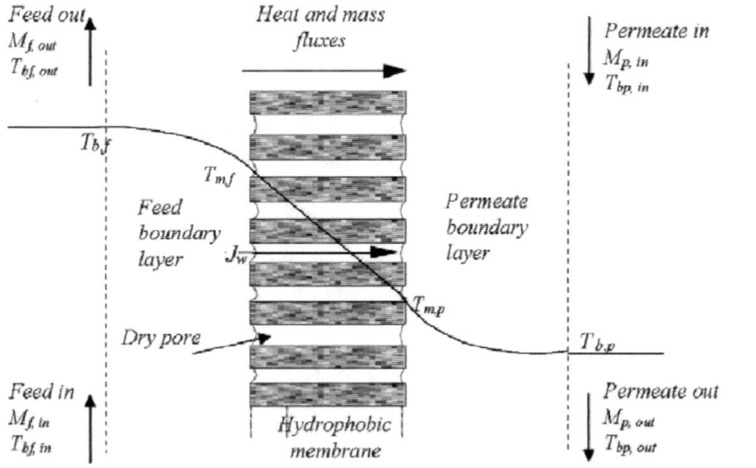

Source: M. Qtaishat et al. / Desalination 219 (2008) 272–292

30

Simulation of DCMD in Aspen Plus

15-20 effects for good approximation

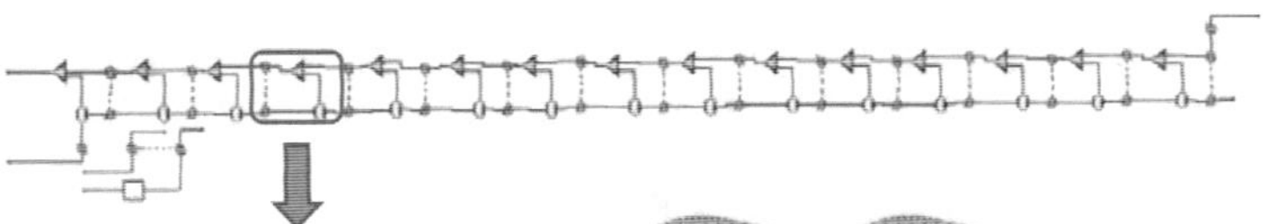

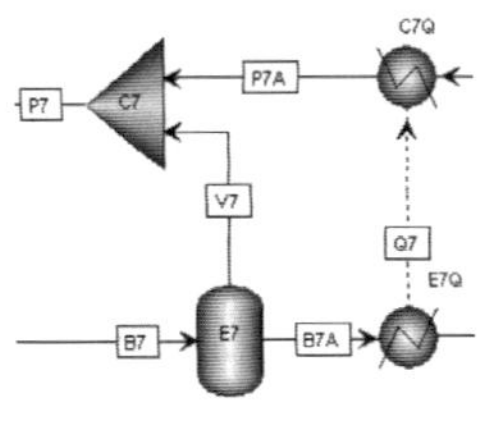

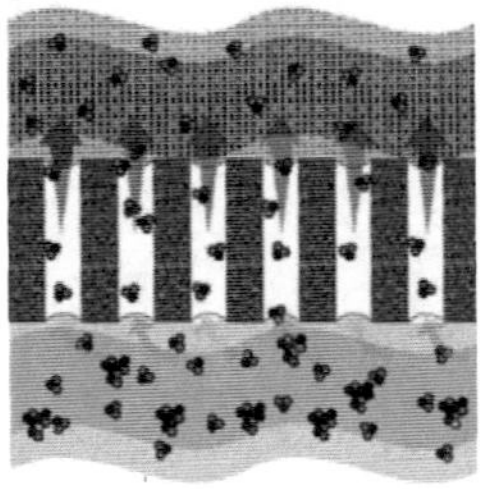

Source: http://www.nap.edu/openbook.php?record_id=12027&page=128

31

Hot Feed (E1, E2)

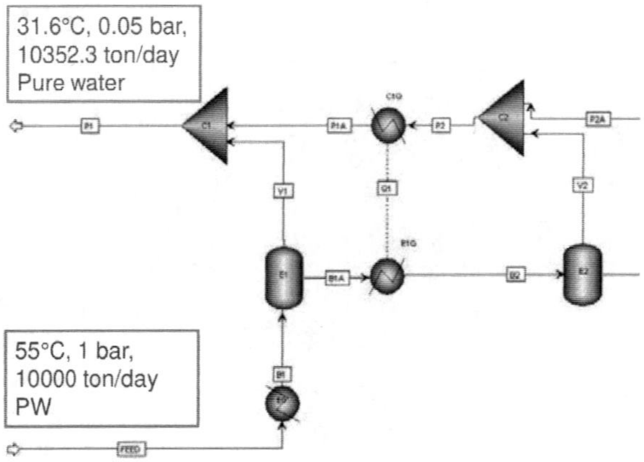

32

Cold Feed (E14, E15)

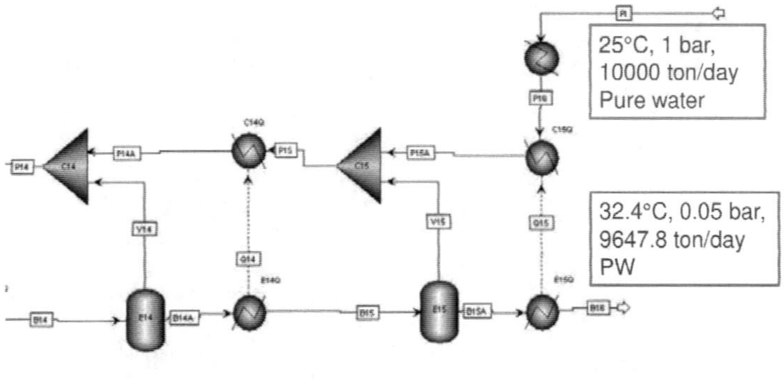

33

Conclusion Remarks

- All 3 desalination technologies (MSF, MED-TVC and MD) have potentials to treat PW with pros and cons;
- Coupling flared gas as heating source is beneficial but challenging;

34

Acknowledgements

- Professor Chau-Chyun Chen
  (Department of Chemical Engineering, Texas Tech University)
- Professor John L. Gossage
  (Department of Chemical Engineering, Lamar University)
- Texas Hazardous Waste Research Center
- Lamar University Graduate Scholarship

35
감사합니다
Natick
Grazie
Danke
Ευχαριστίες
Dalu
Obrigado
Thank You
Köszönöm
Tack
Спасибо
Dank
Gracias
谢谢
Merci
Seé
ありがとう

* DESMAR SOLAR

NATALY AGUDELO
YURANI GUTIÉRREZ
ANDRES OLAYA
FELIPE SASTOQUE

Energetic and exergetic analisys from a seawater solar desalinator

DESMAR SOLAR

*SCHEDULE

- Introduction
- Justification
- Objectives
- Resources available at the university.
- Main processes of water desalination.
- Material and energy balances.
- Preliminary test.
- Environmental considerations.
- Results
- Bibliography.

DESMAR SOLAR

*INTRODUCTION

*Water is the main element that humans use in their daily work. Of all the water available on the planet only 3% is fresh water and is on the surface. The remaining water is either frozen at the poles or distributed in the oceans.

DESMAR SOLAR

*JUSTIFICATION

Desalination processes are quite costly and cause pollution to the environment because their main source of energy is at its most fossil fuels.

Therefore, this investigation is intended to use efficient and clean energy sources to develop the process

DESMAR SOLAR

UNIVERSIDAD LIBRE
SCIENTIA FONS LIBERTATIS
COLOMBIA
*COLOMBIAN SOLAR MAP
DESMAR SOLAR

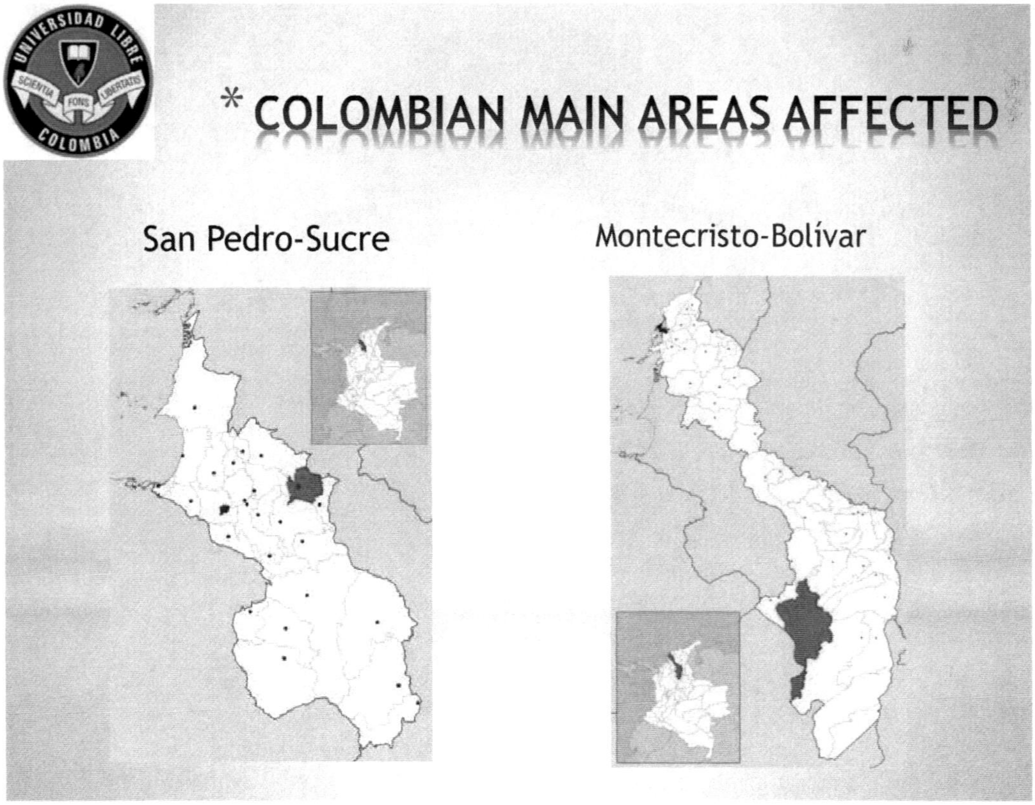
UNIVERSIDAD LIBRE
SCIENTIA FONS LIBERTATIS
COLOMBIA
* COLOMBIAN MAIN AREAS AFFECTED
San Pedro-Sucre
Montecristo-Bolívar

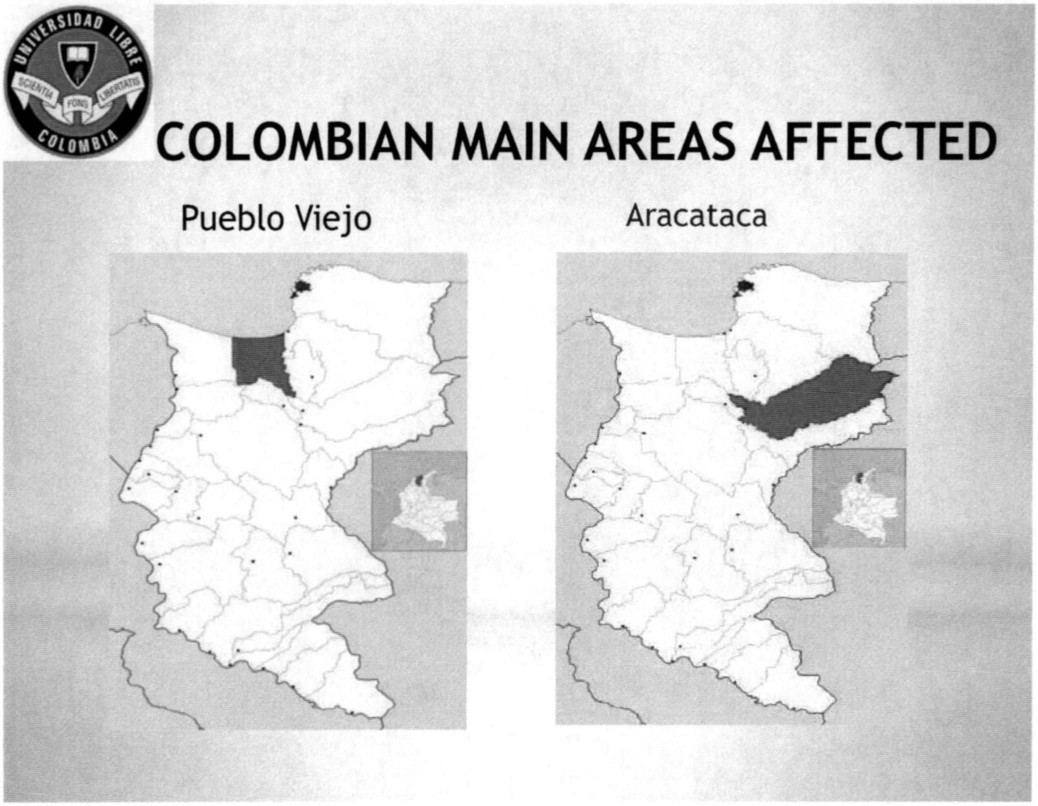
UNIVERSIDAD LIBRE
SCIENTIA FONS LIBERTATIS
COLOMBIA
COLOMBIAN MAIN AREAS AFFECTED
Pueblo Viejo
Aracataca

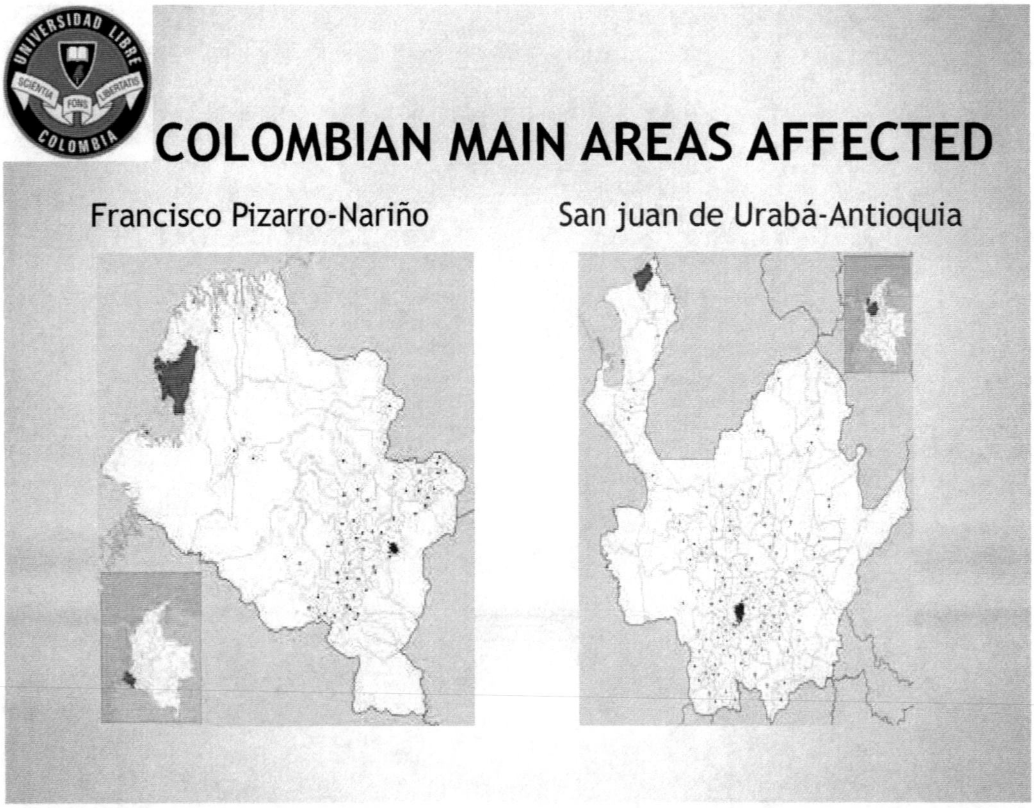
UNIVERSIDAD LIBRE
SCIENTIA FONS LIBERTATIS
COLOMBIA
COLOMBIAN MAIN AREAS AFFECTED
Francisco Pizarro-Nariño
San juan de Urabá-Antioquia

COLOMBIAN MAIN AREAS AFFECTED

Timbiquí

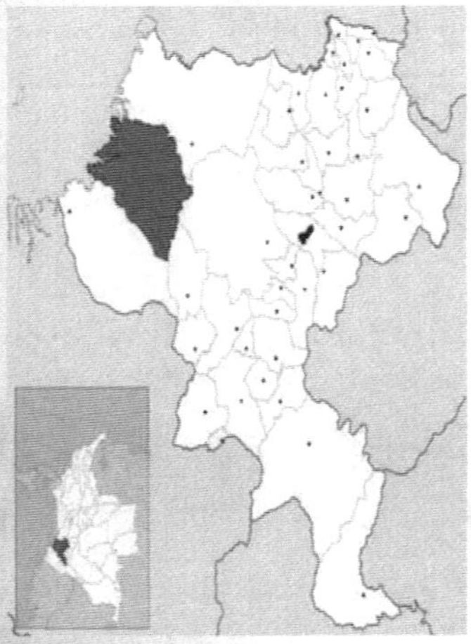

Villa rica

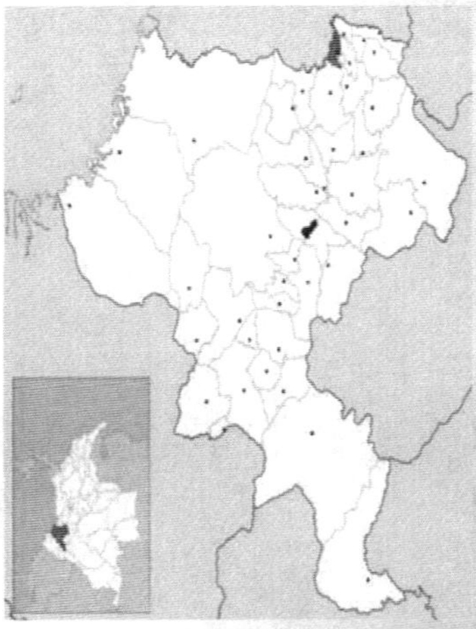

* GENERAL PURPOSE

Establish the energy and exergy analysis of a desalination system using solar energy.

DESMAR SOLAR

*SPECIFIC PURPOSE

* Determine the energy requirements of each of the stages of a solar desalination process.
* Establish the efficiency of different models of solar energy collection system to be used in the desalination process.
* Evaluate the exergy of the process and establish improvements in the design selecting the most suitable model system for the process
* Determine physical properties of the materials.
* Quantify the effect of temperature and the properties of the materials for their selection and use in a solar desalination.
* Establish the lifetime of the material chosen from the tests already performed.

DESALINATION PROCESS

* Desalination by reverse osmosis

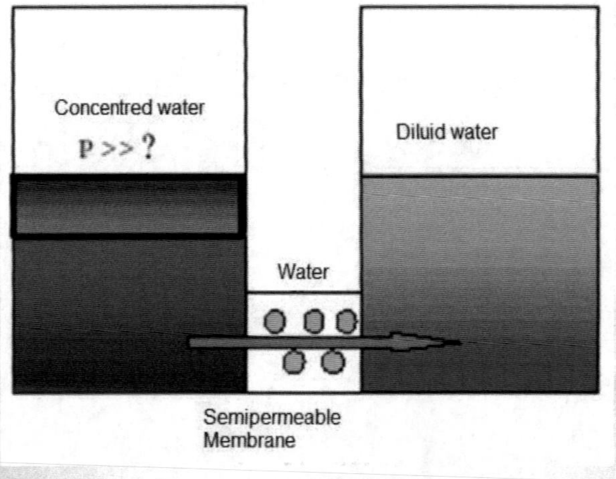

DESMAR SOLAR

*Desalination by Freezing

*Desalination by hydrate formation

*Flash point desalination

DESMAR SOLAR

*Desalination by distillation

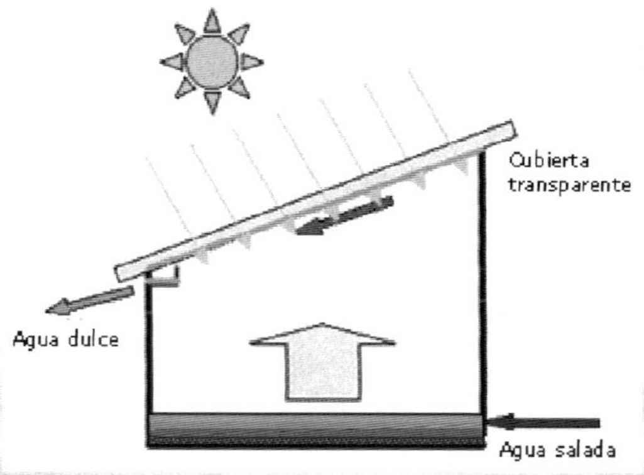

DESMAR SOLAR

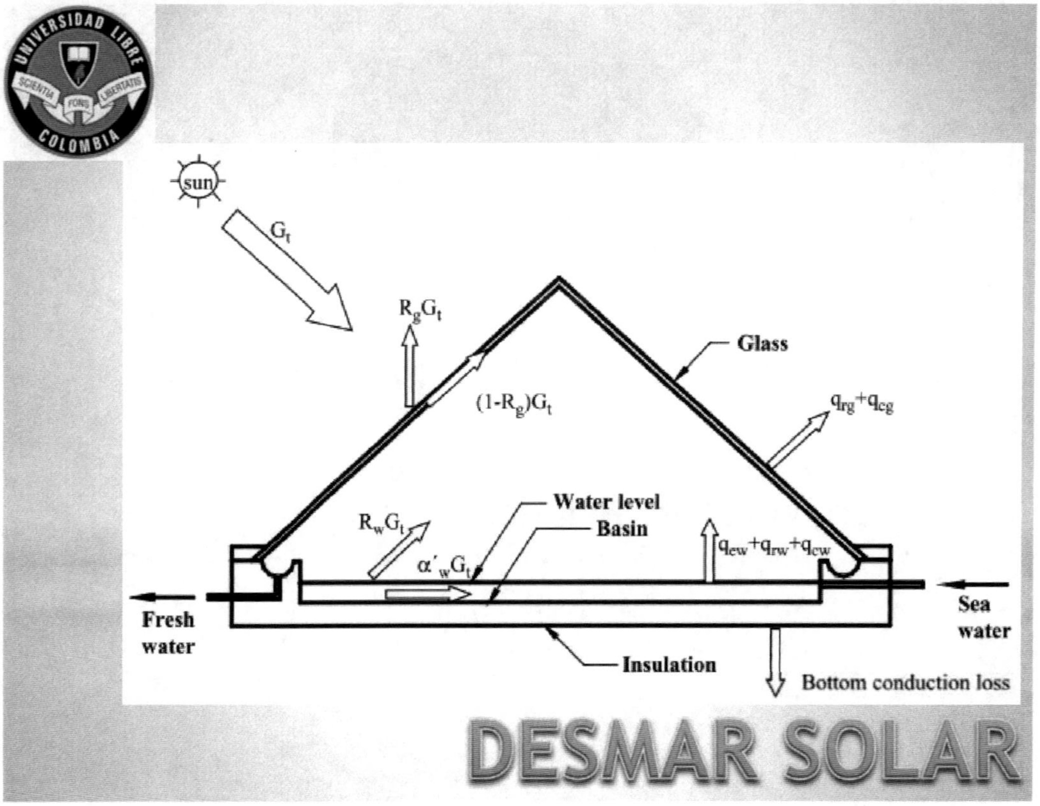
sun
G_t
R_gG_t
Glass
$(1-R_g)G_t$
$q_{rg}+q_{cg}$
Water level
R_wG_t
Basin
$q_{ew}+q_{rw}+q_{cw}$
α'_wG_t
Fresh water
Sea water
Insulation
Bottom conduction loss
DESMAR SOLAR

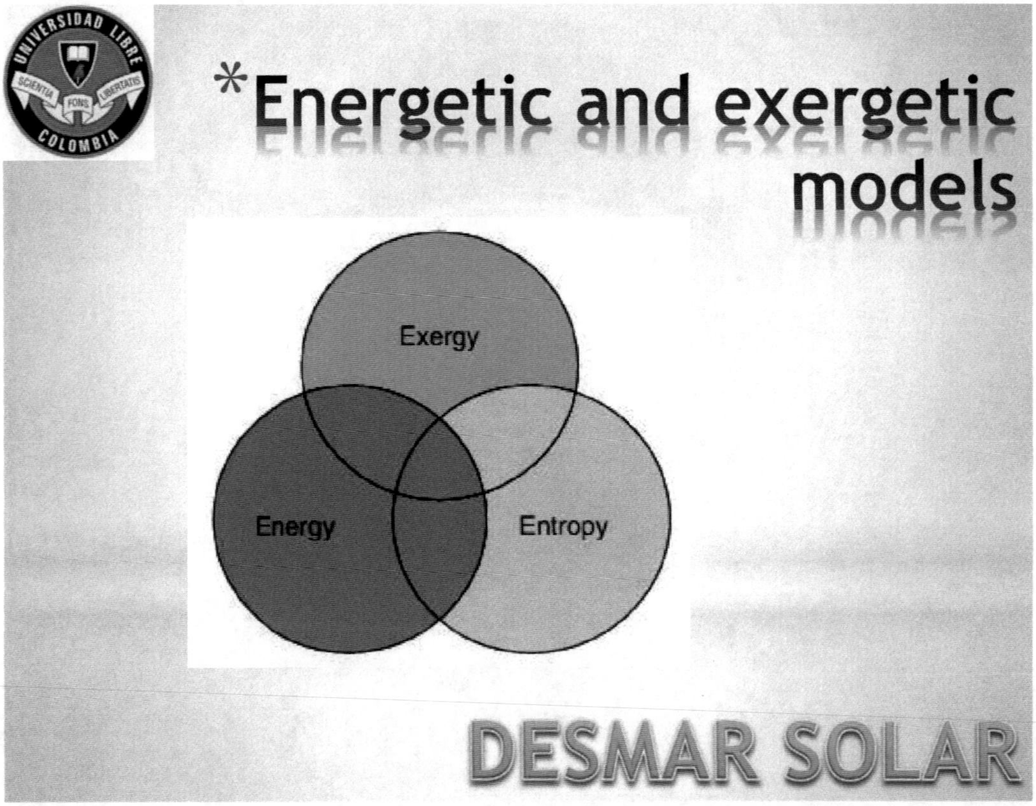
*Energetic and exergetic models
Exergy
Energy
Entropy
DESMAR SOLAR

ENERGY AND EXERGY DEFINITION?

*Energy: The ability that have the matter to produce work in the form of motion, light, heat, etc..

*Exergy: A thermodynamic property that determines the useful work potential of a given amount of energy that can be achieved by the spontaneous interaction between a system and its environment. Reports the system's potential usefulness as a source of labor.

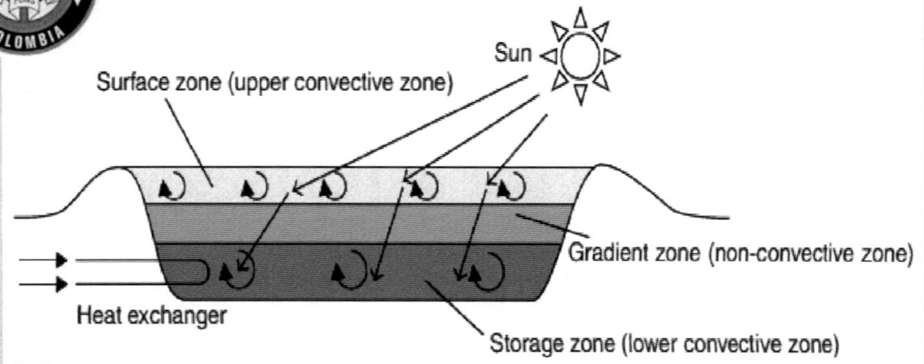

$$n = \frac{m_s h_s + m_w h_w}{G_t}$$

*EXERGY

$$Q_{net} = Q_{HSZ,solar} - Q_{bottom} - Q_{up} - Q_{side}$$

$$n_{HSZ} = 1 - \frac{(Q_{bottom} + Q_{up} + Q_{side})}{Q_{HSZ,solar}}$$

DESMAR SOLAR

*ENTROPY

$$\begin{aligned} s &= x_s s_s + x_w s_w \\ &= x_s [s_{s,\mathrm{pure}}(T,P) - R\ln(x_s)] + x_w [s_{w,\mathrm{pure}}(T,P) - R\ln(x_{\mathbf{w}})] \\ &= x_x s_{s,\mathrm{pure}}(T,P) - R[x_s \ln(x_s) + x_w \ln(x_w)] \end{aligned}$$

DESMAR SOLAR

*EXPERIMENTS

*EXPERIMENT 1.

Distillation through heating plate

DESMAR SOLAR

*HEATING DATA

	TEMPERATURE (°C)	TIME (min)
Evaporated	67	02:46
Finished	91	43:33

FLOW COOLING

Measure	Volume (mL)	Time(s)	Cooling flow(mL/seg)	Cooling flow(L/min)
1	1000	25,57	39,1	2,35
2	1000	25,86	38,7	2,32
3	1000	25,26	39,6	2,38
	PROMEDIO	25,56	39,1	2,35

DESMAR SOLAR

RESULTS

Solution	Temperature (°C)	Volume (mL)	ppm
Initial	20	200	$90x10^3$
Final	20	128	$148x10^3$
Desalinated	20,4	68	0

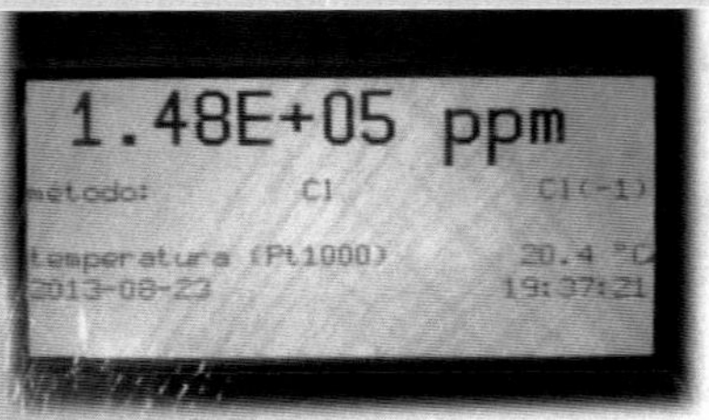

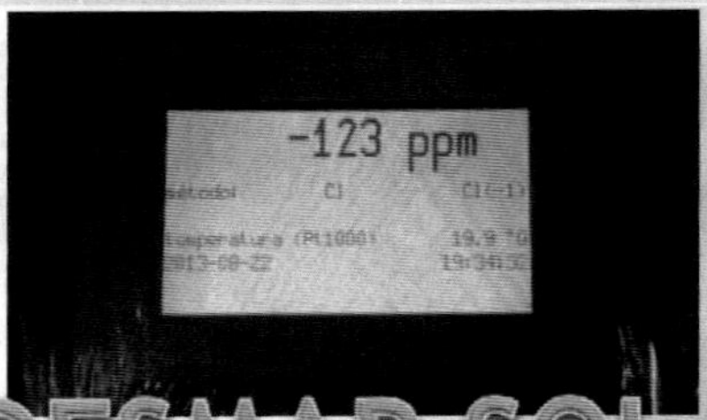

DESMAR SOLAR

BALANCE RESULTS

1

Matter balance		
Input	Output	
200 ml	128 mL	68 mL
90×10^3	148×10^3	0

me = m1+ m2	$90x10^{3} * 0{,}200 = 0 * 0{,}068 + 0{,}128$ *ppm output
ppm output=	140×10^{3}

PERCENTAGE OF ERROR

$$\frac{ppm_{plancha} ppm_{\ salida}}{ppm_{plancha}} * 100 \qquad \frac{148000 - 140000}{148000} * 100 = 5.40\%$$

DESMAR SOLAR

2

Energy Balance		
	w*t	
Input	$2{,}8 \times 10^{7}$	W * seg
	7,8	kWh

$$\rho v C_{H2O} \Delta T = m \Delta H + m C_{H2O} \Delta T$$

ρ	1037,5	kg/m^3
v	0,0002	m^3
C H2O	4184	kJ/(seg * K)
ΔT	70,68	K
m	0,2075	kg
ΔH	173	kJ/(kg * K)
61362	=	61398
	36 J	

DESMAR SOLAR

EXPERIMENT 2

Distillation through incandescent bulbs

DESMAR SOLAR

HEATING DATA SAMPLE

	TEMPERATURE (°C)	TIME(min)
Evaporated	38	46
Finished	90	380

DESMAR SOLAR

BALANCE RESULT

Matter balance		
Input	Output	
200 ml	148mL	52 mL
90×10^3	115×10^3	0

me = m1+ m2	$90 \times 10^3 * 0{,}200 = 0 * 0{,}052 + 0{,}148$ *ppm output
ppm output=	$121{,}62 \times 10^3$

PERCENTAGE OF ERROR

$$\frac{ppm_{plancha} ppm_{\ salida}}{ppm_{plancha}} * 100 \qquad \left|\frac{115000 - 121620}{115000}\right| * 100 = 5{,}76\%$$

DESMAR SOLAR

RESULTS

Solution	Temperature (°C)	Volume (mL)	ppm
Initial	20	200	$90x10^3$
Final	20	148	$121,62x10^3$
Desalinated	54	52	0

DESMAR SOLAR

*MATERIALS FOR THE DESALINATOR

DESMAR SOLAR

*TYPES OF MATERIALS

*Metallic Materials

*Ceramic Materials

*Plastics Materials

DESMAR SOLAR

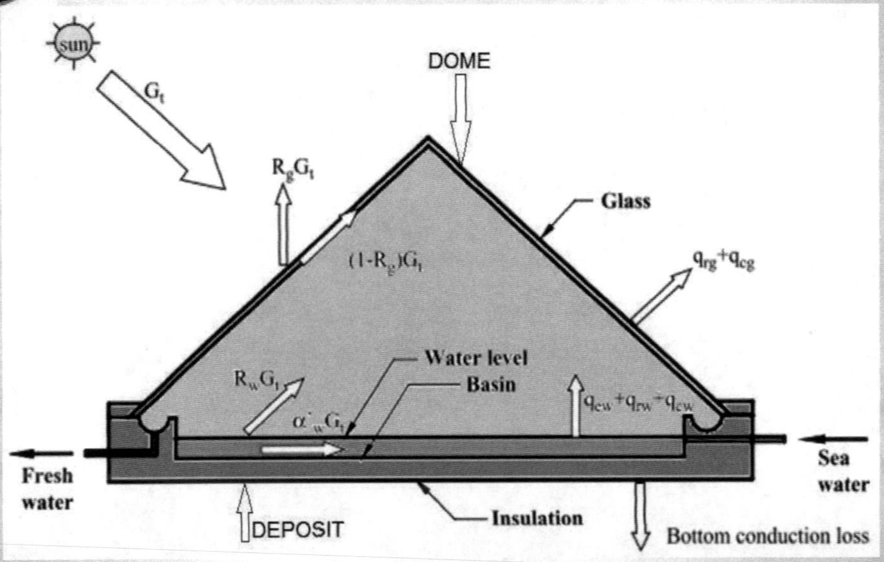

DESMAR SOLAR

UNIVERSIDAD LIBRE
SCIENTIA FONS LIBERTATIS
COLOMBIA
*MATERIALS FOR THE DOME
DESMAR SOLAR

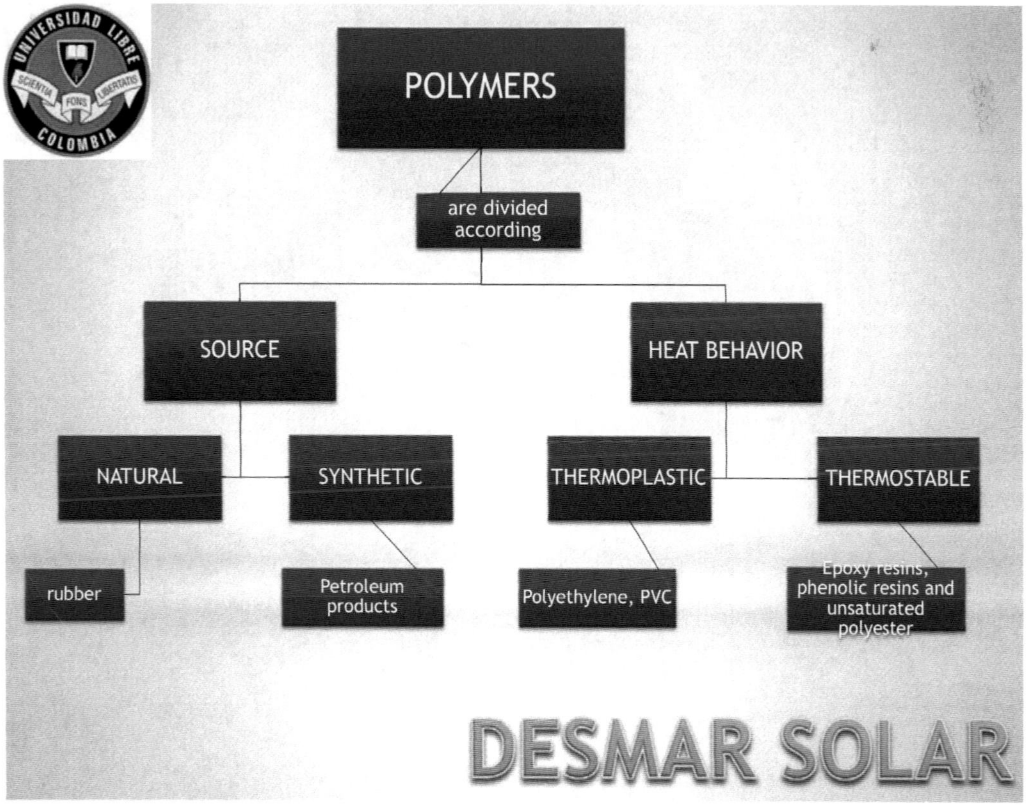
UNIVERSIDAD LIBRE
SCIENTIA FONS LIBERTATIS
COLOMBIA
POLYMERS
are divided according
SOURCE
HEAT BEHAVIOR
NATURAL
SYNTHETIC
THERMOPLASTIC
THERMOSTABLE
rubber
Petroleum products
Polyethylene, PVC
Epoxy resins, phenolic resins and unsaturated polyester
DESMAR SOLAR

*PROPERTIES

*__Elongation:__ Quantity that measures the increase in length of a material

*__Thermal conductivity:__ Ability of a material to conduct heat

*__Specific heat:__ The heat to be supplied to the unit mass of a substance to raise its temperature by one unit

*__Coefficient of thermal expansion:__ Coefficient that express the length change when a body undergoes a temperature change

DESMAR SOLAR

	•Polietileno de baja densidad (LDPE)	•Polietileno de alta densidad (HDPE)	•Polipropileno (PP)	•Cloruro de polivinilo rigido(pvc)	•Cloruro de polivinilo plastico(pvc)
Elongation [%]	90-800	50-1000	200-700	002-40	200-450
Thermal conductivity [kw/mk]	3,35	4,6-5,2	1,17	1,3-2,9	1,3-1,7
Specific heat [kj/kgk]	2,3	2,3	1,9	0,84-1,17	1,26-2,09
Coefficient of linear thermal expansion [K^-1]	1e-4 - 2e-4	1,1e-4 - 1,3e-4	5,8-10,2	5-18,5	7-25
Continuous use temperature [K]	355,37-373,15	394,26	394,26-433,15	338,71-352,59	338,71-352,59
Deflection temperature [K]	100-121	140-190	200-230	135-180	....
Solvents	N/a	N/a	N/a	Cetonas, ésteres,aromaticos,hidrocarburos	Cetonas, ésteres,aromaticos,hidrocarburos
Elasticity in flexure limit [pa]	....	6894,76	41368,56-55158,08	68947,6-110316,16	N/a

*APPROPRIATE MATERIALS

- *Low density polyethylene (LDPE)
- *High density polyethylene (HDPE)
- *Polypropylene (PP)
- *Polyvinyl chloride (PVC)

DESMAR SOLAR

DEPOSIT MATERIALS

DESMAR SOLAR

CERAMIC MATERIALS

* Nonmetallic inorganic solids produced by heat treatment.
* Compared with metals and plastics are hard, non-combustible and non-rusting.
* They can be used in environments with high temperature tribological and corrosive.

DESMAR SOLAR

CERAMIC´S PROPERTIES

* High melting temperature.
* Low density.
* Strength, hardness, structural strength.
* Corrosion resistance.
* Good thermal and electrical insulators.

DESMAR SOLAR

ENVIRONMENTAL CONSIDERATIONS

- By using solar energy production CO2 is removed.?
- The use and advancement of renewable energy is implemented. (In our case solar energy is used).
- Drinkable water in coastal areas.

DESMAR SOLAR

*BIBLIOGRAPHY

- Robert Cook (1986). Computer Simulation of Stress-Strain Behavior in Polymeric Materials. MRS Proceedings, 79, 409 doi:10.1557/PROC-79-409.
- David L. Edwards and Mircea Chipara (2004). Polymeric materials for Solar Sail: The combined effects of polymer thickness, radiation, and temperature. MRS Proceedings, 851, NN1.6 doi:10.1557/PROC-851-NN1.6.
- Alice B. Paterakis (1992). The Consolidation and Desalination of Ceramic Impregnated with Sodium Sulphate. MRS Proceedings, 267, 1071 doi:10.1557/PROC-267-1071.
- T. Y. Wu and D. L. Questad (1992). Mechanical Characterization and Constitutive Description of Thin-Film Polymeric Materials. MRS Proceedings, 264, 143 doi:10.1557/PROC-264-143.(Jr., n.d.)
- Jr., W. D. C. (n.d.). *Materials Science Engineering, An Introduction.* (third edit., p. 785).

DESMAR SOLAR

➢Suzana P. Nunes, Ali Reza Behzad and Klaus-Viktor Peinemann (2013). Self-assembled block copolymer membranes: From basic research to large-scale manufacturing. Journal of Materials Research, 28, pp 2661-2665. doi:10.1557/jmr.2013.253.

➢Paterakis, A. B. (1992). The Consolidation and Desalination of Ceramic Impregnated with Sodium Sulphate. *MRS Proceedings, 267.*

➢Paul C. Painter, M. M. C. (1996). *Fundamentos de ciencia de polimeros: un texto introductorio* (p. 467). Cambridge University Press.

➢Zalba, B. (2003). Review on thermal energy storage with phase change: materials, heat transfer analysis and applications. *Applied Thermal Engineering, 23*(3), 251–283. doi:10.1016/S1359-4311(02)00192-8

DESMAR SOLAR

352632 A Qualitative in Field Test to Characterize Oil-Shale Reservoirs

Abstract:

Shale oil recovery, despite its extraction costs and low yield per shale matrix is becoming a more feasible solution to the rapidly emerging global fuel demands. However, many oil-shale reserves are located several thousand feet below the earth's surface, which makes it extremely difficult and expensive to extract. In addition, the quality of the shale matrix may have a broad range such that the potential yield through and other methods may vary significantly across neighboring reservoirs. And determining the quality of the oil-shale matrix and its potential yield of synthetic/tight oil using laboratory analytical methods can take several weeks. This work presents a rapid and simple qualitative "in-field" test-tube test method that categorizes and prioritizes a given shale matrix based upon its composite. This test method can be a significant cost benefit and a proof positive indicator of where to focus exploration and recovery interest.

Pressure Regulation and Kick Attenuation with Wired Drill Pipe Technology in Managed Pressure Drilling

Hector Perez
Reza Asgharzadeh Shishavan
John Hedengren
Brigham Young University

Outline

- Overview
- Motivation for Pressure Control
- Method and Structure of the Control System
- Results
- Results of Automated BHA Pressure

Motivation for Pressure Control

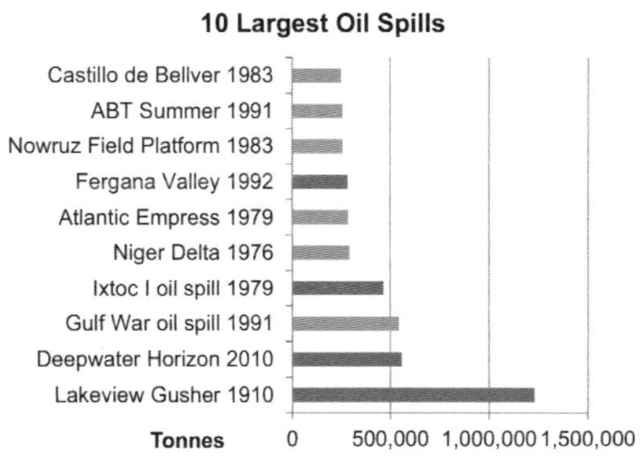

Motivation for Drilling Control

Unsafe Operating Region

Well Fracture or Formation Damage

High BHP limit

Safe Operating Region

Low BHP limit

Reservoir Pore Pressure

Unsafe Operating Region

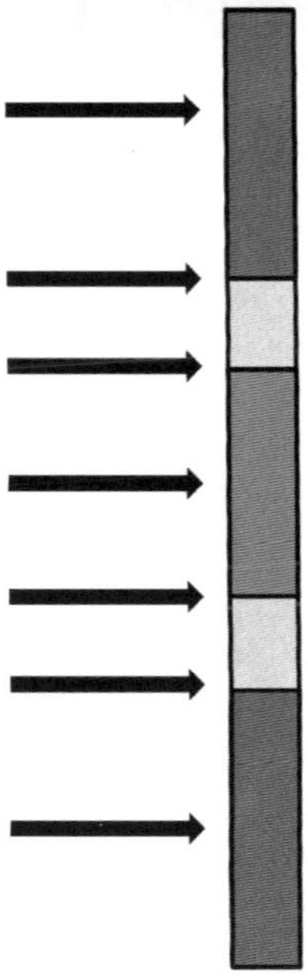

BHP Control Objectives

Pressure Regulation

Kick Attenuation

Managed Pressure Drilling System

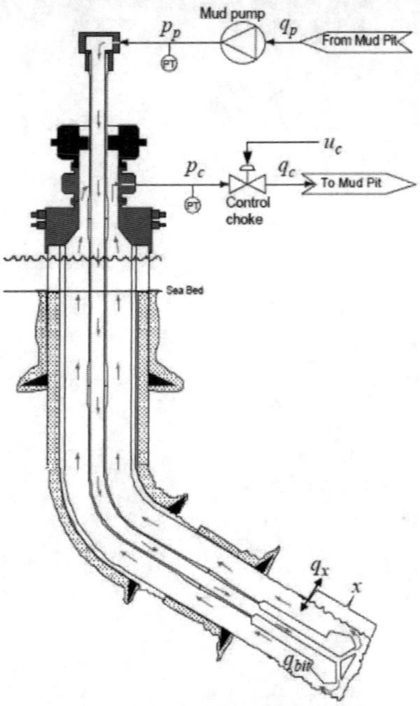

- ❑ Overbalanced Drilling
- ❑ Managed Pressure Drilling (MPD)
- ❑ Underbalanced Drilling

MPD offers:

- ❑ More precisely control of borehole pressure
- ❑ Decrease NPD and mitigate drilling hazards

Manual Control versus Automated Control

Challenges

- Large Time Delays
- Complex Behavior of Multiphase flow
- Sudden Change in Formation Properties

Benefits of BHP control Automation

- Improve Reliability of Pressure control
- Reduce the possibility of human errors

Objective

Designing a comprehensive controller which Simultaneously

- Regulates the BHA Pressure within specified range
- Attenuates Unwanted gas influx when it happens
- Uses latest automation technologies
- Has improved performance in comparison with manual and conventional controllers

Wired Pipe Technology to be used in Control Scheme

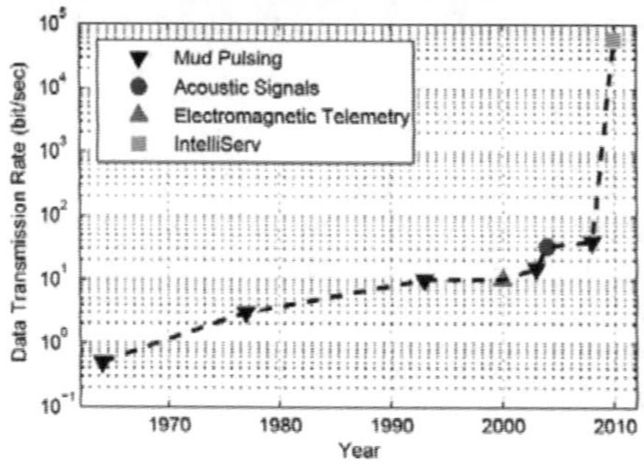

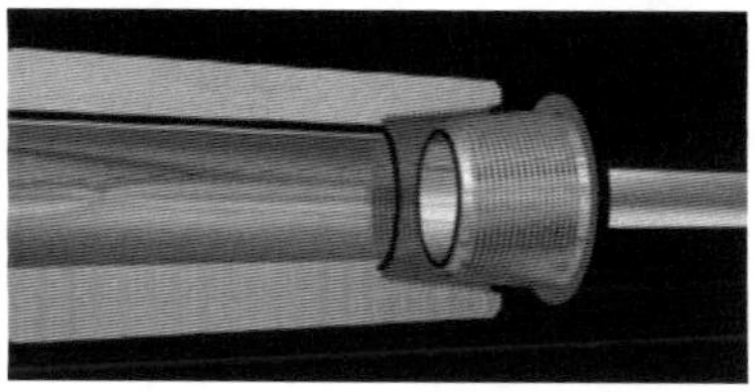

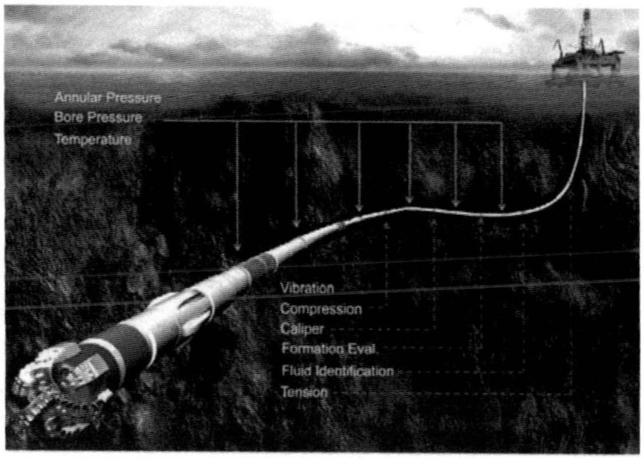

Challenges in the designed controllers

Mud Pulse Telemetry Based

- ❑ Indirect control of BHP
- ❑ High uncertainty in estimated variables values
- ❑ Still Large time delay in kick detection

Wired Pipe Telemetry Based

- ❑ Directly control of BHP and annular pressures ✔
- ❑ Increased precision in estimating the unmeasured process variables ✔
- ❑ Improved kick detection ✔

Schematic of the MPD Controller

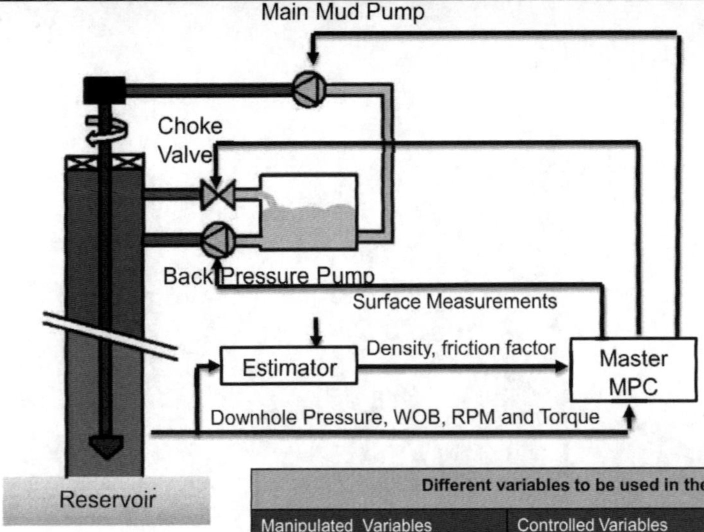

Different variables to be used in the current research		
Manipulated Variables	Controlled Variables	Estimated Variables
Rig Pump Flow Rate	BHP	Annulus Drill Fluid Density
Choke Valve Opening	Annular Pressures	Annulus Friction Factor
		Reservoir Pore Pressure

Controller Structure

Multivariate Nonlinear Model Predictive Controller (MPC)

What are advantages over conventional controllers?

Nonlinear Feature of the Controller

Multivariate Nonlinear Model Predictive Controller (MPC)

- Drilling is a non linear process
- Current controllers use linear (simplified) models
- A nonlinear controller can capture the dynamics better
- Designing nonlinear controllers are more challenging

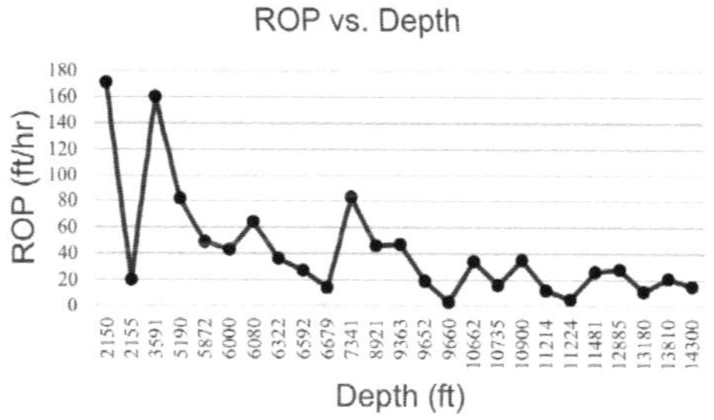

Figure 1. Change of ROP with drilling depth. Field data from (Kinabalu East-1, 1990)

Model Predictive Control and Moving Horizon Estimator

Controller

Multivariate Nonlinear Model Predictive Controller (MPC)

- Future prediction capability – Improved control performance
- MPC vs. Conventional Controllers

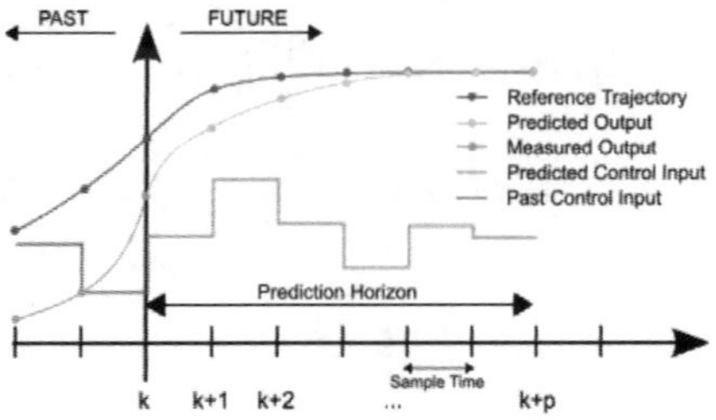

Estimator

- Moving Horizon Estimator
- Uses both downhole and surface measurements
- More precise and faster parameter and state estimation

Software

- APMonitor Optimization Suite
- Designed Especially for Dynamic Nonlinear Processes
- Using the Most Efficient Solvers Available
- Solution Time: Fraction of a Second

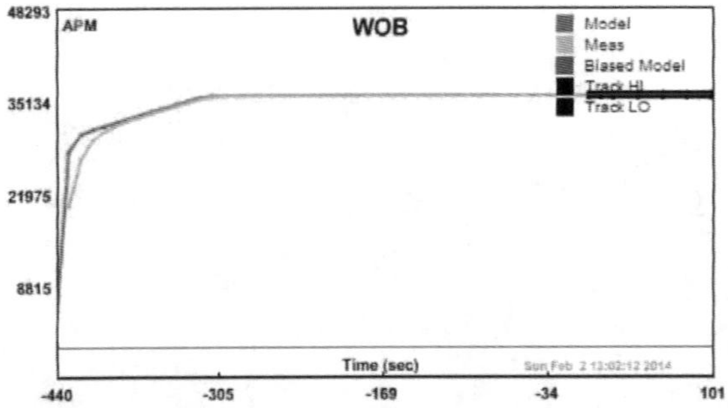

Lower Order Model

$$\dot{p}_p = \frac{\beta_d}{V_d}(q_{pump} - q_{bit})$$

$$\dot{p}_c = \frac{\beta_a}{V_a}(q_{bit} + q_{back} - q_{choke} + q_{res})$$

$$\dot{q}_{bit} = \frac{1}{M}(p_p - F_d|q_{bit}|q_{bit} + \rho_d g h_{bit} - p_{bit})$$

$$\dot{\mathbf{h}}_{bit} = \boldsymbol{ROP}$$

Balance Equations

$$M_a = \rho_a \int_0^{l_w} \frac{1}{A_a(x)} dx$$

$$M_d = \rho_d \int_0^{L_{dN}} \frac{1}{A_d(x)} dx$$

$$M = M_a + M_d$$

$$p_{bit} = p_c + \rho_a \times F_a|q_{bit} + q_{res}|(q_{bit} + q_{res})h + \rho_a g(h_{bit} - (n-1)h)$$

$$q_{choke} = K_c z_{choke}\sqrt{\rho_{an}(p_c - p_0)}$$

Closure Relations

Zhou, Kaasa, et al. (2008

Kick Detection and Attenuation

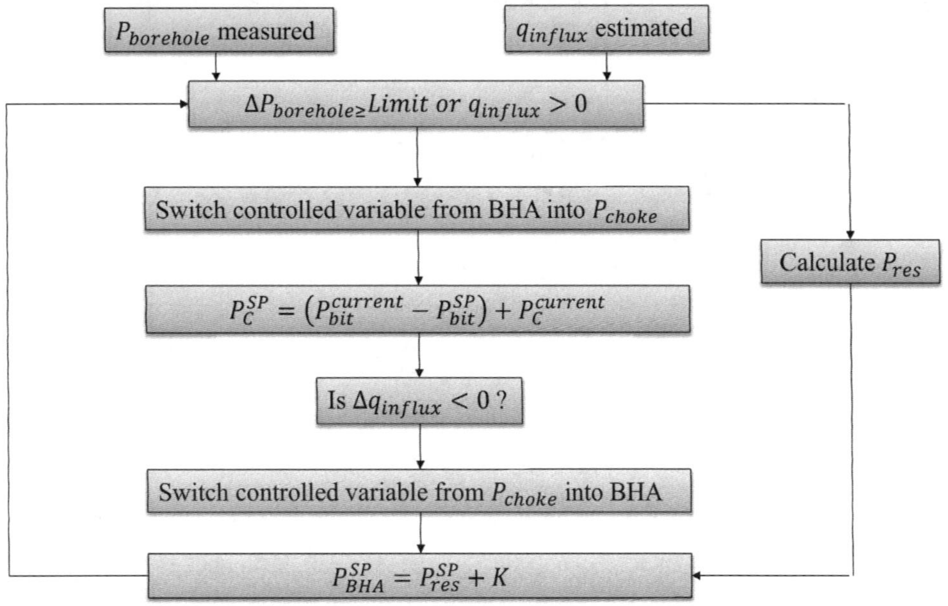

Results

Apr 2, 2014 Pressure Regulation and Kick Attenuation with Wired Drillpipe in MPD

Automated BHA pressure control

Table 1. Initial well conditions used for simulations

Variable	Value (SI units)	Value (US units)
Distance between two successive pressure sensors h_a	150 m	492.1 ft
ROP	0.01 m/s	0.033 ft/s
Fluid density in the drill string - ρ_d	0.012 x 10^{-5} kg/m^3	74.9 lb/ft^3
Fluid density in the annulus - ρ_a	0.014^a x 10^{-5} kg/m^3	87.4 lb/ft^3
Bulk modulus in the annulus β_d	14000 bar	2.03 x 10^5 psi
Bulk modulus in the annulus β_a	14000 bar	2.03 x 10^5 psi
Annulus cross sectional area	0.021 m^2	0.226 ft^2
Friction factor in the drill string F_d	10000 bar m^{-6} s^2	116.3 psi ft^{-6} s^2
Friction factor in the annulus F_a	1250^a m^{-5}	3.29 ft^{-5}
Lumped density per length parameter for the drill string - M_d	5700 x 10^{-5} kg/m^4	108.5 x 10^{-5} lb/ft^4
Lumped density per length parameter for the annulus – M_a	800 x 10^{-5} kg/m^4	15.22 x 10^{-5} lb/ft^4
Well depth (h_{bit})	2000 m	6562 ft
^{a}For connection procedures, the values used for simulation were 0.01 (density) and 1000 (friction factor)		

Apr 2, 2014 Pressure Regulation and Kick Attenuation with Wired Drillpipe in MPD

MHE Estimator

Within 30 seconds of operation, the estimator has accurately determined the friction factor and has determined the annulus density within 1.5% of the actual value.

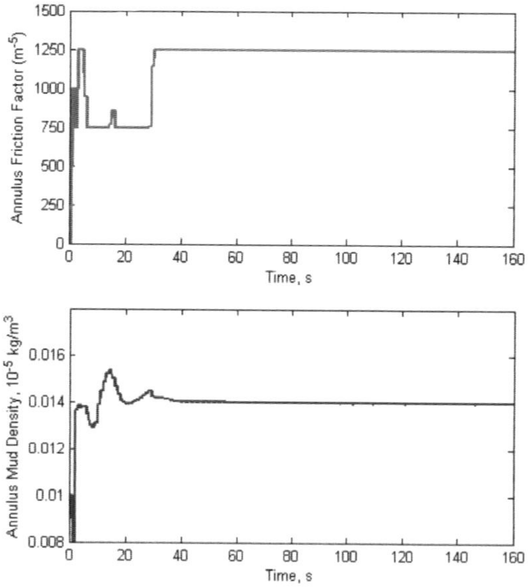

Apr 2, 2014 Pressure Regulation and Kick Attenuation with Wired Drillpipe in MPD

Case 1: Normal Operation

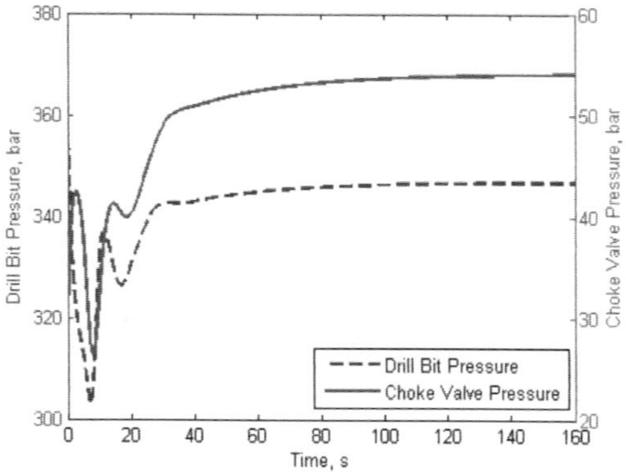

Within 30 seconds, the drill bit pressure has been controlled within 0.72% of its set point of 345 bar

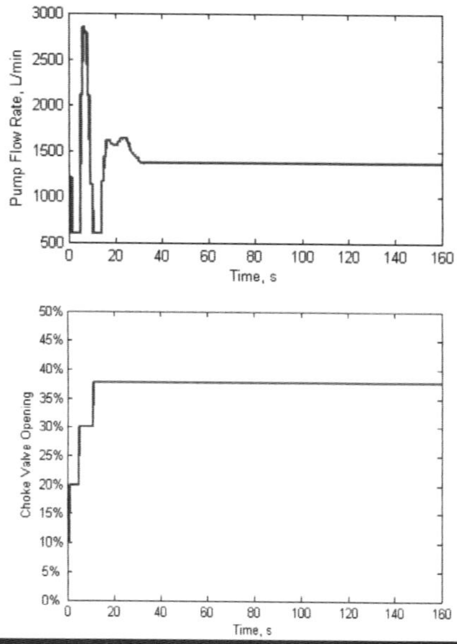

Apr 2, 2014 Pressure Regulation and Kick Attenuation with Wired Drillpipe in MPD

Case 2: Connection Procedure

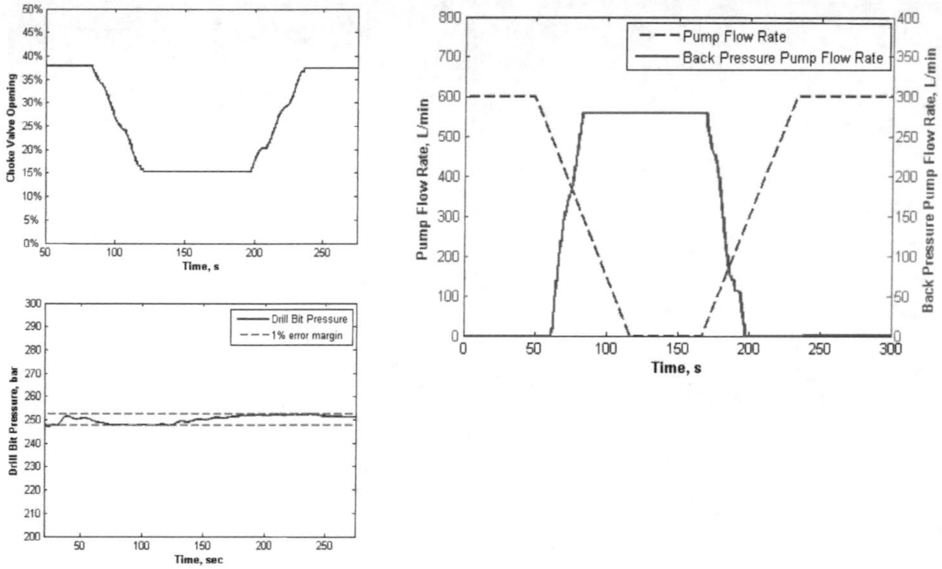

Apr 2, 2014 Pressure Regulation and Kick Attenuation with Wired Drillpipe in MPD

Case 3: Unwanted Gas Influx

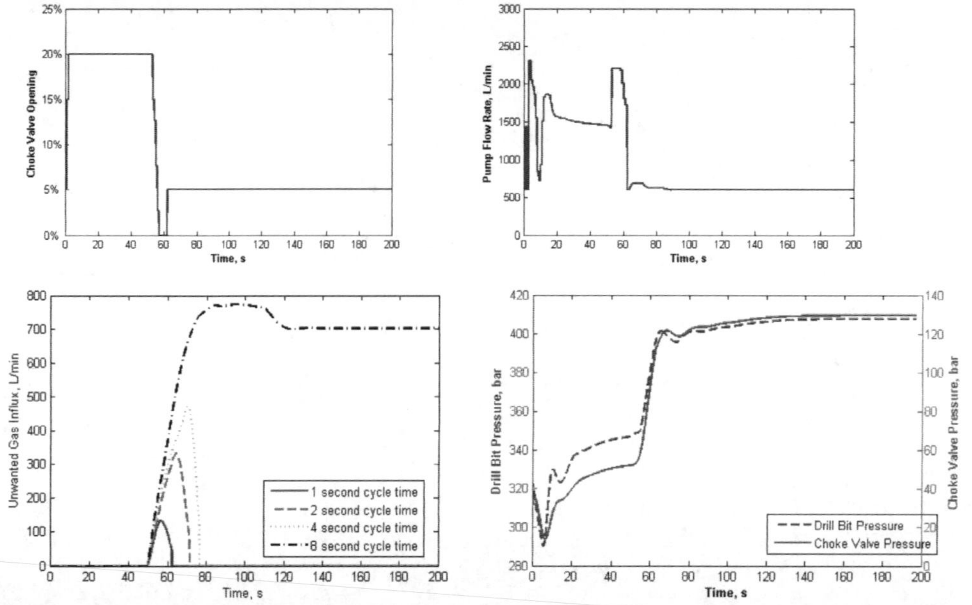

Apr 2, 2014 Pressure Regulation and Kick Attenuation with Wired Drillpipe in MPD

Thank You for Your Attention

14AIChE®
Spring Meeting, New Orleans, LA

Risks from Unconventional Wells Gasification

Dennis B. Butts
Americas Risk Practice
Environmental Resources Management (ERM)
CityCentre Four
840 West Sam Houston Parkway North, Suite 600, Houston, TX 77024
dennis.butts@erm.com

Prepared for Presentation at
American Institute of Chemical Engineers
2014 Spring Meeting
Topical 2: Shale Gas and Tight Oil
New Orleans, LA
March 30 – April 2, 2014

UNPUBLISHED

Risks from Unconventional Wells Gasification

Dennis B. Butts
Americas Risk Practice
Environmental Resources Management (ERM)
CityCentre Four
840 West Sam Houston Parkway North, Suite 600, Houston, TX 77024
dennis.butts@erm.com

Keywords: LNG, Risk Management, Process Safety, Unconventional Wells, Shale

Abstract

With the continued differential between diesel and natural gas costs, onshore well operators have been working to convert both drilling rigs and hydraulic fracturing pumps to utilize natural gas as fuel. Associated with this conversion is a change in the risk profile that may not be fully recognized or understood in the field. Risk management philosophies and assumptions need to adjust to the difference in hazard characteristics between the fuels, as well as the variety of delivery methods.

Specific topics will include:

- Motivations for fuel conversion and the differences in scale for rig power, natural gas substitution on diesel hydraulic fracturing pumps, and dedicated natural gas powered hydraulic fracturing pumps. Risk mitigation and field training requirements may not fully accounted for in investment decisions.
- Scale of conversion and location specific availability drives natural gas sourcing, which can then drive personnel risk and well pad layout issues depending on whether the gas is sourced from local wellheads, pipeline, Compressed Natural Gas (CNG), or Liquefied Natural Gas (LNG). Separate Simultaneous Operations (SIMOPS) issues can arise based on the specifics of each sourcing method.
- Associated risks need to be appropriately managed in a manner that involves the personnel directly exposed. And there is a need to ensure that well sites which did not previously present a threat to surrounding activities do not become a threat due to either direct risk from the fuel conversion itself or from escalation risks not previously of concern.

1. Introduction

Over the past few years there has been a significant increase in efforts to substitute natural gas for diesel to provide fuel for drilling and fracturing onshore wells. This is an expansion of the conversion efforts that were already underway to convert transportation fleets away from diesel fuel. While the reasons for this may vary, and do include environmental benefits, the most significant reason is cost savings.

Natural gas has been averaging about $4.50/mmbtu [1] over the past 6 months, which is about $0.59 per diesel gallon equivalent (DGE) [1]. Even with added distribution costs, if pipeline gas is available, the economics are very attractive. Wellhead gas is even more attractive. And if LNG is the only alternative at a well site, at $4.50/mmbtu, the natural gas component of LNG is about $0.34 per gallon [1], with a markup of about $0.15 per gallon for liquefaction and storage [3] plus transport costs of about $0.15 per gallon (up to 100 miles) [4], giving a breakeven price of $0.64 per gallon or $1.10 per DGE. With typical supplier pricing to bulk customers as a discount to diesel typically around $1.50 per gallon, and more substantial discounts for large scale contracts, there are significant financial benefits for both LNG suppliers and drillers when diesel fuel is in the $4 per gallon range.

And as a bonus to the economic justification, drillers can tout emissions reductions of as much as 20 – 30% relative to diesel [5]. Rarely in these discussions is the subject of added costs or effort for process and technical safety brought up.

Substitution of natural gas for diesel has been underway for some time in heavy vehicle transportation fleets. With added competition in engine conversion markets and among storage tank suppliers, conversion costs have been reduced to the point where heavy vehicles can recover costs in less than 2 years [6]. Efforts to convert drill rigs originated in the Marcellus region, due to environmental concerns, but the economics have driven natural gas rigs into most of the major markets in the US, as well as into western Canada. Since hydraulic fracturing pumps use even more diesel than rigs, recent efforts have been in this direction, though technical challenges regarding power delivery have resulted in two types of conversion. One is the substitution of gas for diesel in existing engine drivers to ensure sufficient horsepower. The other is to use high powered turbine pumps, which are very well suited to gas power. Large scale conversion of fuel sources is not unprecedented, as the US heavy truck fleet went from less than 10% diesel powered to nearly 100% diesel powered from 1950 to 1980 [6].

So what does this mean in terms of risk management? Aside from the business and financial risk of converting equipment to a fuel in natural gas that is historically more volatile in pricing than diesel fuel at a time of historically high oil to gas price differentials, which is a subject for another paper, there are practical technical safety considerations that come into play as drilling sites convert from diesel to natural gas. These can result in increases to the risk profile of a well pad that is not generally accounted for, as well as changes in the risk profile within the drilling operation itself.

This paper will discuss some of the issues and options regarding conversion and highlight specific risk issues associated with each.

2. Sources of Natural Gas

2.1 Typical Volume Requirements

A heavy truck may use 20,000 gallons of diesel per year [4], with significant potential savings being the result of a conversion to natural gas fuel. Conventional fluid drill rigs use an average of 425,000 gallons of diesel per year [5], creating a fuel cost savings equivalent to converting a fleet of over 20 heavy trucks, but at an investment cost that is much lower. But even these volumes pale in comparison to the amounts of diesel used in hydraulic fracturing. Fracking crews can use 5,000 to 10,000 gallons of diesel per day [7], and it is common for the crews to be in operation for as many as 300 days per year, accounting for mobilization and demobilization time. That means 1.5 to 3 million gallons of diesel per year can easily be used by a single crew, often going well to well for the same operator. An unfortunate outcome of this is that conversions become a foregone conclusion, without due consideration to the resulting process safety risks.

2.2 Options for Sourcing and Delivery

Natural gas for a well site can be sourced as gas directly from the wellhead or gathering system, or it can be piped in from a commercial distribution line, when those are available. In more remote locations, distribution lines may not be available, and if the gas is too wet, wellhead or gathering gas may also be unsuitable for use. In those situations the options are to utilize either CNG or LNG. CNG would require considerably more transportation than LNG due to the lesser amount of gas that can be moved and comes with higher costs for storage and handling equipment. Therefore, CNG has limited utility for well sites, though its lower fuel cost makes it useful in some narrow applications. The use of LNG can be cost effective when other options are not available, as it is more available than in the past and although containing less energy than diesel per gallon, an onsite tank containing 15,000 gallons of LNG is still equivalent to over 8600 gallons of diesel [2] and is about 2.4 times the energy density of CNG [9].

2.3 Issues Associated with Each Option

Use of wellhead gas or gas from gathering lines requires either very dry gas or extensive treatment, as most engines converted to utilize natural gas are designed for gas approximating domestic gas. This is in contrast to offshore facilities where engines are often optimized to take advantage of fuel gas diverted from separation trains. Pipeline gas can be the best option, for those sites with ready access to a distribution line, and will provide sufficient flow and pressure for most natural gas drilling rigs. However, pipeline gas may have to have compression added for certain types of fracking pumps, and available flow may not be sufficient for some fracking operations. CNG can be useful in areas that do not have dry gas or ready pipeline access and may not have easy access to LNG. In such areas, CNG facilities can be set up to provide the product to users in the surrounding region much more readily than LNG facilities. To utilize CNG does require more transportation of the gas and more onsite storage than using LNG. With LNG being the last supply option, but becoming very common in the field, this is an option that carries significant process and technical safety risks. LNG is a material than can bring many benefits, but needs to be treated with considerable respect, often more than is witnessed at well sites.

3. Associated Risks

Associated process and technical safety risks arise in several stages of the conversion from diesel to natural gas. In most cases there is an increase in those risk levels relative to diesel fuel, though this is not always fully contemplated by decision makers or well communicated to the field level staff. Whereas diesel fuel is classified as a combustible liquid, NFPA Class II [8], is difficult to ignite if spilled and primarily a localized threat, natural gas is highly flammable vapor and depending on delivery method could have the potential to disperse considerable distances if released.

In use in equipment, natural gas can create the need for hazardous area class or hot work permit zones that were not required around diesel equipment. This issue has been missed on well sites and often is not included in discussions about conversion costs until after being identified in PHA studies or other risk assessments. The same consideration extends to gas monitoring and shutdown systems. System risks are manageable, but need to be given due consideration and treated with the proper respect. Some expansion of this thought is provided below.

3.1 Source of Gas

Wellhead gas primarily introduces risk through the separation and dehydration process. Even relatively dry gas needs to be treated prior to use in most of the engine drivers used for onshore production. To utilize it during drilling usually means that at least one well has already been put into operation on a multi-well pad. The risks associated with production while drilling and completing adjacent wells have to be assessed. Controls that may need to be in place during certain phases of drilling could interfere with the supply of gas. An alternative that may be available within a field of well sites under one operator is to use gas from the gathering system. While treatment would also be required for that gas, the potential SIMOPS issues of a producing well on a pad with drilling activity. In either case the gas being treated will be a relatively high pressure and is likely to result in some amount of condensate or natural gas liquids (NGL) production and handling. This can increase the fire and explosion risks above those normally found during drilling activity.

Pipeline gas is only an option when a distribution line is nearby. But if available, this can provide a supply of natural gas that is already treated and can be adjusted to the necessary pressure at the takeoff location from the distribution network. In analysis of fire and explosion risks, pipeline gas supply most often has the smallest consequence zones and the least amount of equipment susceptible to failures. This is even the case if high horsepower fracking pumps are in use that requires some degree of compression of the natural gas to reach necessary operating pressures.

CNG introduces risk due to its high pressure, typically near or above 3000 psig [9]. This can both drive the spread of a gas cloud in the event of a release as well as provide the means through depressurization to cool the gas to a low enough temperature that it will disperse for some distance along the ground, further lengthening dispersion distances. Storage under pressure also introduces a finite risk of boiling liquid expanding vapor explosion (BLEVE) that could spread debris.

LNG usage at well sites is capable of imposing a larger potential impact zone than a well blowout scenario. And this is excluding a catastrophic failure of the storage medium, which is low likelihood event due to the dual tank construction of trailers that handle LNG. When released, LNG will initially disperse in a manner very similar to liquefied petroleum gas (LPG), but as the edges of the dispersing cloud warm, flammable vapors will loft upward. Assumptions regarding layout and spacing that are based on LPG can address the flammable breadth of a possible LNG cloud, an example of which is shown in Figure 1 from PHAST [10], going out about 850 feet (260 meters) to the 50% lower flammable limit (LFL). Maximum distance to a 1 psi overpressure, good for window breakage, trailer damage, and upsetting neighbors, would extend another 300+ feet (100 meters) beyond the 50% LFL distance. In general, the height of the 50% LFL section of the cloud will remain under 15 feet (5 meters), but this also needs to be confirmed relative to the drill rig flare height. And the failure of a LNG loading hose would impose an even larger impact zone. Though many operators end up using gas detection, this is only effective if appropriate emergency response procedures are in place.

Figure 1: 50% LFL Dispersion Distance for 1-inch LNG Leak [10]

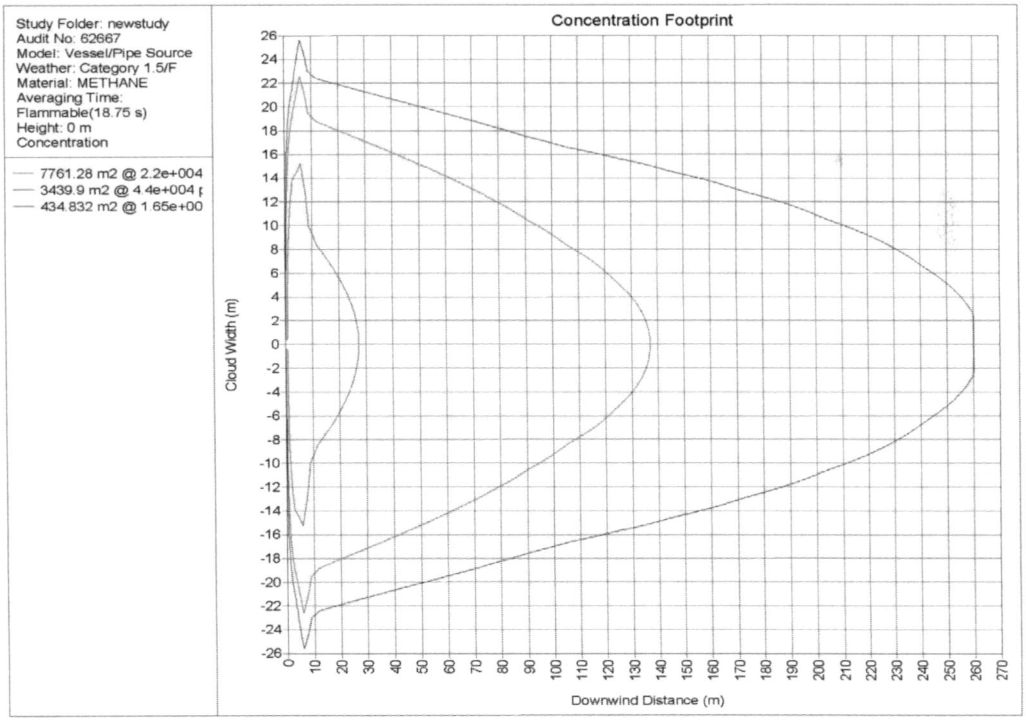

For LNG most releases behave as impinged releases, due to the low pressure environment. For CNG, the best case is for a free field release, which provide dispersion distances of less than half that of LNG. However, horizontal releases of CNG are capable of dispersion distances beyond those of LNG, in particular if they impact other equipment and become partially impinged, as shown for dispersion to 50% LFL for a 1-inch CNG release in **Figure 2**. An interesting phenomenon is how the LNG dispersion slumps outward from the beginning, due to release at temperatures low enough for the methane to be denser than air, but the CNG is initially released at warmer temperatures, then settles into a much lower temperature, resulting in a slumping spread.

Figure 2: 50% LFL Dispersion Distance for 1-inch CNG Horizontal Leak [10]

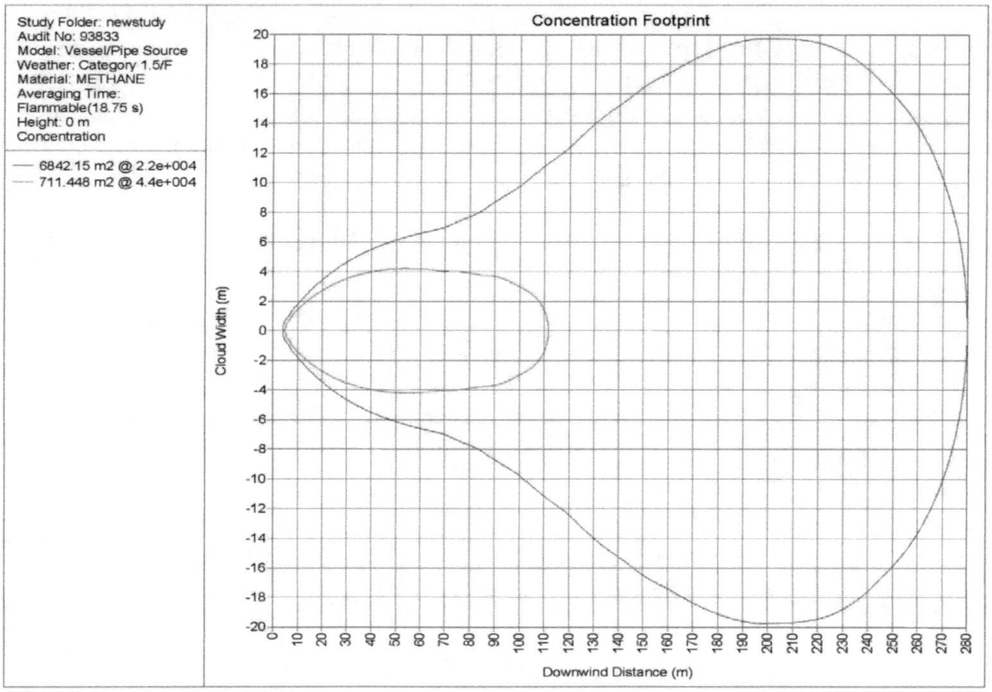

As a reference, an ignited well blowout, traditionally considered a worst case event at a well pad, will typically have an effect distance to 1500 Btu/hr-ft^2 (4.7 kW/m^2) of about 500 feet (150 meters).

3.2 Refueling at Site

When utilizing CNG or LNG as the source of gas, there are additional considerations for storage at the site and for refueling deliveries. Both require delivery by truck, just like diesel fuel, but the release of diesel from an overturned truck is not the same as a release of compressed or liquefied natural gas. This is why there are stringent design requirements on CNG and LNG transport vehicles. However, when assessing risks it is wise to keep in mind that those standards were developed for paved highway transportation, not gravel access roads to well sites. In addition to possible visibility issues, there are the impacts such roads could have on the expected behavior of safety features on the vehicles.

3.3 Compression and Delivery of Gas to Equipment

Further issues arise in situations where either pipeline gas or LNG is used to supply equipment such as hydraulic fracturing pumps that require higher pressure feed gas. In these circumstances, arrangements have to be made to provide compression at the site. A site that is typically already very crowded with equipment. In terms of risk from the equipment it is normally a localized threat, however the extra equipment adds a level of complexity that can complicate operations. The added complexity can add a human factors risk dimension that is rarely assessed in drilling situations.

4. Conclusion

Considerable financial and environmental benefits can be had by converting drilling operations from diesel fuel to natural gas, as long as the current differential in crude oil versus natural gas pricing exists. To achieve those benefits does entail some risk, with this paper having focused on the process and technical safety risks. In some instances, the benefits and risks can pull in opposing directions creating a challenge to find equilibrium between the two.

Depending on the source of natural gas for the drilling and completions operations, due consideration needs to be given to:

- Training needs for field staff that are used to dealing with combustible diesel fuel and not highly flammable natural gas;
- Flammable gas dispersion distances as well as resulting maximum potential zones for 1 psi overpressures, in particular for LNG or CNG applications;
- Review of drill rig flare location relative to natural gas supply point(s) to account for potential height of dispersion;
- Road conditions and their potential impact on transportation of fuels;
- Hazardous area classification zones and resulting equipment alterations, layout changes, or work permit requirements;
- Gas detection needs and the emergency response planning to accompany the detection; and
- Hazard assessment requirements and when they are activated to address changes made to drilling operations or well pad layouts once in service.

As we further explore this question, it is necessary to keep in mind not only the benefits of the ongoing process of converting away from diesel fuel to natural gas at well sites, but also the risk management challenges that can arise.

5. References

[1] US Energy Information Administration, 2014

[2] US Department of Energy, Fuel Energy Spreadsheet, 2014

[3] Quine, Tom, "Fundamentals in LNG", ASGMT White Papers, 2010

[4] Center for Energy Economics, University of Texas at Austin

[5] Oil & Gas Technology, "Gas Powers Fracking Plays", April 10, 2013

[6] Blomerus, Paul and Oulette, Patric, "LNG As a Fuel for Demanding High Horsepower Engine Applications: Technology and Approaches",

[7] Schutt, Bryan, "Diesel to Gas Switching Trend Takes Off in Drilling Sector", SNL Financial, September 3, 2013

[8] NFPA 30, National Fire Protection Association, 2012

[9] Envocare Ltd., "Liquefied Petroleum Gas (LPG), Liquefied Natural Gas (LNG), and Compressed Natural Gas (CNG)", 2007

[10] PHAST 6.7, DNV GL AS

353108 Supercritical CO_2 / Brine / Hydrocarbon Interactions in Shale Rock Fracture: A Geo-Microfluidics Approach

Abstract:

Shale oil and gas is a critical unconventional resource for the energy sector in the US and worldwide. The booming of its production in the recent decade, due to the hydraulic fracturing technique, has led to serious environmental concerns such as huge water demands and toxic additives into fracking water which may pollute groundwater. A proposed solution is to use supercritical CO_2 ($scCO_2$) as a more environment-friendly alternative fracking fluid to replace water for enhanced oil recovery (EOR) in oil shale. Many challenges lie in the way of this scheme such as to understand multiphase flow of $scCO_2$ at injection sites, its interactions with brines, hydrocarbons, and surrounding geologic matrix. We present here a microfluidic approach to investigate the micro-scale multiphase fluid flow and reactive transport within low-permeability shale rock. Unlike the conventional microfluidic techniques, the geo-microfluidic chip is directly fabricated from a thin-section of shale rock, hence provides the genuine surface properties. Simple fracture-like patterns are created on the shale rock surface by machining and femto-second laser etching to simulate the natural structures. Flow experiments in the geo-microfluidic chip are conducted within a high-pressure, high-temperature apparatus to simulate the in situ reservoir conditions. The interactions between $scCO_2$/brine/hydrocarbon will be quantified and the hydrocarbon extraction efficiency of using $scCO_2$ and using fracking water will be compared. The displacement patterns, the dissolution rate of $scCO_2$ in brine/hydrocarbon, and the surface wettability of $scCO_2$ and shale rock will also be investigated. These micro-scale experiments results will help to reveal the underlying mechanisms of multi-phase fluid flow under in situ temperatures and pressures.

352468 Model Shale Gas Properties from a Statistical Mechanical Perturbed Hard-Sphere Chain Cubic Equation of State

Abstract:

Shale gas is an increasingly important energy source in North America. Hence, an accurate description of the physical, transport and thermodynamic properties of shale gas mixtures are a key prerequisite for process design. A particularly advantageous feature is that multiple properties can be obtained from a single equation of state (EOS) for a model system. Traditional EOS's are cubic in molar volume and are derivatives of the basic van der Waals (1873) EOS. Redlich and Kwong (1949), Soave (1973) and Peng and Robinson (1976) are attempts to improve the accuracy of the basic vdW EOS; however, as is shown early in this talk, these equations of state overcorrect the attractive term to compensate for deficiencies in the repulsive term. On the other hand these EOS's have the advantages of a computationally simple form and *three well-defined molar volume roots*. Nonetheless, while these EOS's often give a reasonably accurate representation of the phase composition of simple binary and ternary alkane mixtures (model shale gas systems); poor results are obtained for other properties including liquid density, mixture second virial coefficient and diffusion coefficient.

On the other hand, a class of accurate molecularly-based but complex and computationally intense equations of state: the statistical associating fluid theory (SAFT) equations (Huang and Radosz, 1990; Chapman et. al., 1990), which account for individual molecular level effects (i.e., hard-sphere repulsion, London dispersion, covalent bond chain formation, multipolar forces and hydrogen bonding) have successfully emerged in the past few decades to account for multiple effects in pure fluids and mixtures including associating and/or polydisperse components. These EOS's capture the major molecular level effects; however, convergence difficulties and the existence of several molar volume roots complicate implementation of these EOS routines.

Consequently, considering the above difficulties, we have developed an approach which incorporates the rich molecular description of the SAFT equations into a cubic in molar volume form. Specifically, molecular simulation data for hard-sphere compressibility factor is nearly exactly fit to a simple empirical function in packing fraction, which ensures an overall cubic in molar volume equation of state and from which two universal parameters (applicable to all fluids and fluid mixtures) are obtained. Furthermore, London dispersion attractive forces are described by an extremely simple polynomial refit of data generated for Chen and Kreglewski's (1977) *thirty-six term* double sum in reciprocal absolute temperature and packing fraction. Finally, chain formation (covalent bonding) is described by a modified form of Wertheim's (1984) thermodynamic first-order perturbation theory (TPT1) for segment covalent bonding, in which a diimer reference radial distribution function is employed to obtain more accurate results. Conventional forms of the hard-dimer radial distribution function (Chiew, 1991; Yethiraj and Hall, 1992; Ghonasgi and Chapman, 1994; Sadus, 1999) were refit to a novel empirical functional form which results in an equation of state which is overall cubic in molar volume.

The resulting theoretically-based EOS was applied to predict several model shale gas properties including compressibility factor versus packing fraction isotherms, the vapor-liquid coexistence curves, the second-virial coefficient versus temperature and the self-diffusion coefficient versus pressure isotherms for the constituent pure fluids; and vapor-liquid equilibria and binary diffusion coefficients versus pressure for binary systems. The observed performance of the molecularly-based *cubic* EOS presented in this paper is generally better than the performance of conventional cubic equations of state and about equal to the performance of the more complex non-cubic SAFT equations of state for the same fluids but with much less computational effort. Ongoing work involves development of *cubic* EOS terms which account for multipolar and hydrogen bonding effect to include water and other species in the shale gas mixtures.

REFERENCES

Chapman, W. G., K. E. Gubbins, G. Jackson and M. Radosz, *New Reference Equation of State for Associating Liquids*, Ind. Eng. Chem. Res. **29**, 1709-1721 (1990).

Chen, S. S. and A. Kreglewski, *Applications of the Augment van der Waals Theory of Fluids: I. Pure Fluids*, Berichte der Bunsengesellschaft fur Physikalische Chemie, **81**, 1048-1052 (1977).

Chiew, Y. C., *Percus-Yevick Integral Equation Theory for Athermal Hard-Sphere Chains. II. Average Intermolecular Correlation Functions*. Molec. Phys. **73**, 359-373 (1991).

Ghonasgi, D. and W. G. Chapman, *A New Equation of State for Hard-Chain Molecules*, J. Chem. Phys. **100**, 6633-6639 (1994).

Huang, S. H. and M. Radosz, *Equation of State for Small, Large, Polydisperse and Associating Molecules*, Ind. Eng. Chem. Res., **29**, 2284-2294 (1990).

Peng, D.-Y. and D. B. Robinson, *A New Two-Constant Equation of State*, Ind. Eng. Chem., **15**, 59-64 (1976).

Redlich, O. and J. N. S. Kwong, *On the Thermodynamics of Solutions. V. An Equation of State. Fugacities of Gaseous Solutions*, Chem. Rev. **44**, 233-244 (1949).

Sadus, R. J., *A Simple Equation of State for Hard-Sphere Chains*, AIChE J., **45**, 2454-2457 (1999).

Soave, G. *Equilibrium Constants from a Modified Redlich-Kwong Equation of State*. Chem. Eng. Sci., **27**, 1197-1203 (1972).

van der Waals, J. D., *Over de Continuiteit van den Gas- en Vloeistoftoestand (On the Continuity of the Gas and Liquid State)*. Ph.D. thesis, Leiden, The Netherlands, (1873).

Wertheim, M. S., *Fluids with Highly Directional Attractive Forces I. Statistical Thermodynamics*, J. Stat. Phys., **35**,

19-34 (1984).

Yethiraj, A. and C. K. Hall, *Monte Carlo Simulations and Integral Equation Theory for Microscopic Correlations in Polymeric Fluids*, J. Chem. Phys., **96**, 797-809 (1992).

353985 Resolving Sampling Inaccuracies Through Numerical Study of Liquid-Liquid Vertical Dispersed Flows

Abstract:

In the post-production, downstream processing of heavy oil and bitumen, surface treating facilities often involve vertical flows of mixtures containing droplets of oil, fine solid particles and water. In some cases, ISOLOK™ (piston) samplers are used to withdraw samples from the main flow for composition analysis (i.e. per cent of oil, water and solids in the mixture). The results are used for mass balancing and for process trouble-shooting. Plant operators, on numerous occasions, found that other measurements indicated that the oil composition in these samples was lower than expected, but there was no way to verify this with field measurements. As a result, numerical simulations of liquid-liquid dispersed flow in a vertical pipe were carried out using the two-fluid approach implemented in ANSYS CFX. A dispersion of oil in water (where water is the continuous phase) was studied. A comparison of the present numerical results with previous experimental results showed that, depending on the droplet size and droplet size distribution, the volume concentration of oil at the pipe radius where the ISOLOK™ sample is collected could be 50% lower than the bulk (average) concentration. The results of this study have the industry looking for improved method of sample collection as well as on-line, real-time composition measurement techniques.

HYDRODYNAMICS OF POLYOLEFIN SLURRY LOOP REACTOR: CFD ANALYSIS

(Paper no: 140b)

Rupesh K. Reddy [1] and **Nandakumar K** [1]

[1]*Department of Chemical Engineering, Louisiana State University, Baton Rouge, LA 70803*

Erno Elovainio [2]

[2]*Borealis Polymers Oy, P.O.Box 330, FI-06101 Porvoo, Finland*

Sameer Vijay and [3] and **Christof Wurnitsch** [3]

[3]*Borealis Polyolefine GmbH, Linz, Austria,*

Introduction

Use of slurry loop reactors is quite common in polyolefin production industry and, large amounts of polyolefins such as polypropylene and polyethylene are produced in such reactors (Total length in the range of 200-400 m, diameter in the range of 0.3-0.6 m). In this technique solid (catalyst particles) and liquid (monomer) are circulated in array of vertical tubes (4-8 legs) interconnected with elbows, using one or two internal axial pumps (circulation velocity 6 – 10 m/s). For efficient design and scale-up of these reactors, understanding the hydrodynamics of the slurry mixture is essential.

Using the Two Fluid Model (TFM) we are investigating the hydrodynamic interactions of solid and liquid to make the reactor more energy efficient and productive. The goal is to understand the phase distributions near the impeller exit zones and near the bends and their dispersion characteristics. During this investigation, no polymerization reaction and related effects were actually simulated.

CFD Model Formulation

In the present work, an Eulerian multi-fluid model, which considers the conservation of mass and momentum for the solid and liquid phases, has been adopted. The kinetic theory of granular flow, which considers the conservation of solid fluctuation energy, has been used for closure. The unsteady state equations of continuity for solid and liquid phases are given by:

$$\frac{\partial(\rho_L \epsilon_L)}{\partial t} + \nabla.(\rho_L \epsilon_L \langle u_{Li}\rangle) = 0 \tag{1}$$

$$\frac{\partial(\rho_S \epsilon_S)}{\partial t} + \nabla.(\rho_S \epsilon_S \langle u_{Si}\rangle) = 0 \tag{2}$$

The momentum balance equations for each phase are :

$$\frac{\partial(\rho_L \epsilon_L \langle u_{Li}\rangle)}{\partial t} + \nabla.(\rho_L \epsilon_L \langle u_{Li}\rangle\langle u_{Li}\rangle) = -\epsilon_L \nabla\langle p\rangle + \nabla.\left(\epsilon_L \mu_{eff,L}(\nabla\langle u_{Li}\rangle + (\nabla\langle u_{Li}\rangle)^T)\right) + \rho_L \epsilon_L \vec{g} - K_{LS}? \tag{3}$$

$$\frac{\partial(\rho_S \epsilon_S \langle u_{Si}\rangle)}{\partial t} + \nabla.(\rho_S \epsilon_S \langle u_{Si}\rangle\langle u_{Si}\rangle) = -\epsilon_S \nabla\langle p\rangle - \nabla\langle p_S\rangle + \nabla.\begin{pmatrix} \epsilon_S \mu_{eff,S}(\nabla\langle u_{Si}\rangle + (\nabla\langle u_{Si}\rangle)^T) \\ +(\epsilon_S \lambda_S)(\nabla.\langle u_{Si}\rangle I) \end{pmatrix} + \rho_S \epsilon_S \vec{g} + K_{LS}? \tag{4}$$

Where, μ_{eff} is the effective viscosity, g is the gravitational acceleration, and K_{LS} is the interphase momentum transfer force. The phasic volume fractions satisfy the following condition:

$$\in_L + \in_S = 1 \tag{5}$$

Constitutive equations are required to close these governing equations. Interactions between the phases involve various momentum exchange mechanisms such as the drag, the lift and the added mass force, etc. However, the contribution of drag force has been considered because it is the dominant interaction force, while the effect of the other forces has been considered relatively small (Cheng and Zhu, 2005; Cornelissen et al., 2007).

Drag law: In the present work the drag coefficient (C_D) has been modeled by using the equations proposed by Gidaspow (1994).

$$\text{For } \in_L > 0.8 \quad K_{LS} = 3C_D \frac{\in_S \in_L \rho_L V_S^2}{d_P} \in_L^{-2.65} \tag{6}$$

$$\text{For } \in_L \leq 0.8 \quad K_{LS} = 150 \frac{\in_S^2 \mu_L V_S}{\in_L d_P^2} + 1.75 \frac{\in_S \rho_L V_S^2}{d_P} \tag{7}$$

Where,

$$C_D = \frac{24}{\in_L \mathrm{Re}_S}[1 + 0.15(\in_L \mathrm{Re}_S)^{0.687}] \tag{8}$$

$$\mathrm{Re}_S = \frac{d_P V_S \rho_L}{\mu_L} \tag{9}$$

These drag equations are based on the total surface area of the particle.

For the turbulent regime, *k-ε* model along with granular kinetic theory has been used. The detailed model and the other constitutive equations for the solid phase are discussed in Reddy and Joshi (2009). Multiple Reference Frame (MRF) methodology was used to model the pump dynamics

All terms of the governing equations for unsteady state were discretized using the second-order upwind differencing scheme. The SIMPLE algorithm has been employed for the pressure-velocity coupling. The convergence criterion (sum of normalized residuals) is set at 5 X 10^{-4} for all the equations. The simulations have been performed on LONI cluster Queenbee, HPC machine Supermike and XSEDE machine Stampede. These are 5000 processor machines and each processor has a clock speed of 2.4GHz and 2GB RAM (the number of processers used for each simulation was varied and given in next sections). The present study employs commercial CFD software ANSYS FLUENT. As we study different cases, the specific inflow boundary conditions relevant to the cases will be identified. No slip boundary conditions are always employed at all the walls.

Results and Discussion

Preliminary 3D simulations of single loop reactor (2 leg) and 4 leg loop reactor:

Initially the 3D CFD model was developed and validated in a single loop reactor. Both steady state and unsteady simulations were performed and the pump impeller was a simple 3-blade pitched turbine. The details of the simulations are given in Table 1. The steady state simulations in 4-leg loop were also performed with the 6 blade impeller. The geometric models were shown in Figure 1.

Separate simulations were performed to examine the effect of the stator vanes and results are shown in Figure 2. It was observed that immediately following the downstream of the pump, the particles are forced radially outwards towards the wall due to the high centrifugal force generated by the impeller. This leads to accumulation of the solids in the rising leg (Figure 2A). Moreover, the flow generated by the pump is also expected to be low. We need more axial flow than the rotational/tangential flow for uniform distribution of the particles.

Table 4.1: Simulation details for the 3D single loop reactor

Diameter	0.6 m
Total length	39 m
Initial solid Concentration	15 % (by Volume)
Density of the Liquid	400 kg/m^3
Density of the Solids	800 kg/m^3
Particle Size	1.8 mm
Impeller	Pitch Blade (150 Rad/s)
No of cells	1,200,000
No of processors	8
CPU speed	3 GHz
Memory per processer	4 Gb
Time taken for steady state	Less than 24 hr
Time take for unsteady for simulating 4 sec	3- 4 days

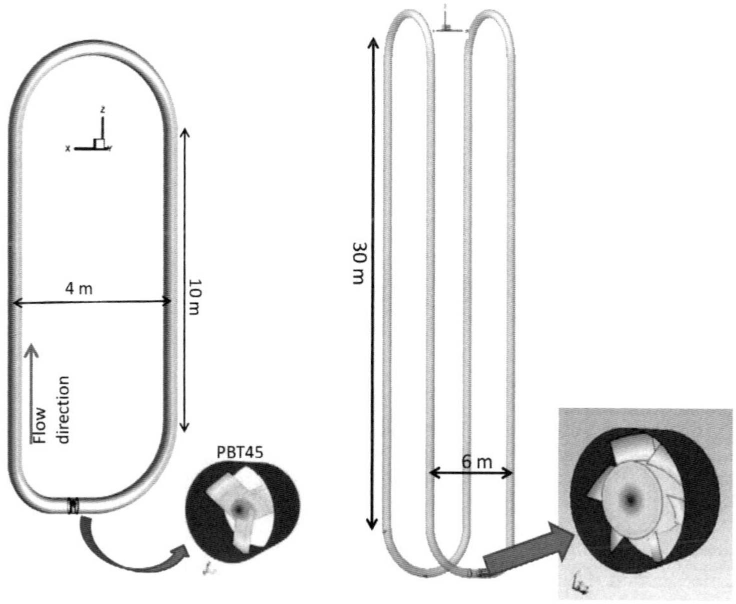

Figure 1: Single loop reactor and 4-leg loop reactor with pump

In the presence of the stator vanes (Figure 2B) immediate on the downstream of the impeller, this accumulation of the solids was not observed. It is because the orientations of these stator vanes are in such a way that the rotational flow generated by the impeller was converted effectively into the axial flow. It was also observed that the flow rate also increased in the presence of the stator vanes.

One unsteady state simulation (with single loop) was also performed with the initial condition from the steady state result. It was observed that when the slug is in the rising leg the pressure drop across the pump is higher than that for the slug is in the down flow leg (Figure 3).

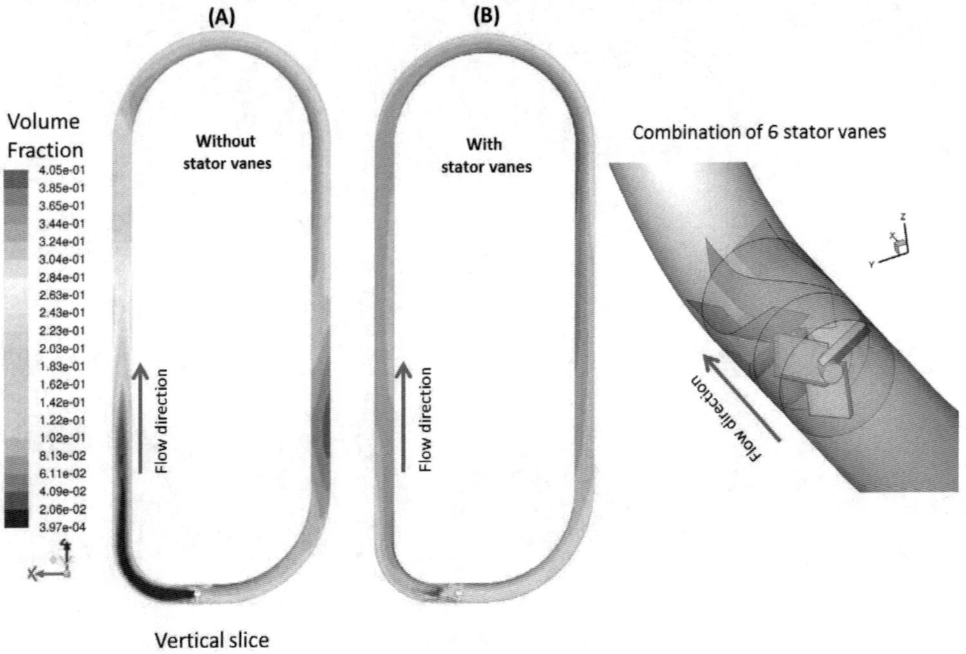

Figure 2: Solid volume fraction of single loop with and without stator vanes

Validation of the CFD Model: Simulations of RTD in single loop:

One of the important objectives of this study is the validation of the present multiphase-CFD model by comparing the simulated results with the tracer experimental results in single loop (Figure 3). The particle tracer is patched at the downstream side of the impeller. For exact quantitative comparison and validation, we must have the same initial width particle tracer in the simulations as in the experiments. For this few trial simulations were performed.

The comparison of simulated results with the experimental results in a small loop reactor after 13th cycle is shown in Figure 4. The main quantitative results available from the measurement are the pulse width at the half maximum for initial tracer pulse (0.16 s) and after 13th cycle (0.55 s). The simulated value of the pulse width at the half maximum after 13th cycle is 0.52 s, which is well in agreement with the experimental value of 0.55 s.

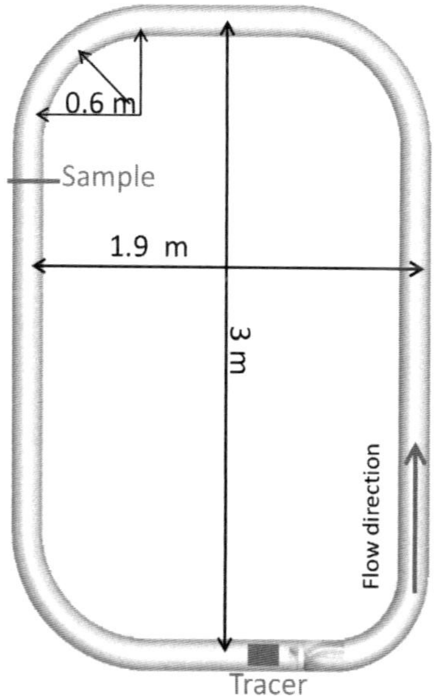

Figure 3: Schematic diagram of the single loop used for CFD model validation

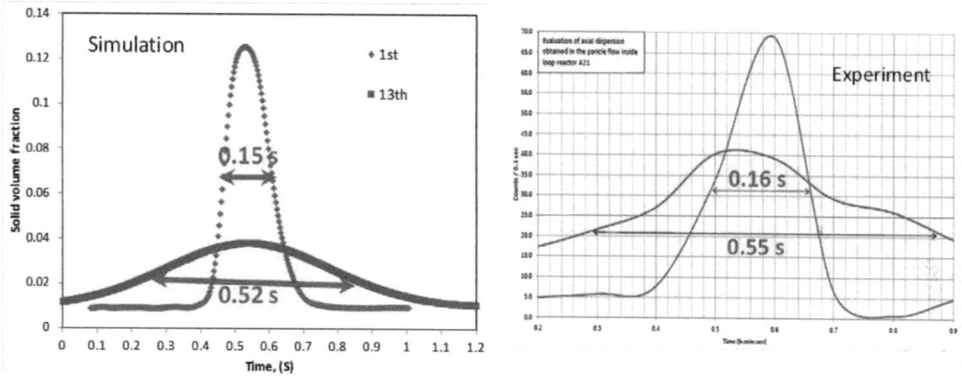

Figure 4: Comparison between the simulated results with the experimental results after 13th cycle

Simulation in the 4 leg loop reactor:

The simulations in the 4 leg loop reactor also demonstrate the importance of the stator vanes in the uniform distribution of particles at the downstream end of the pump. Another important observation made in this 4 leg loop simulation was that at the elbows the particles were accumulated at the outer walls due to centrifugal forces (Figure 4 at elbows 2 and 3).

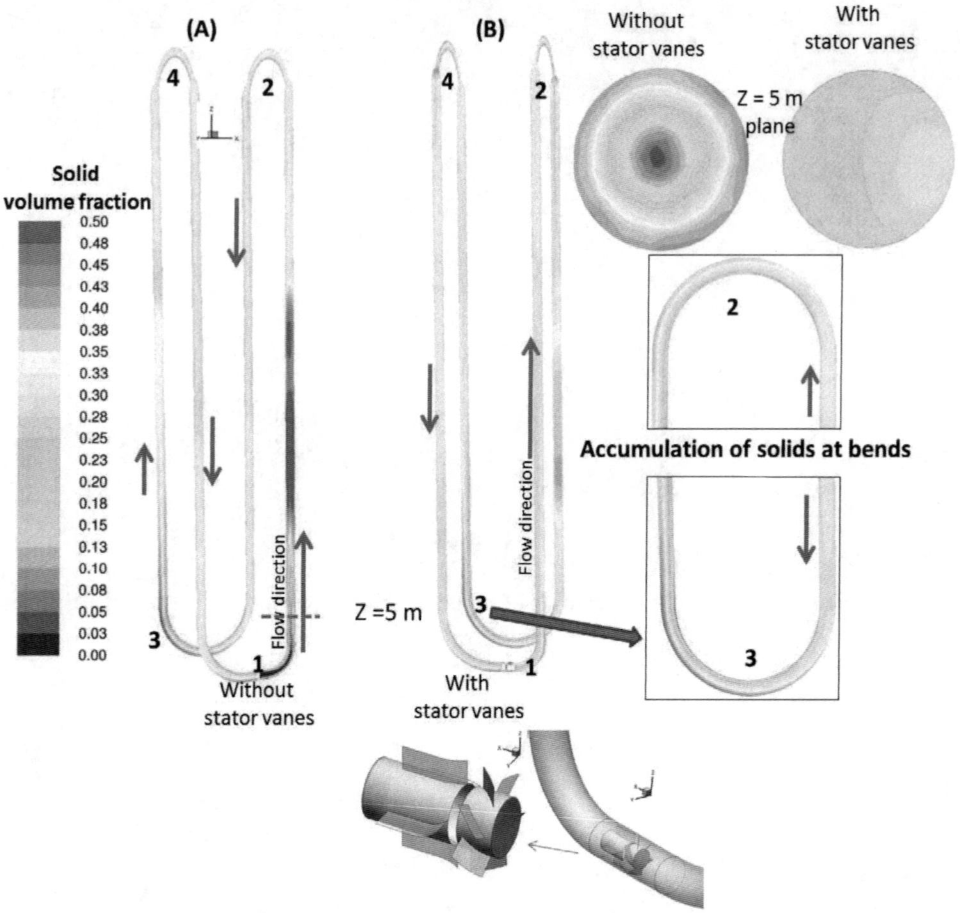

Figure 4: Solid volume fraction of 4-leg loop with and without stator vanes

The fate of these accumulated solids in the elbows was examined with the help of single elbow simulations. Several case studies focused on understanding the segregation mechanism *around bends* for *different particle sizes*. 2.5mm size particles segregate more than 1.0 mm or 0.5 mm and 30% initial solids concentration results in greater segregation than 15% initial solids concentration.

Acknowledgements

We would like to acknowledge the High Performance Computing Center (HPC) at Louisiana State University, Louisiana Optical Network Initiative (LONI) and XSEDE for providing access to computational resources.

References:

Cheng, Y., Zhu, J., 2005. CFD modeling and simulation of hydrodynamics in liquid–solid circulating fluidized beds. Canadian Journal of Chemical Engineering 83, 177–185.

Cornelissen, J. T., Taghipour, F., Escudie, R., Ellis, N., Grace, J. R., 2007. CFD modeling of liquid-solid fluidized bed. Chemical Engineering Science 62, 6334 -6348.

Gidaspow, D., 1994. Multiphase flow and fluidization: continuum and kinetic theory descriptions. Academic Press, San Diego.

Reddy, R. K., Joshi, J. B., 2009. CFD modeling of solid liquid fluidized beds of mono and binary particle mixtures. Forwarded to Chemical Engineering Science.

353987 Thermodynamics Coupled CFD Simulation of Multiphase Flows

Abstract:

CFD modeling of multiphase flows with phase change has long been a big challenge in the oil and gas industry. Understanding both hydrodynamic and thermodynamic behaviors of the oil-water-gas system is critical for many processes from upstream to downstream such as oil and gas recovery from shale, transport along pipelines and distillation or cracking in a refinery. The hydrodynamics determines the multiphase flow or the local flow conditions such as pressure and temperature and hence affects the thermodynamic behavior of the mixture components. On the other side, the thermodynamic behavior such as phase change will greatly modify the mixture properties and thus changes the flow dynamics. Unlike the pure component multiphase flows where phase change is described by a simple pressure-temperature diagram, multi-component multiphase systems have different dew and bubble points at which phase change takes place in different directions. The additional complexity is that the state where phase change occurs strongly depends on the mole/mass fraction of each component in the mixture. This talk will discuss the development of a thermodynamic coupled CFD model addressing the above complexity under the Eulerian-Eulerian multiphase flow modeling framework. The potential application of this modeling approach in process simulations will be highlighted.

353988 A Numerical Model to Simulate the Foaming Process and Address the Challenges of a Rapidly Changing Industry

Abstract:

As environmental regulations and demand for high efficiency insulations increase, the thermoplastic foams industry is looking for improved polymer properties and processing conditions which lead to small cell size (100-1000nm) and high porosity (>95%) foams using CO_2 as the blowing agent. A modeling tool that can help in the design of experiments for selecting the right polymer and processing conditions can reduce the product development time and speed up the process of commercialization.

In this presentation, we focus on the on-going work to develop a numerical model for simulating simultaneous nucleation and bubble growth in thermoplastic foams. Being based on the "Influence Volume Approach" the model differentiates between nucleation time and depressurization time as is the case in real foaming processes. Classical nucleation theory is used to predict the rate of nucleation of bubbles. By solving the mass, momentum and species conservation equations for each bubble, the model are capable of predicting average bubble size, bubble size distribution and bulk porosity.

Comparison of the model predictions to experimental data and the current model limitations are discussed. The effect of depressurization rate, blowing agent diffusivity, polymer viscosity and surface tension are also delineated. Enhancements to the model in the form of improved equation-of-state and its effect on the model predictions will also be presented.

353989 Effects on FCC Catalyst Residence Time Distributions Through Simulations of Circulating Reactors

Abstract:

Introduction

Catalytic upgrading[3,10] of biomass pyrolysis vapors prior to condensation has recently received attention as a promising strategy for improving the compatibility of bio-oil with conventional processing infrastructure. Riser reactors, staples of the petroleum processing industry, are commonly used for FCC (fuel catalytic cracking) and other processes that share characteristics with the envisioned biomass to liquid fuel pathway.[2] We are currently undertaking a computational study to assess the operational feasibility of using such reactors with pyrolysis vapors.

Specifically, we examine the effects of mass flow rates and boundary conditions on the solid catalyst residence times. The flow rates include the ratios of pyrolysis vapor and fluidizing gases (steam, N_2) to solid catalyst. The boundary conditions include inlet positions, diameters, and the outlet configuration. We additionally examine the effects on clustering, temperature, and velocity profiles.

Starting with non-reactive two- and three-dimensional simulations, this study will ultimately incorporate kinetic mechanisms for catalyzed secondary gas phase reactions to study the overall performance of the reactors. Experimental measurements of bulk quantities (temperature, pressure, steady state catalyst weight %) from a pilot scale reactor will be used for validation of the model fidelity. This work forms a baseline for future studies where we will model both the primary pyrolysis as well as catalyzed secondary gas phase reactions to study and improve the overall performance of the reactors.

Goals of Research

The objective of this project is to simulate and study the catalytic upgrading of pyrolysis vapors through validation with the pilot scale FCC DCR (Davison circulating riser) reactor that is arriving at NREL in Feb. 2014. Through validation of experiments with the DCR, we will better understand and improve the conversion of biomass to liquid transportation fuels, specifically gasoline and diesel precursors. This conversion is highly dependent on the contact time of the pyrolysis vapor with the FCC catalyst, which is typically of the zeolite type. Too short of a contact time leads to under conversion and too long leads to excessive coking, deactivation of the catalyst, and/or undesirable products. In an actual riser, there is a distribution of residence times and unsteady creation, destruction, and flow of particle clusters, particularly prevalent at the walls. The distribution's average as well as its shape (skew preferring slower times) is predicted to have an effect on the resulting biofuel products.

Numerical Approach

Gas-particle flows in FCC reactors occur over a wide range of scales. In order to have a simulation that can run for experimental time and length scales, a continuum and filtered equation approach is warranted. The two-fluid model (TFM) equations for gas-particle flows, which have been developed and analyzed extensively over the past five decades, are able to capture their existence in a robust manner.[1] However, to resolve the FCC catalyst particle clusters and bubble-like voids at all length scales, extremely fine spatial resolution is necessary, but impractical for three-dimensional simulations. An engineering tool, the filtering of the two-fluid equations[4,5,6] is used to probe the macro-scale flow characteristics directly, without having to resolve the smaller scale structures. In this approach, the influence of the small-scale structures appears as residual correlations, called subgrid models. These subgrid modifications of the equations have been constructed with constitutive or closure models that are designed to model macro-scale gas-solids interactions, for example particle drag, as well as the partial slip wall effects of the boundary conditions.

With the MFIX TFM (two-fluid) method[8,9], using solids with a particle size distribution of solid catalyst particles (Geldart A type, size 20-140 micrometers with mean of 75), the clustering and residence times of solids and vapors will be determined. Peregrine, the NREL supercomputer with ~15,000 available cores is utilized for these simulations. The use of a hybrid model using distributed (domain decomposition) and shared memory (threading of loops) parallelism is important for efficient, three dimensional, multi-component simulations.

Results and Initial Conclusions

Preliminary simulations for two and three dimensional riser models have been conducted to determine the effect of initial and boundary conditions on the catalyst and vapor residences times. A preliminary simulation result for a catalyst to feedstock ratio equal to three, with 3 and 1 weight percent respectively (relative to feed) of steam and additional fluidizing nitrogen has been conducted for the cylindrical riser of ~1cm diameter and height ~3m. Using a Cartesian stretched grid of size 96x2440 cells, the simulation was conducted for 50 seconds of operation using 768 cores for one week on Peregrine at NREL. Lower resolution simulations were also conducted up to 150 seconds. These baseline simulations were first conducted with out the subgrid scale models such that a highly resolved result could be used for a comparison to subsequent filtered results with the subgrid clustering and wall model modifications. By finding the optimal filter size and wall subgrid parameters, coarse grid simulations can be used to obtain the desired bulk properties at a fraction of the computational cost.

MISSOURI S&T | MISSOURI UNIVERSITY OF SCIENCE AND TECHNOLOGY | Founded 1870 | Rolla, Missouri

DETAILED PROCESS MODELING OF A SUSTAINABLE COAL, WIND, AND NUCLEAR HYBRID ENERGY SYSTEM

140f – Enabling Process Innovation Through Computation

Kyle L. Buchheit

Joseph D. Smith, Ph.D.

Energy Research & Development Center

Contact: (573) 225-2277

kyle.buchheit@mst.edu

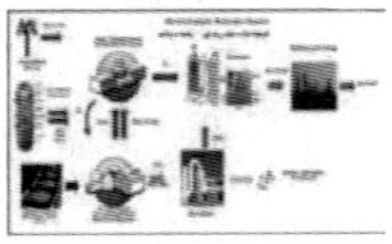

April 2, 2014

MISSOURI S&T | MISSOURI UNIVERSITY OF SCIENCE AND TECHNOLOGY | Founded 1870 | Rolla, Missouri

Outline

- Global Energy Production
- Hybrid Energy System
- Aspen Plus Model
- Stream Results
- Efficiencies
- Future Work

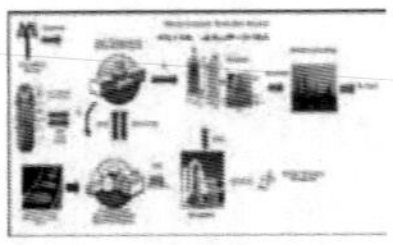

Yearly Global Energy Consumption

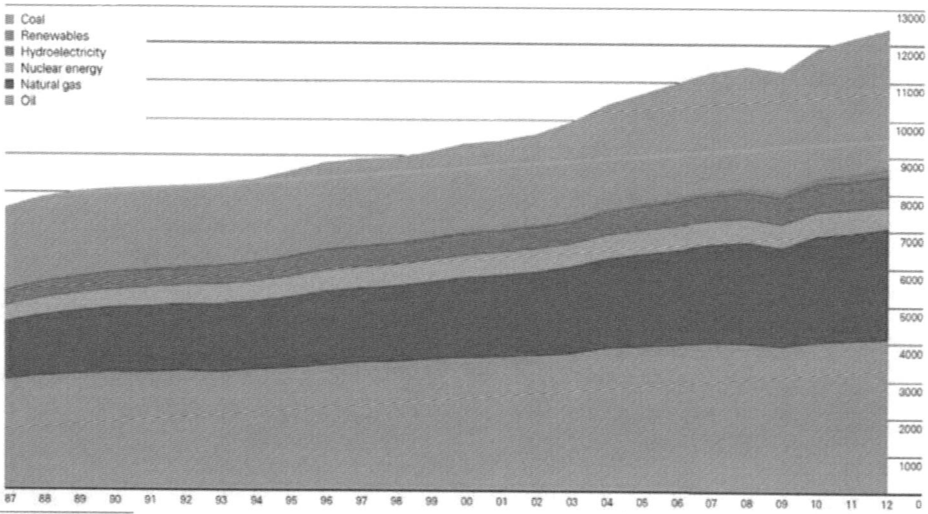

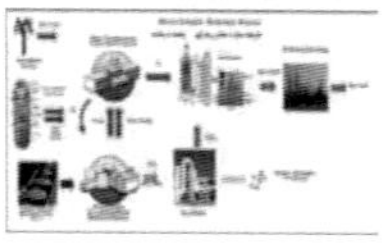

BP Statistical Review of World Energy – June 2013

MISSOURI S&T Missouri University of Science and Technology Founded 1870 | Rolla, Missouri

Global Man-made Carbon Dioxide Emissions

CO_2 emissions: Sectoral Approach

million tonnes of CO_2

	1971	1975	1980	1985	1990	1995	2000	2005	2008	2009	2010	% change 90-10
World *	14 064.8	15 668.5	18 042.2	18 623.5	20 973.9	21 843.8	23 509.1	27 187.4	29 483.0	28 946.7	30 276.1	44.4%

* Total world includes non-OECD total, OECD total as well as international marine bunkers and international aviation bunkers.

International Energy Agency – 2012

"We can't have an energy strategy for the last century that traps us in the past. We need an energy strategy for the future – an all-of-the-above strategy for the 21st century that develops every source of American-made energy." - President Barack Obama, March 15, 2012

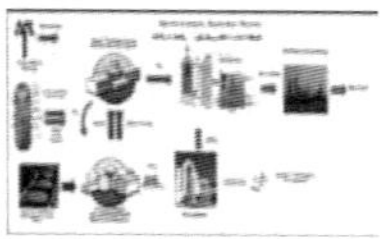

Hybrid Energy System

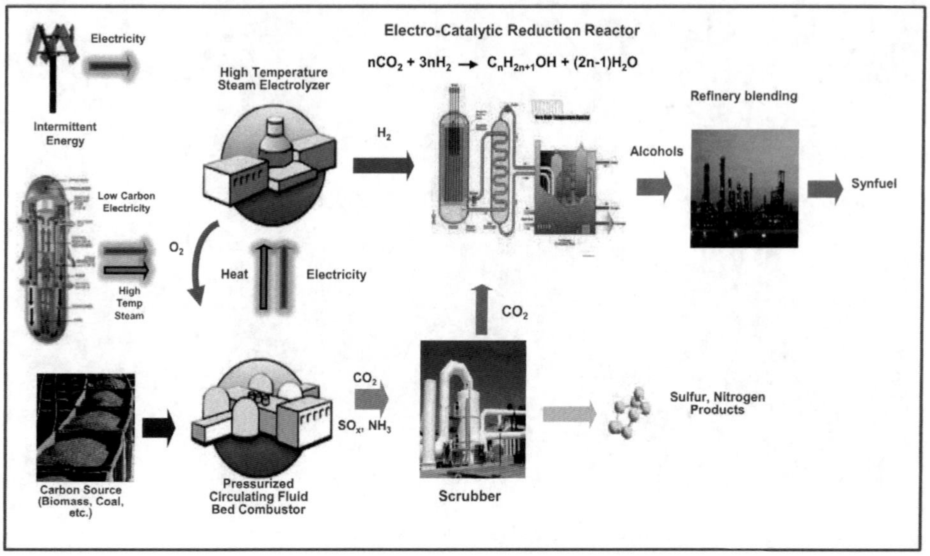

Hybrid Energy System

- PCFB Coal Combustor
  - Super-Critical Combustor
  - Fuel Selectivity
- Nuclear SMR
  - Light-Water Reactors
  - Safety Features
- Wind
  - Geography
  - Technology

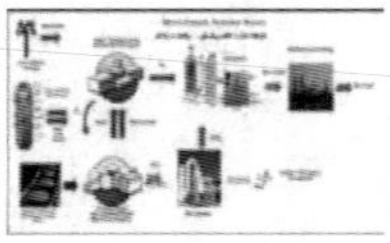

Missouri University of Science and Technology Founded 1870 | Rolla, Missouri

Pressurized Circulating Fluidized Bed (PCFB) Combustor

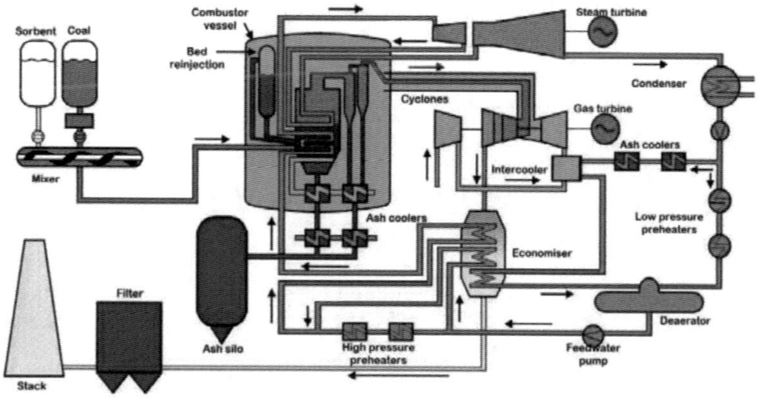

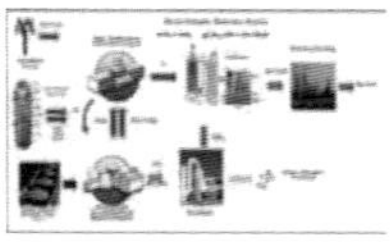

7

Missouri University of Science and Technology Founded 1870 | Rolla, Missouri

Aspen Plus overall PFD

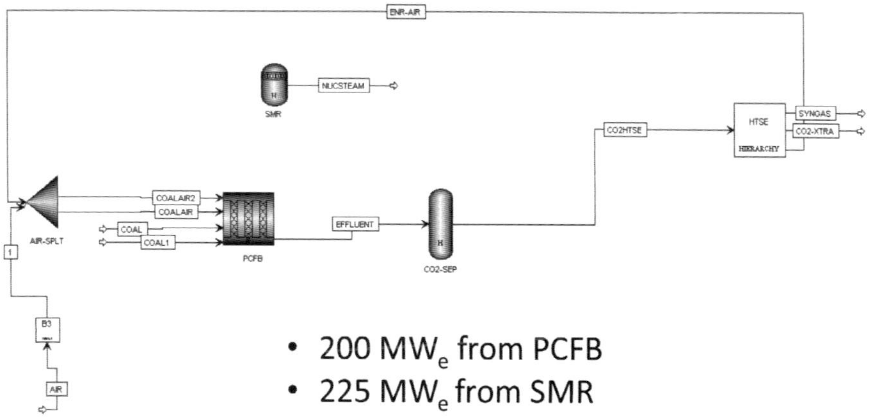

- 200 MW_e from PCFB
- 225 MW_e from SMR

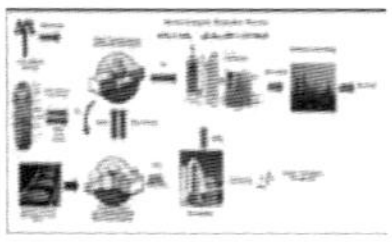

8

Missouri University of Science and Technology — Founded 1870 | Rolla, Missouri

PCFB Model Parameters

Feature (100 MW unit block)	*Value*	*Comments:*
Overall plant efficiency (HHV) - %	39.30	Not accounting for CO_2 compression
Fuel supply kg/s	7.5	Dry coal. HHV equals 33.927 MJ/kg
Unit power - MW_e	100	Full plant 400 MW_e requiring 4 unit blocks
CO_2 generated – kg/s	26.47	
CO_2 captured – kg/s	23.92	Thus emitting 2.5518 kg/s CO_2
CO_2 capture rate - %	90.36	

Alstom P200 PCFB / Sargas Technology targets

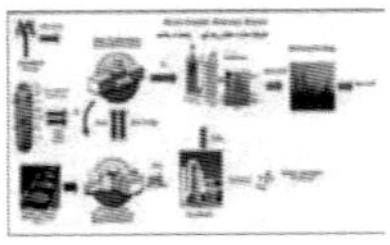

9

Missouri University of Science and Technology — Founded 1870 | Rolla, Missouri

PCFB Hierarchy

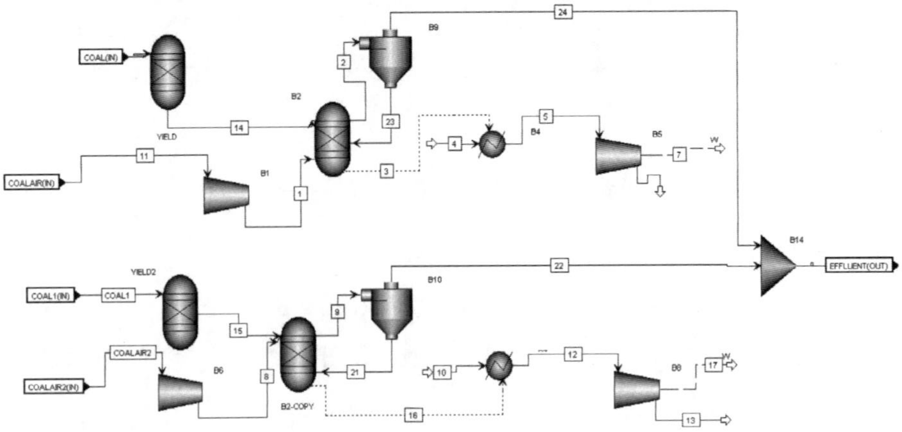

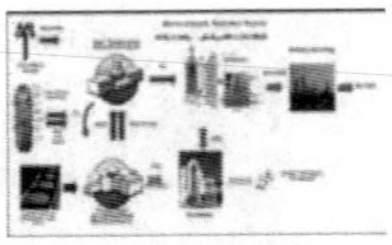

10

Missouri University of Science and Technology Founded 1870 | Rolla, Missouri

SMR Hierarchy

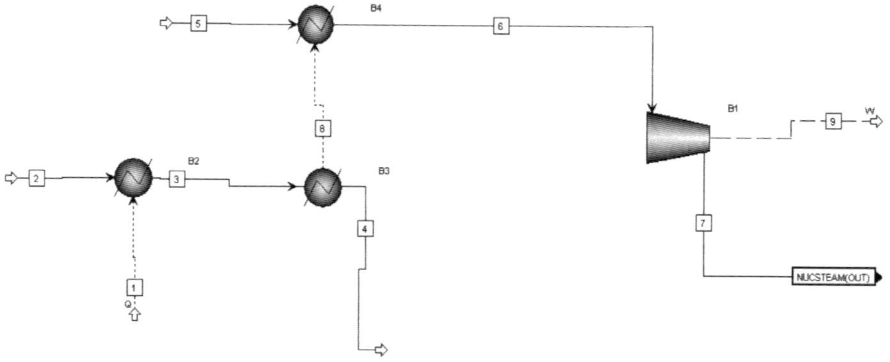

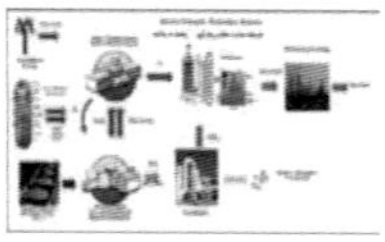

11

Missouri University of Science and Technology Founded 1870 | Rolla, Missouri

Benfield Hierarchy

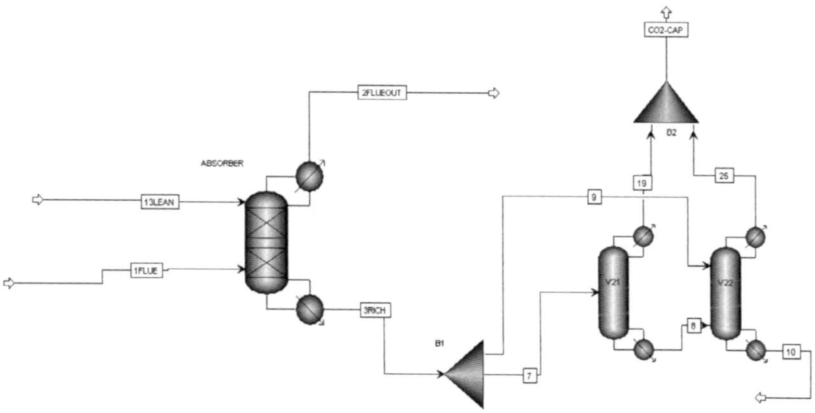

- 5 m K^+ / 2.5 m Piperazine Aqueous solution
- 90% Base carbon dioxide absorption

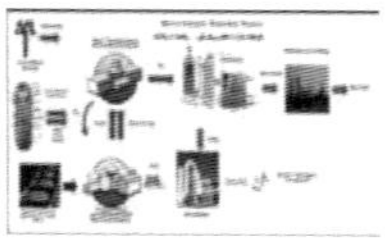

12

MISSOURI S&T Missouri University of Science and Technology Founded 1870 | Rolla, Missouri

High Temperature Steam Electrolysis

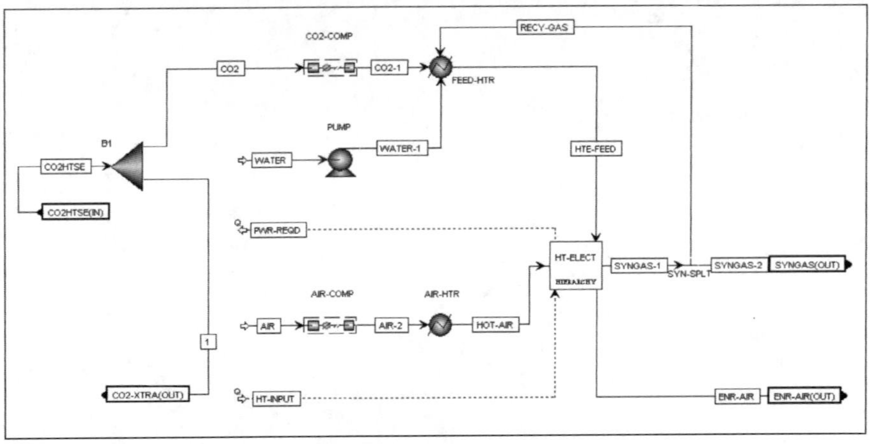

- Outer Hierarchy
    - Prepare stream conditions
    - Recover heat

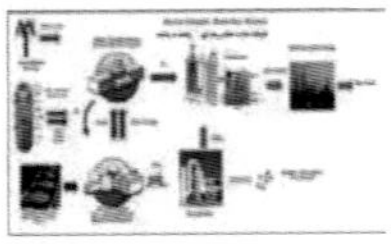

13

MISSOURI S&T Missouri University of Science and Technology Founded 1870 | Rolla, Missouri

High Temperature Steam Electrolysis

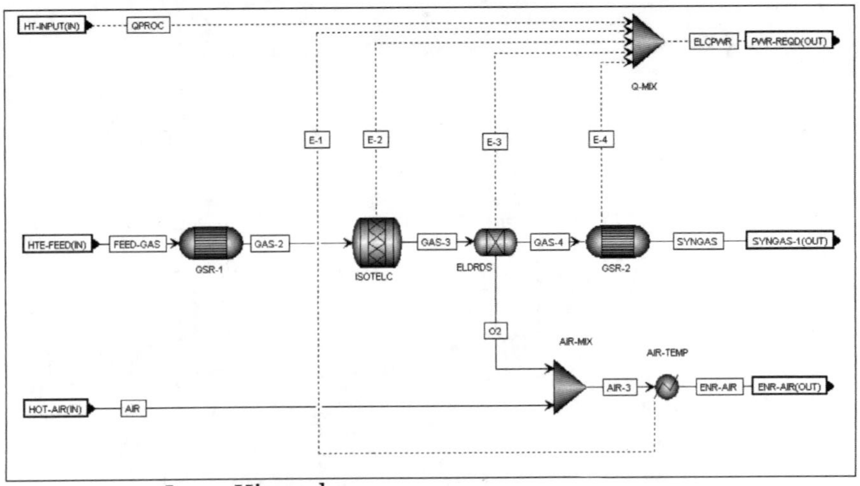

- Inner Hierarchy
    - Water-gas-shift reaction
    - Electrolysis model
    - In-line FORTRAN for Energy required

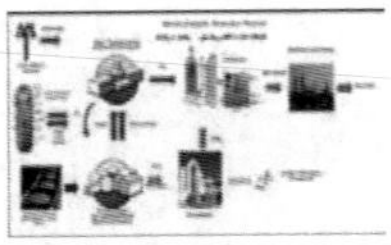

14

Simulation Stream Results

Stream Name	PCFB Flue Gas	CO_2 to HTSE	Enriched Air	Unused CO_2	Syngas	Coal Air
Total Flow (lbmol/hr)	36,134	7,742	11,071	4,749	8,515	32,471
Component Flows (lbmol/hr)						
CO2	8,602	7,742	-	4,749	393	-
CO	949	-	-	-	2,599	-
N2	22,510	-	5,536	-	-	22,442
H2	343	-	-	-	5,522	-
O2	TRACE	-	5,535	-	-	10,029
H2O	3,657	-	-	-	-	-
Mole Fractions						
CO2	0.2381	1.0000	-	1.0000	0.0462	-
CO	0.0263	-	-	-	0.3053	-
N2	0.6229	-	0.5001	-	-	0.6911
H2	0.0095	-	-	-	0.6485	-
O2	TRACE	-	0.4999	-	-	0.3089
H2O	0.1012	-	-	-	-	-

- 40% CO_2 Utilization

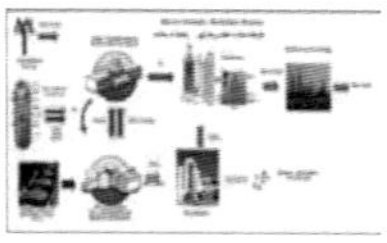

Efficiencies

Individual Efficiencies	
Power Plant Type	Efficiency
PCFB	43%
SMR	28%
Wind Turbines	34%
Global Average Coal	33%
Subcritical coal	>38%
Supercritical Coal	>42%
Ultra-supercritical coal	45-50%

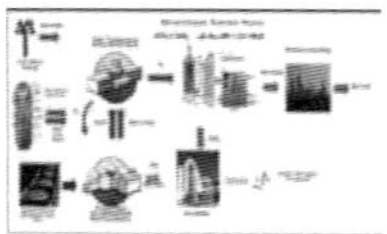

MISSOURI S&T — MISSOURI UNIVERSITY OF SCIENCE AND TECHNOLOGY — Founded 1870 | Rolla, Missouri

Dynamic Wind

- Aspen Plus is a steady-state process simulator
  - o Wind is highly dynamic / time dependent
- Update current Aspen Plus model
  - o Integrate with Wind Model into Aspen Dynamics

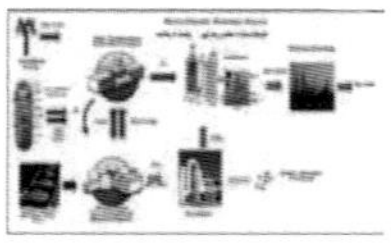

17

MISSOURI S&T — MISSOURI UNIVERSITY OF SCIENCE AND TECHNOLOGY — Founded 1870 | Rolla, Missouri

Testing and Model Validation

- Missouri S&T's Energy Research and Development Center
  - o Design and build Hybrid Energy System components
  - o Campus wind and solar data
  - o Test Hybrid system for model comparison

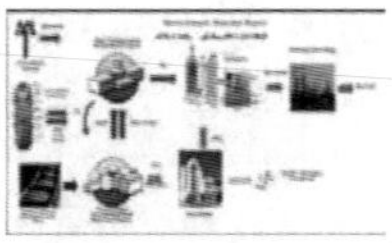

18

Future Work

- Integrate SMR / PCFB water cycles
- Electro-Catalytic / Fischer-Tropsch reactor for synfuel production
- Total system optimization based on overall cost of electricity
- Bio feed stock for PCFB
- Solar Integration for renewable energy

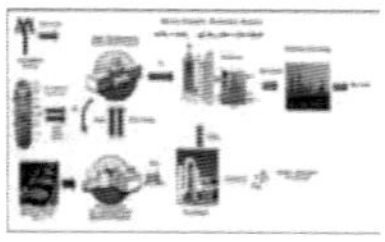

19

MISSOURI S&T Missouri University of Science and Technology Founded 1870 | Rolla, Missouri

References

BP plc: "BP Statistical Review of World Energy 2013"

International Energy Agency: "CO2 Emissions From Fuel Combustion: Highlights" Second Edition March 2013

Obama,B.: "Remarks by the President on Energy" March 15, 2012.
Accessed October 25th 2013
http://www.whitehouse.gov/the-press-office/2012/03/15/remarks-president-energy

PFBC Environmental Energy Technology, Inc.: "PFBC: Competitive Clean Coal Power Utilizing Pressurized Fluidized Bed Combined-Cycle Technology"

Westinghouse Electric Company LLC.: "The Westinghouse Small Modular Reactor: Building on 125 Years of Innovation" 2013

Ryan Blinn: "Westinghouse Small Modular Reactor Design and Application" June 2012

Rochelle, Gary T., Seibert, Frank et al.: "CO2 Capture by Absorption with Potassium Carbonate" Final Report December 2007

McKellar, M. G., O'Brien, J. E., et al.: "Optimized Flow Sheet for a Reference Commercial-Scale Nuclear-Driven High-Temperature Electrolysis Hydrogen Production Plant" Idaho National Labs, November 2007

Hetland, J., Christensen, T.: "Assessment of a fully integrated SARGAS process operating on coal with near zero emissions" Sargas

U.S. Energy Information Administration: "Levelized Cost of New Generation Resources in the Annual Energy Outlook 2013" January 2013

U.S. Energy Information Administration: "Average Operating Heat Rate for Selected Energy Sources",
http://www.eia.gov/electricity/annual/html/epa_08_01.html Last Accessed October 24th 2013

International Energy Agency: "Technology Roadmap: High-Efficiency, Low-Emissions Coal-Fired Power Generation" 2012

20

MISSOURI S&T | MISSOURI UNIVERSITY OF SCIENCE AND TECHNOLOGY | Founded 1870 | Rolla, Missouri

- Questions?

Kyle L. Buchheit
Joseph D. Smith, Ph.D.
Energy Research & Development Center

Contact: (573) 225-2277
kyle.buchheit@mst.edu

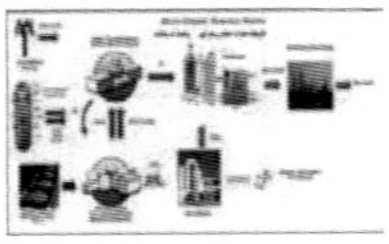

21

AUTHOR INDEX

AUTHOR INDEX

American Institute of Chemical Engineers
3 Park Avenue
New York, New York 10016-5991

CURRAN ASSOCIATES INC
proceedings.com